# Developing Multi-Agent Systems with JADE

**Wiley Series in Agent Technology**

Series Editor: Michael Wooldridge, *Liverpool University, UK*

The 'Wiley Series in Agent Technology' is a series of comprehensive practical guides and cutting-edge research titles on new developments in agent technologies. The series focuses on all aspects of developing agent-based applications, drawing from the Internet, telecommunications, and Artificial Intelligence communities with a strong applications/technologies focus.

The books will provide timely, accurate and reliable information about the state of the art to researchers and developers in the Telecommunications and Computing sectors.

Titles in the series:

| | |
|---|---|
| Padgham/Winikoff: Developing Intelligent Agent Systems | 0470861207 (June 2004) |
| Pitt (ed.): Open Agent Societies | 047148668X (August 2004) |

# Developing Multi-Agent Systems with JADE

**Fabio Bellifemine,** *Telecom Italia, Italy*
**Giovanni Caire,** *Telecom Italia, Italy*
**Dominic Greenwood,** *Whitestein Technologies AG, Switzerland*

John Wiley & Sons, Ltd

Telephone (+44) 1243 779777

Email (for orders and customer service enquiries): cs-books@wiley.co.uk
Visit our Home Page on www.wiley.com



***Other Wiley Editorial Offices***

John Wiley & Sons Inc., 111 River Street, Hoboken, NJ 07030, USA

Jossey-Bass, 989 Market Street, San Francisco, CA 94103-1741, USA

Wiley-VCH Verlag GmbH, Boschstr. 12, D-69469 Weinheim, Germany

John Wiley & Sons Australia Ltd, 42 McDougall Street, Milton, Queensland 4064, Australia

John Wiley & Sons (Asia) Pte Ltd, 2 Clementi Loop #02-01, Jin Xing Distripark, Singapore 129809

John Wiley & Sons Canada Ltd, 6045 Freemont Blvd, Mississauga, ONT, L5R 4J3, Canada

Wiley also publishes its books in a variety of electronic formats. Some content that appears in print may not be available in electronic books.

Anniversary Logo Design: Richard J. Pacifico

***Library of Congress Cataloging-in-Publication Data:***

Bellifemine, Fabio.
Developing multi-agent systems with JADE / Fabio Bellifemine,
Giovanni Caire, Dominic Greenwood.
p. cm.
Includes bibliographical references and index.
ISBN-13: 978-0-470-05747-6 (cloth : alk. paper)
1. Intelligent agents (Computer software) 2. Object-oriented programming (Computer science) 3. Java (Computer program language) I. Caire, Giovanni. II. Greenwood, Dominic. III. Title.
QA76.76.I58B45 2007
006.3–dc22

2006036200

***British Library Cataloguing in Publication Data***

A catalogue record for this book is available from the British Library

ISBN: 978-0-470-05747-6 (HB)

Typeset in 9/11pt Times by Laserwords Private Limited, Chennai, India
Printed and bound in Great Britain by Antony Rowe Ltd, Chippenham, Wiltshire
This book is printed on acid-free paper responsibly manufactured from sustainable forestry in which at least two trees are planted for each one used for paper production.

# Contents

# The Authors

**Fabio Bellifemine** is a senior project manager at the Research Labs of Telecom Italia. He inspired, originated and led the JADE Project and, later, he launched and presided at the JADE Governing Board. He graduated summa cum laude from the Computer Science University of Torino in 1988 and, before joining Telecom Italia (formerly CSELT), until 1994 he was a researcher at the Italian National Research Council (CNR) working in the field of digital video coding. In 1996 he joined the FIPA standardization body and he was attracted by the software agent technology. In 2000, on behalf of the Work Package 1 of the FACTS Project, he was awarded the first prize in the FIPA competition for the best agent-based application: 'New generation TV entertainment system'. He served as chairman of the FIPA Architecture Board and, next, of the X2S Technical Committee with the challenging goal of promoting to standard the core specifications of FIPA. The effort and the passion for this emerging technology were recognized by appreciation certificates of FIPA for his outstanding contributions. He is IEEE Senior Member, member of the Technical Committee on Distributed Intelligent Systems of the IEEE Systems, Man, and Cybernetics Society, and member of the Program Committee of some scientific conferences in this research field. He is also member of the Expert Group of JSR 271 which is going to specify the third generation of Java MIDP specifications. His research interests mainly focus on agent technologies and on their adoption to develop better and more efficient systems and applications; recently, he launched in Telecom Italia a new research program in the field of Wireless Sensor Networks.

**Giovanni Caire** graduated summa cum laude from the Politecnico di Torino in 1992 and, after that, joined the Research Labs of Telecom Italia (formerly CSELT) where he is now a senior project manager. He started working in the Multimedia and Video Services department where he was involved in several international collaborative projects. In 1995 he led the ATMAN ACTS European Project dealing with advanced techniques of audio-visual content trading. Since 1998 his interest has been in the field of distributed applications with particular focus on the Java technology. He started working in the JADE Project in 2000 when he led the working group that ported the platform on the Java Micro Edition within the scope of the LEAP IST European Project. In 2004 he represented Telecom Italia in the Java Workstream of the OMTP international initiative whose goal was to define an open platform on mobile terminals enabling service providers to develop a uniform and simplified customer experience. In 2005 he actively participated in the Expert Group of the JSR 232 Mobile Operational Management which addressed the customization of the OSGi specification for hand-held devices. Currently he is working in the OSS Innovation department of Telecom Italia where he leads the software development of an important project employing agent technology and in particular the JADE platform in the field of network management. At the same time he is actively involved in the JADE Board with the role of Technical Leader.

**Dominic P. A. Greenwood** is Head of Research at Whitestein Technologies AG in Zúrich, Switzerland. He has been active in the field of multi-agent software and systems for several years from both a research and commercial perspective, including participation in FIPA and involvement with JADE as a member of the JADE Board and as designer of the JADE Web Services Integration Gateway add-on. He received a Ph.D. for his work on 'adaptive strategies for controlling non-linear behaviour in communication networks' from Staffordshire University in 1997, after which he took a position as group leader of distributed network management research at Fujitsu Telecommunications

Europe in the UK. In 2000 he joined the Network Agent Research group of Fujitsu Laboratories of America in Sunnyvale, California, where he became involved in several aspects of agent-related research and standardization. It was during this period that he became actively involved with FIPA, serving as contributor to various activities including the FIPA Abstract Architecture and as co-chair of the FIPA Ontology technical committee. He also served as technical coordinator of JSR 87, Java Agent Services, which addressed the creation of a core API for agent systems based on a reification of the FIPA Abstract Architecture. In 2003 he joined Whitestein Technologies AG in Zúrich, becoming involved in a broad variety of projects related to the commercial application of agent technology and several European-wide collaborative projects, including the JADE Board. He is a member of several program committees and has a number of academic publications, with his current research interests including distributed and complex adaptive systems, proactive computing, ubiquitous systems, service-oriented computing and commercial applications of software agent technology and autonomic computing.

# Contributors

The following contributed to various parts of the book:

Agostino Poggi, Dipartimento di Ingegneria dell'Informazione, Università degli Studi di Parma (Section 2.1).

The membership of FIPA (Section 2.2) Jordi Cucurull-Juan, Joan Ametller-Esquerra and Ramon Marti, Universitat Autònoma de Barcelona (Chapter 6).

Alois Reitbauer, Profactor Research, (Chapter 7).

Roland Mungenast, Profactor Research (Section 9.2).

Dirk Bade, University of Hamburg, Department of Computer Science, Distributed and Information Systems (Chapter 11).

Karl-Heinz Krempels, Sven Lilienthal and Ananda Sumedha Markus Widyadharma, RWTH University of Aachen, Department of Computer Science, Communication and Distributed Systems (Chapter 11).

Vincent Louis and Thierry Martinez, France Telecom, Research & Development (Chapter 12).

Chris Van Aart, Y'All BV (Section 13.1).

David B. Bernstein, Caboodle Networks, Inc. (Section 13.2).

Pavel Tichy, Rockwell Automation Research Center (Section 13.3).

Alexander Pokahr, Lars Braubach, Andrzej Walczak and Winfried Lamersdorf, University of Hamburg Department of Informatics, Distributed Systems and Information Systems (Section 13.4).

# Preface

In life, many important undertakings often begin with some degree of happenstance, appearing as both hesitant and uncertain as they start to take on a distinct form. JADE was such an undertaking, with its true nature only able to emerge through a period of growth and evolution. From this quiet genesis, JADE has now become a major open source software project with a worldwide scope and user base. It is arguably the most popular software agent technology platform available today, a fact that never ceases to delight and amaze the project's initiators given that it was neither planned nor anticipated.

This remarkable evolution is of course the result of myriad contributions from an inspired and self-motivated grass-roots open source community. Contributing groups and individual JADE users and developers are often silently masked behind an email address, but have nevertheless been able to participate and collaborate at many different levels with suggestions, ideas, contributions, and, in several cases, entire software modules, add-ons and new derivative projects. It is our hope that this book will assist with yet further spreading knowledge of JADE, helping to unify, inspire, motivate and grow this community.

We would like to make a very special thank you and acknowledgement to all of our international colleagues and friends who have contributed to JADE over the years. This book would not be possible without you. Naturally we must also reserve an additional and very warm thank you to everybody able to directly contribute material to this book.

# 1

# Introduction

Agent-Oriented Programming (AOP) is a relatively new software paradigm that brings concepts from the theories of artificial intelligence into the mainstream realm of distributed systems. AOP essentially models an application as a collection of components called agents that are characterized by, among other things, autonomy, proactivity and an ability to communicate. Being autonomous they can independently carry out complex, and often long-term, tasks. Being proactive they can take the initiative to perform a given task even without an explicit stimulus from a user. Being communicative they can interact with other entities to assist with achieving their own and others' goals. The architectural model of an agent-oriented application is intrinsically peer to peer, as any agent is able to initiate communication with any other agent or be the subject of an incoming communication at any time.

Agent technology has been the subject of extensive discussion and investigation within the scientific community for several years, but it is perhaps only recently that it has seen any significant degree of exploitation in commercial applications. Multi-agent systems are being used in an increasingly wide variety of applications, ranging from comparatively small systems for personal assistance to open, complex, mission-critical systems for industrial applications. Examples of industrial domains where multi-agent systems have been fruitfully employed include process control, system diagnostics, manufacturing, transportation logistics and network management.

When adopting an agent-oriented approach to solving a problem, there are a number of domain-independent issues that must always be solved, such as how to allow agents to communicate. Rather than expecting developers to develop this core infrastructure themselves, it is convenient to build multi-agent systems on top of an agent-oriented middleware that provides the domain-independent infrastructure, allowing the developers to focus on the production of the key business logic.

This book describes JADE (Java Agent DEvelopment framework), probably the most widespread agent-oriented middleware in use today. JADE is a completely distributed middleware system with a flexible infrastructure allowing easy extension with add-on modules. The framework facilitates the development of complete agent-based applications by means of a run-time environment implementing the life-cycle support features required by agents, the core logic of agents themselves, and a rich suite of graphical tools. As JADE is written completely in Java, it benefits from the huge set of language features and third-party libraries on offer, and thus offers a rich set of programming abstractions allowing developers to construct JADE multi-agent systems with relatively minimal expertise in agent theory. JADE was initially developed by the Research & Development department of Telecom Italia s.p.a., but is now a community project and distributed as open source under the LGPL licence.

**The JADE website is http://jade.tilab.com.** In addition, a book companion website from Wiley is available at http://wiley.com/go/bellifemine_jade.

---

*Developing Multi-Agent Systems with JADE* Fabio Bellifemine, Giovanni Caire, Dominic Greenwood

The book's intended audience is primarily application developers with goals to both provide a comprehensive explanation of the features provided by JADE and also to serve as a handbook for programmers. Many of the features discussed are supported with exemplary code, either ad hoc or positioned with the context of a pervasive 'book trading' example. All code snippets and application examples provided in the book relate to version 3.4.1 of JADE released on November 2006; although we expect this material will remain consistent with future versions and is, in the most part, backward compatible with previous versions of JADE.

The book is structured as follows: Chapter 2 presents an overview of agent technology discussing the most relevant agent-related concepts, architectures and tools. Moreover, it provides a summary of the FIPA specifications that represent the most important standardization activity conducted in the field of agent technology. Chapter 3 presents the core architecture of JADE, its components and supported features. It also shows how to start the platform, launch agents and use the graphical administration tools. Chapters 4, 5 and 6 focus on the main Application Programming Interfaces (APIs) that JADE provides to access its features. Chapter 4 describes the basic features, Chapter 5 the advanced features and Chapter 6 is dedicated to agent mobility. Chapter 7 presents the internal architecture of the JADE kernel and explains how to modify and extend its behaviour. Chapter 8 addresses the development and deployment of JADE-based applications in the mobile and wireless environment. Chapters 9, 10 and 11 present additional configurations and tools that can be used to solve issues that must typically be taken into account when deploying real-world applications. Chapter 12 then discusses the JADE Semantic Framework, an important and recently released module that exploits the formal semantics of messages exchanged by agents. Finally, Chapter 13 gives an overview of other relevant tools that can be used with JADE.

# 2

# Agent Technology Overview

The first part of this chapter presents an overview of agent technology, including a brief discussion on agent architectures, programming languages, and tools. It also provides a set of relevant bibliographic references to scientific papers and applications of multi-agent systems.

The second part of the chapter describes the FIPA specifications, today the most widespread and accepted set of standards for multi-agent platforms and applications. JADE is compliant with the FIPA specifications; it has also in some ways extended the FIPA model in several areas, but in all aspects related to interoperability, the core purpose of FIPA, JADE fully respects the FIPA specifications.

## 2.1 ABOUT AGENTS

Agents are considered one of the most important paradigms that on the one hand may improve on current methods for conceptualizing, designing and implementing software systems, and on the other hand may be the solution to the legacy software integration problem.

### *2.1.1 WHAT IS AN AGENT?*

The term 'agent', or software agent, has found its way into a number of technologies and has been widely used, for example, in artificial intelligence, databases, operating systems and computer networks literature. Although there is no single definition of an agent (see, for example, Genesereth and Ketchpel, 1994; Wooldridge and Jennings, 1995; Russell and Norvig, 2003), all definitions agree that an agent is essentially a special software component that has autonomy that provides an interoperable interface to an arbitrary system and/or behaves like a human agent, working for some clients in pursuit of its own agenda. Even if an agent system can be based on a solitary agent working within an environment and if necessary interacting with its users, usually they consist of multiple agents. These multi-agent systems (MAS) can model complex systems and introduce the possibility of agents having common or conflicting goals. These agents may interact with each other both indirectly (by acting on the environment) or directly (via communication and negotiation). Agents may decide to cooperate for mutual benefit or may compete to serve their own interests.

Therefore, an agent is *autonomous*, because it operates without the direct intervention of humans or others and has control over its actions and internal state. An agent is *social*, because it cooperates with humans or other agents in order to achieve its tasks. An agent is *reactive*, because it perceives its environment and responds in a timely fashion to changes that occur in the environment. And

*Developing Multi-Agent Systems with JADE* Fabio Bellifemine, Giovanni Caire, Dominic Greenwood

an agent is *proactive*, because it does not simply act in response to its environment but is able to exhibit goal-directed behavior by taking initiative.

Moreover, if necessary an agent can be *mobile*, with the ability to travel between different nodes in a computer network. It can be *truthful*, providing the certainty that it will not deliberately communicate false information. It can be *benevolent*, always trying to perform what is asked of it. It can be *rational*, always acting in order to achieve its goals and never to prevent its goals being achieved, and it can *learn*, adapting itself to fit its environment and to the desires of its users.

### *2.1.2 ARCHITECTURES*

Agent architectures are the fundamental mechanisms underlying the autonomous components that support effective behaviour in real-world, dynamic and open environments. In fact, initial efforts in the field of agent-based computing focused on the development of intelligent agent architectures, and the early years established several lasting styles of architecture. These range from purely reactive (or behavioural) architectures that operate in a simple stimulus–response fashion, such as those based on the *subsumption architecture* of Brooks (1991) at one extreme, to more deliberative architectures that reason about their actions, such as those based on the belief desire intention (BDI) model (Rao and Georgeff, 1995), at the other extreme. In between the two lie hybrid combinations of both, or layered architectures, which attempt to involve both reaction and deliberation in an effort to adopt the best of each approach. Thus agent architectures can be divided into four main groups: logic based, reactive, BDI and layered architectures.

*Logic-based* (symbolic) architectures draw their foundation from traditional knowledge-based systems techniques in which an environment is symbolically represented and manipulated using reasoning mechanisms. The advantage of this approach is that human knowledge is symbolic so encoding is easier, and they can be constructed to be computationally complete, which makes it easier for humans to understand the logic. The disadvantages are that it is difficult to translate the real world into an accurate, adequate symbolic description, and that symbolic representation and manipulation can take considerable time to execute with results are often available too late to be useful.

*Reactive* architectures implement decision-making as a direct mapping of situation to action and are based on a stimulus–response mechanism triggered by sensor data. Unlike logic-based architectures, they do not have any central symbolic model and therefore do not utilize any complex symbolic reasoning. Probably the best-known reactive architecture is Brooks's subsumption architecture (Brooks, 1991). The key ideas on which Brooks realized this architecture are that an intelligent behaviour can be generated without explicit representations and abstract reasoning provided by symbolic artificial intelligence techniques and that intelligence is an emergent property of certain complex systems. The subsumption architecture defines layers of finite state machines that are connected to sensors that transmit real-time information (an example of subsumption architecture is shown in Figure 2.1). These layers form a hierarchy of behaviours in which the lowest

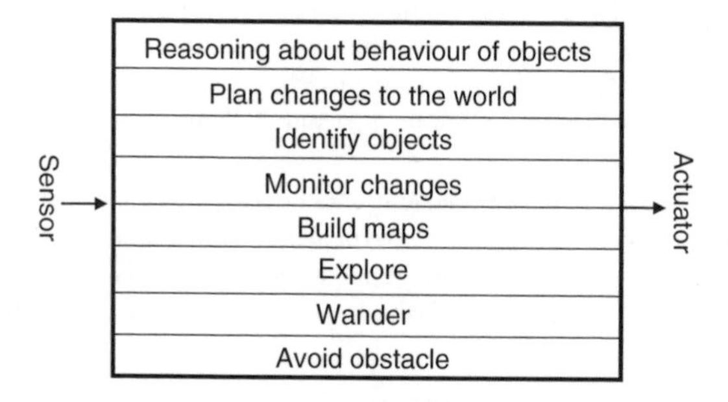


**Figure 2.1** A subsumption architecture for robot navigation

levels have less control than higher levels of the stack, thus decision-making is achieved through goal-directed behaviours. Subsumption-designed agents perceive conditions and act, but do not plan. The advantage of this approach is that it will perform better (i.e. respond faster but not reason better) in dynamic environments, as well as that they are often simpler in design than logic-based agents. However, the fact that reactive agents do not employ models of their environment results in some disadvantages. In fact, sensor data may not be sufficient to determine an appropriate action and the lack of agent state makes it almost impossible to design agents that learn from experience. Moreover, given that the agent acts on the basis of the interaction among the different behaviours, it is very hard to engineer reactive agents to fulfil specific tasks, in particular, when such agents must be realized through a large number of behaviours.

*BDI* (Belief, desire, intention) architectures are probably the most popular agent architectures (Rao and Georgeff, 1995). They have their roots in philosophy and offer a logical theory which defines the mental attitudes of belief, desire and intention using a modal logic. Many different agent-based systems have been realized that implement BDI (e.g. PRS (Georgeff and Lansky, 1987), JAM (Huber, 1999), JACK (Howden *et al.*, 2001), dMARS (d'Inverno *et al.*, 1998 and JADEX (see Section 13.4)) with a wide range of applications demonstrating the viability of the model. One of the most well-known BDI architectures is the Procedural Reasoning System (PRS) (Georgeff and Lansky, 1987). This architecture is based on four key data structures: beliefs, desires, intentions and plans, and an interpreter (see Figure 2.2).

In the PRS system, beliefs represent the information an agent has about its environment, which may be incomplete or incorrect. Desires represent the tasks allocated to the agent and so correspond to the objectives, or goals, it should accomplish. Intentions represent desires that the agent has committed to achieving. Finally, plans specify some courses of action that may be followed by an agent in order to achieve its intentions. These four data structures are managed by the agent interpreter which is responsible for updating beliefs from observations made of the environment, generating new desires (tasks) on the basis of new beliefs, and selecting from the set of currently active desires some subset to act as intentions. Finally, the interpreter must select an action to perform on the basis of the agent's current intentions and procedural knowledge.

*Layered* (hybrid) architectures allow both reactive and deliberative agent behaviour. To enable this flexibility, subsystems arranged as the layers of a hierarchy are utilized to accommodate both types of agent behaviour. There are two types of control flows within a layered architecture: horizontal (Ferguson, 1991) and vertical layering (Muller *et al.*, 1995). In horizontal layering, the layers are directly connected to the sensory input and action output (see Figure 2.3), which essentially has each layer acting like an agent. The main advantage of this is the simplicity of design since if the agent needs $n$ different types of behaviours, then the architecture only requires $n$ layers. However, since each layer is in effect an agent, their actions could be inconsistent prompting the need for a mediator function to control the actions. Another complexity is the large number of possible interactions between horizontal layers–$m^n$ (where $m$ is the number of actions per layer). A vertical layer architecture eliminates some of these issues as the sensory input and action output are each

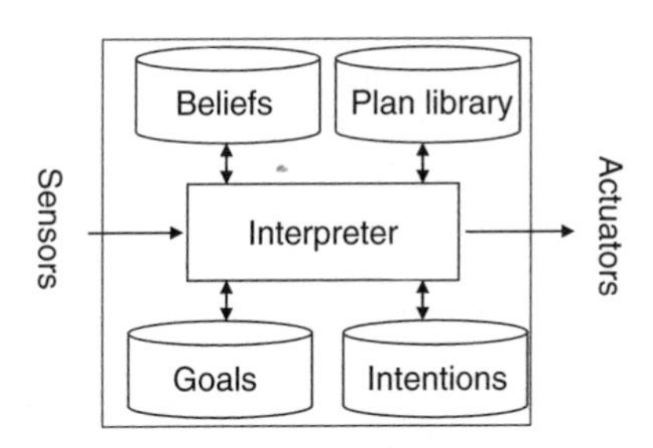


**Figure 2.2** The PRS agent architecture

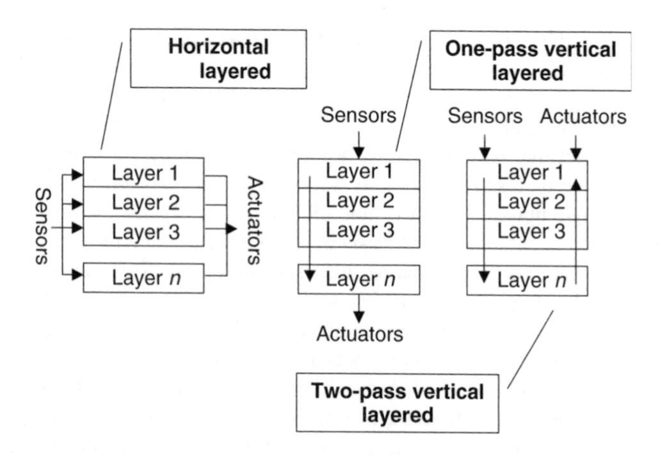


**Figure 2.3** Data and control flows in the layered architectures

dealt with by at most one layer each (creating no inconsistent action suggestions). The vertical layered architecture can be subdivided into one-pass and two-pass control architectures. In one-pass architectures, control flows from the initial layer that gets data from sensors to the final layer that generates action output (see Figure 2.3). In two-pass architectures, data flows up the sequence of layers and control then flows back down (again, see Figure 2.3). The main advantage of vertical layered architecture is the interaction between layers is reduced significantly to $m2(n-1)$. The main disadvantage is that the architecture depends on all layers and is not fault tolerant, so if one layer fails, the entire system fails.

### *2.1.3 COMMUNICATION AND COORDINATION*

One of the key components of multi-agent systems is communication. In fact, agents need to be able to communicate with users, with system resources, and with each other if they need to cooperate, collaborate, negotiate and so on. In particular, agents interact with each other by using some special communication languages, called agent communication languages, that rely on speech act theory (Searle, 1969) and that provide a separation between the communicative acts and the content language. The first agent communication language with a broad uptake was KQML (Mayfield *et al.*, 1996). KQML was developed in the early 1990s as part of the US government's ARPA Knowledge Sharing Effort. It is a language and protocol for exchanging information and knowledge that defines a number of performative verbs and allows message content to be represented in a first-order logic-like language called KIF (Genesereth and Ketchpel, 1994).

Currently the most used and studied agent communication language is the FIPA ACL (see Section 2.2.2), which incorporates many aspects of KQML (Labrou *et al.*, 1999). The primary features of FIPA ACL are the possibility of using different content languages and the management of conversations through predefined interaction protocols. Coordination is a process in which agents engage to help ensure that a community of individual agents acts in a coherent manner (Nwana *et al.*, 1996). There are several reasons why multiple agents need to be coordinated including: (1) agents' goals may cause conflicts among agents' actions, (2) agents' goals may be interdependent, (3) agents may have different capabilities and different knowledge, and (4) agents' goals may be more rapidly achieved if different agents work on each of them. Coordination among agents can be handled with a variety of approaches including organizational structuring, contracting, multi-agent planning and negotiation.

*Organizational structuring* provides a framework for activity and interaction through the definition of roles, communication paths and authority relationships (Durfee, 1999). The easiest way of ensuring coherent behaviour and resolving conflicts seems to consist of providing the group with an agent which has a wider perspective of the system, thereby exploiting an organizational or hierarchical structure. This is the simplest coordination technique and yields a classic master/slave or client/server architecture for task and resource allocation among slave agents by a master agent. The master controller can gather information from the agents in the group, create plans and assign tasks to individual agents in order to ensure global coherence. However, such an approach is impractical in realistic applications because it is very difficult to create such a central controller, and in any case, centralized control, as in the master/slave technique, is contrary to the decentralized nature of multi-agent systems.

An important coordination technique for task and resource allocation among agents and determining organizational structure is the *contract net protocol* (Smith and Davis, 1980). This approach is based on a decentralized market structure where agents can take on two roles, a manager and contractor. The basic premise of this form of coordination is that if an agent cannot solve an assigned problem using local resources/expertise, it will decompose the problem into sub-problems and try to find other willing agents with the necessary resources/expertise to solve these sub-problems. The problem of assigning the sub-problems is solved by a contracting mechanism consisting of: (1) contract announcement by the manager agent, (2) submission of bids by contracting agents in response to the announcement, and (3) the evaluation of the submitted bids by the contractor, which leads to awarding a sub-problem contract to the contractor(s) with the most appropriate bids (see Figure 2.4).

Another approach is to view the problem of coordinating agents as a planning problem. In order to avoid inconsistent or conflicting actions and interactions, agents can build a multi-agent plan that details all the future actions and interactions required to achieve their goals and interleave execution with additional planning and replanning. Multi-agent planning can be either centralized or distributed. In *centralized multi-agent planning*, there is usually a coordinating agent that, on receipt of all partial or local plans from individual agents, analyses them to identify potential inconsistencies and conflicting interactions (e.g. conflicts between agents over limited resources). The coordinating agent then attempts to modify these partial plans and combines them into a multi-agent plan where conflicting interactions are eliminated (Georgeff, 1983). In *distributed multi-agent planning*, the idea is to provide each agent with a model of other agents' plans. Agents communicate in order to build and update their individual plans and their models of other agents until all conflicts are removed (Georgeff, 1984).

Partial global planning integrates the strengths of the organizational, planning, and contracting approaches by uniting them into a single approach (Durfee and Victor, 1987). The goal of this approach is to gain the multi-agent planning benefits of detailed, situation-specific coordination

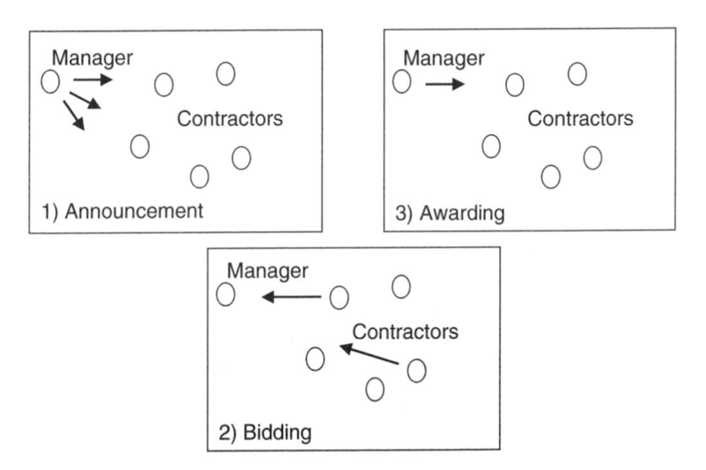


**Figure 2.4** Phases of the contract net protocol

while avoiding excessive computation and communication costs. This is possible because the jointly known organizational structures effectively prune the space of possible plans to keep the problem tractable. In addition, partial global planning views contracts as jointly held plans that specify future exchanges of tasks and results among agents. Hence, within partial global planning, coordination involves both sharing tasks and sharing results; both adhering to long-term organizational roles and reactively planning to achieve short-term objectives.

Negotiation is probably the most relied upon technique for coordinating agents. In particular, negotiation is the communication process of a group of agents in order to reach a mutually accepted agreement on some matter (Bussmann and Muller, 1992). Negotiation can be competitive or cooperative depending on the behaviour of the agents involved. *Competitive negotiation* is used in situations where agents have independent goals that interact with each other; they are not a priori cooperative, share information or willing to back down for the greater good. *Cooperative negotiation* is used in situations where agents have a common goal to achieve or a single task to execute. In this case, the multi-agent system has been centrally designed to pursue a single global goal.

### *2.1.4 PROGRAMMING LANGUAGES AND TOOLS*

Multi-agent systems programming languages, platforms and development tools are important components that can affect the diffusion and use of agent technologies across different application domains. In fact, the success of multi-agent systems is largely dependent on the availability of appropriate technology (i.e. programming languages, software libraries and development tools) that allows relatively straightforward implementation of the concepts and techniques that form the basis of multi-agent systems.

Multi-agent systems can be realized by using any kind of programming language. In particular, object-oriented languages are considered a suitable means because the concept of *agent* is not too distant from the concept of *object*. In fact, agents share many properties with objects such as encapsulation, and frequently, inheritance and message passing. However, agents also differ from objects in several key ways; they are autonomous (i.e. they decide for themselves whether or not to perform an action on request from another agent); they are capable of a flexible behaviour; and each agent of a system has its own thread of control.

Agent-oriented programming languages are a new class of programming languages that focus on taking into account the main characteristics of multi-agent systems. Minimally, an agent-oriented programming language must include some structure corresponding to an agent, but many also provide mechanisms for supporting additional attributes of agency such as beliefs, goals, plans, roles and norms. Today, several agent-oriented languages are available (Bordini *et al.*, 2006). Some are designed from scratch, directly encoding some theory of agency, while others extend existing languages to suit the peculiarities of the paradigm. Moreover, some take either a purely declarative or a purely imperative programming approach (FLUX (Thielscher, 2005) and JACK Agent Language (Winikoff, 2005) are respective examples). Other hybrid languages combine the declarative and imperative features of other languages (3APL (Dastani *et al.*, 2005) and Jason (Bordini *et al.*, 2005) are examples of this class). Moreover, some of these languages (e.g. 3APL and JACK Agent Language) provide an integrated development environment to simplify the realization of agent-based systems by automating some tedious coding tasks such as refactoring and testing.

Software platforms and frameworks are the other key means enabling the development of multi-agent systems. Most provide a means to deploy multi-agent systems on different types of hardware and operating systems, usually providing a middleware to support their execution and essential operations such as communication and coordination. Some of these platforms and frameworks have the common goal of providing FIPA-compliant functionalities to support interoperation between different multi-agent systems. Moreover, some also have the goal of supporting different kinds of hardware, communication networks and agent architectures (e.g. JADE (Bellifemine *et al.*, 2001)), and others of uniquely supporting special kinds of agents (e.g. mobile agents (Lange and Oshima, 1998)).

An important characteristic that multi-agent systems should provide is the capability to support interoperability among legacy software systems. Therefore, the availability of software tools for their integration with other common technologies can be a key to their success. The Internet is one the most important application domains and the most important communication means that multi-agent systems can use to provide interoperability among legacy software systems; therefore, a lot of current research and development work is oriented towards providing suitable techniques and software tools for the integration of multi-agent systems with Web technologies such as, for example, Web services and Semantic Web technologies.

Web services are a technology that is rapidly changing the Internet through the provision of a language-neutral, environment-neutral programming model fostering use of the Web for application integration both inside and outside an enterprise (Tsalgatidou and Pilioura, 2002; Weikum, 2001). Different important works have proposed integration between agent and Web services technologies as an ideal means of both supporting software interoperability and for providing more sophisticated services. In particular, agents have proven useful when directly acting as Web services, providing agent-based services for consumption by Web services, and dynamically coordinating the execution of a set of Web services by providing new services as a composition of other services (Buhler and Vidal, 2005; Greenwood *et al.*, 2005; Negri *et al.*, 2006) (see also Chapter 10 of this book).

The Semantic Web is an extension of the current Web in which information is given well-defined meaning, thereby better enabling computers and people to work in cooperation. In particular, the Semantic Web provides an infrastructure and a set of technologies that enable not just Web pages, but databases, services, programs, sensors, personal devices and even household appliances to both consume and produce data on the Web (Hendler *et al.*, 2002). Many different organizations are working in the realization of multi-agent systems where Semantic Web technologies are used to support agents both in the searching, filtering and manipulation of information and in the composition of processes (Labrou, 2002; Sycara *et al.*, 2003; Silva *et al.*, 2003).

### *2.1.5 APPLICATIONS OF MULTI-AGENT SYSTEMS*

Multi-agent systems are being used in an increasingly wide variety of applications, ranging from comparatively small systems for personal assistance to open, complex, mission critical systems for industrial applications (Jennings and Wooldridge, 1998).

Industrial applications are very important for multi-agent systems because they are where the first multi-agent system techniques were experimented with and demonstrated their initial potential. Examples of the application of multi-agent systems in industrial settings include process control (Jennings, 1994), system diagnostics (Albert *et al.*, 2003), manufacturing (Parunak, 1987), transportation logistics (Neagu *et al.*, 2006), and network management (Greenwood *et al.*, 2006).

One of the most important application fields of multi-agent systems is information management (Decker and Sycara, 1997). In particular, the Internet has been shown as an ideal domain for multi-agent systems due to its intrinsically distributed nature and the sheer volume of information available. Agents can be used, for example, for searching and filtering this mass of information (Klusch, 2001). The Internet has also pushed the use of agent technologies in the commerce and business process management fields. In fact, before the spread of Internet commerce, business process management was almost entirely driven by human interactions: humans deciding when to buy goods, how much they are willing to pay, and so on. Now electronic commerce and automated business processes are increasingly assuming a pivotal role in many organizations because they offer opportunities to significantly improve the way in which the different entities involved in the business process interact. In this scenario multi-agent systems have been shown to be both suitable for the modelling and the design of business process management systems (Camarinha-Matos and Afsarmanesh, 2001) and as key components for the automation of some or all the steps of these processes (Jennings *et al.*, 1996).

Traffic and transportation is also an important field, where the distributed nature of traffic and transport processes and the strong independence among the entities involved in such processes make multi-agent systems a valuable tool for realizing genuinely effective commercial solutions (Neagu *et al.*, 2006). Several domains have been addressed, with some such as OASIS (Ljungberg and Lucas, 1992) providing strong evidence that multi-agent systems are an ideal means for open, complex, mission-critical systems. OASIS is a sophisticated agent air traffic control system based on the BDI agent model, deployed and used with success at Sydney airport in Australia.

Telecommunication systems are another application field where multi-agent systems have been used with success. In fact, telecommunication systems are large, distributed networks of interconnected components which need to be monitored and managed in real time, and form the basis of a competitive market where telecommunication companies and service providers aim to distinguish themselves from their competitors by providing better, quicker or more reliable services. Therefore, multi-agent systems are used both for the management of such distributed networks and for the realization of advanced telecommunication services (Fricke *et al.*, 2001; Hayzelden and Bourne, 2001; Greenwood *et al.*, 2006).

Many multi-robotic systems also use multi-agent and distributed planning techniques for the coordination among the different robots. FIRE coordinates the actions of multiple robots at several layers of abstraction (Goldberg *et al.*, 2002). The top planning layer uses a market-based strategy to distribute tasks among robots, where robot travel time is the primary measure of cost. MISUS combines techniques from planning and scheduling with machine learning to perform autonomous scientific exploration with cooperating rovers (Estlin *et al.*, 2005). Distributed planning and scheduling techniques are used to generate efficient, multi-rover coordination plans, monitor plan execution and perform replanning when necessary. Machine learning clustering components are used to deduce geological relationships among collected data and select new science activities. Moreover, this system is able to reason about interdependent goals to perform plan optimization and to increase the value of collected data.

Other interesting multi-agent system applications can be found in health care (Moreno and Nealon, 2003). In fact, multi-agent systems have already been proposed to deal with many different kinds of problems in the health care domain, including patient scheduling and management, senior and community care, medical information access and management, and decision support. Several realized applications have shown that multi-agent systems can be the right solution for building medical decision support systems (Hudson and Cohen, 2002) and improve the coordination between the different professionals involved in the health care processes (Lanzola and Boley, 2002).

## 2.2 THE FOUNDATION FOR INTELLIGENT, PHYSICAL AGENTS (FIPA)

The complete FIPA specification set is listed at the end of the bibliography and it is publicly available on the FIPA website (FIPA). In this section we provide some history and context for FIPA and select a few of the specifications of particular relevance to JADE. Naturally, as JADE is largely an implementation of the FIPA specifications, it is highly dependent on the ideas generated through the specification process and expressed in the documents themselves. Because JADE has in some ways extended the FIPA model in several areas, the specifications do not provide complete coverage. However, the singular fact remains that in all aspects relating to interoperability, the core purpose of FIPA, JADE is compliant.

### *2.2.1 FIPA HISTORY AND GOALS*

FIPA was established in 1996 as an international non-profit association to develop a collection of standards relating to software agent technology. The initial membership, a collection of academic and industrial organizations, drew up a set of statutes guiding the production of a set of *de jure* standard specifications for software agent technologies. At that time software agents were already very well known in the academic community but have to date received only limited attention from commercial enterprises beyond an exploratory perspective. The consortium agreed to produce standards that would form the bedrock of a new industry by being usable across a vast number of applications.

At the core of FIPA is the following set of principles:

1. Agent technologies provide a new paradigm to solve old and new problems;
2. Some agent technologies have reached a considerable degree of maturity;
3. To be of use some agent technologies require standardization;
4. Standardization of generic technologies has been shown to be possible and to provide effective results by other standardization fora;
5. The standardization of the internal mechanics of agents themselves is not the primary concern, but rather the infrastructure and language required for open interoperation.

FIPA was initially set up with a five-year mandate to specify selected aspects of multi-agent systems; this mandate was indefinitely extended in 2001. Until the end of 2005 FIPA was governed by a member-elected Board of Directors responsible for providing strategic guidance and managing formal administrative duties. Decisions relating to the creation of technical groups to produce specifications and oversee the life cycle of in-development specifications were managed by the FIPA Architecture Board (FAB). Members of the FAB were elected by the Board of Directors. Technical work oriented toward the production of specifications was conducted in Technical Committees (TCs) that were created when a new proposal of work was accepted and dissolved when work was completed or abandoned. In addition Work Groups (WGs) were formed as fora to discuss technical issues and establish the groundwork required before forming a TC. Finally, Special Interest Groups (SIGs) were occasionally formed to discuss work related to FIPA that was not intended to lead to technical specification.

At its peak FIPA counted more than 60 members from more than 20 different countries worldwide among its members, and through an iterative process over several years developed a core set of specifications that went through three cycles of review: FIPA'97, FIPA'98 and FIPA2000. Accompanying the last of these iterations was the FIPA Abstract Architecture that abstracted the implementation agnostic principles expressed in the FIPA2000 specification set to produce a specification defining all core architectural elements and their relationships, guidelines for the specification of agent systems in terms of particular software and communications technologies and specifications governing the interoperability and conformance of agents and agent systems. This work and other selected specifications are discussed in Section 2.2.3.

A brief history of FIPA follows:

***1996***: FIPA established its first call for proposals used to seek different application areas of interest to the community and which would form the basis of a FIPA'97 specification set. Of the 12 responses received, four were selected by consensus: personal assistant, personal travel assistance, audio-visual entertainment broadcasting, and network provisioning and management. A list of the agent technologies needed to create these four applications was identified.

***1997***: FIPA'97, the first set of specifications, was identified as a collection of seven parts. The first three would be normative and specify the core middleware technologies of agent management, agent communication and agent/software interaction. The remaining four parts would be informative applications consisting of personal assistant, personal travel assistance, audio-visual entertainment, and broadcasting network provisioning and management.

From the communication perspective, FIPA decided to adopt ARCOL (Sadek, 1991) from France Télécom as the basis of an agent communication language, soon to become known as FIPA-ACL, or just ACL. The decision to adopt ARCOL emerged from an intense and controversial debate concerning the merits of ARCOL over KQML (Labrou *et al.*, 1999). Ultimately ARCOL won the debate because it is underpinned by formal semantics. Following this, FIPA also decided to adopt the SL language as an informative standard for the expression of message content and several cooperation protocols, also provided by France Télécom.

With regard to agent management, FIPA'97 defined the initial FIPA Agent Management Ontology including the concept of an agent platform consisting of an Agent Communication Channel (ACC), Agent Management System (AMS) and Directory Facilitator (DF). IIOP (1999) was selected as the baseline protocol for platform inter- and intra-operability.

***1998–9***: The core specifications were revised with extensions to the basic agent management mechanisms including mobility management and to the communication work including new interaction protocols and human–agent interaction extensions. In addition preliminary work was conducted on agent security management and an ontology service to host and serve domain ontologies.

In late 1998 plans were made for a FIPA-compliant agent software interoperability test for early 1999. Other than JADE several other early platforms were showcased at the event including FIPA-OS (Buckle *et al.*, 2002) and the ComTec Agent Platform (Suguri, 1998).

Also in early 1999 the delayed FIPA'98 specification set was released, consisting of many improvements on, and clarifications of, the FIPA'97 set. Two new TCs were formed, the first to develop the FIPA Abstract Architecture and the second to develop specifications for nomadic application support.

***2000–2***: FIPA issued a new mission statement: 'To promote technologies and interoperability specifications that facilitate the end-to-end inter networking of intelligent agent systems in modern commercial and industrial settings.' This is accompanied by a renewed focus on high-level semantic-based communications, interoperability between agents (rather than platforms), and agent agreements and interactions over extended periods of time.

The preliminary development of the FIPA Abstract Architecture was adopted as the new FIPA overall architectural model. It has well-defined abstractions that will not break as technology changes, contains mappings to commonly used technologies (e.g. CORBA, JINI), and offers support for alternate mechanisms including message transports, content encodings and explicit definition of implicitly used agent terms.

A new life-cycle model for standards was also adopted consisting of the three major phases: preliminary, experimental and standard. The experimental phase implies that the specification must undergo implementation-based proof of concept. The additional classifications of obsolete and deprecated were also introduced.

Two new TCs were formed. The first was to address gateway issues, support for mobile devices, and the collection of interaction protocols into a new library format. The second was to develop a semantic framework to address the relationship between external signals and internal states and account for communicative acts, interaction protocols, contracts, policies, service models and ontologies.

During 2000 and in early 2001 several of the mature specifications were promoted to experimental status and released at the FIPA2000 specification set. Platform implementations of these documents underwent fresh interoperability trials, termed a 'bake-off', including JADE and some of the other leading platforms available at the time – FIPA-OS (Buckle *et al.*, 2002) and Zeus (Nwana *et al.*, 1999).

In 2002 a special TC was created called 'X2S', mandated to harmonize all existing experimental specifications to prepare them for final promotion to standard status. In late 2002, 25 of the FIPA specifications were finally promoted to standard status, 56% of the entire specification set.

***2003–4***: With industrial support waning as standardization of the core specifications was accomplished, FIPA now focused on areas of ad hoc communication, semantics, security, services, modelling and methodologies. The latter of these attempt to clarify the definition of AUML (the Agent Unified Modelling Language) and develop a fragment library and method base, respectively.

In late 2004 the original FIPA organization was discontinued due to lack of continued support.

***2005–current***: Subsequent to a discussion phase and some renewed interest, FIPA was reincorporated in mid-2005 as a standards activity of the IEEE, called FIPA-IEEE. Several workgroups were set up to focus on the areas of agent and Web service interoperability, human–agent communication, mobile agents and peer-to-peer nomadic agents.

To date some of the key achievements of FIPA are as follows:

- A set of standard specifications supporting inter-agent communication and key middleware services.
- An abstract architecture providing an encompassing view across the entire FIPA2000 standards. This architecture underwent an incomplete reification as a Java Community Project known as the Java Agent Services (JAS) (JSR82).
- A well-specified and much-used agent communication language (FIPA-ACL), accompanied by a selection of content languages (e.g. FIPA-SL) and a set of key interaction protocols ranging from single message exchange to complex transactions.
- Several open source and commercial agent tool-kits with JADE generally considered as the leading FIPA-compliant open source technology available today.
- Several projects outside FIPA such as the completed Agentcities project that created a global network of FIPA-compliant platforms and agent application services.
- An agent-specific extension of UML, known as AUML or Agent.

### 2.2.2 *THE CORE CONCEPTS OF FIPA*

During the evolution of FIPA many agent-related ideas have been proposed. Some have reached fruition by promotion to standard status, several have been developed but remain incomplete, and others have fallen by the wayside for one reason or another. Of all these ideas, those of most central importance to the work are agent communication, agent management and agent architecture. This section discusses some of the key concepts associated with each of these areas.

#### 2.2.2.1 Agent Communication

Agents are fundamentally a form of distributed code processes and thus comply with the classic notion of a distributed computing model comprising two parts: components and connectors. Components are consumers, producers and mediators of communication messages exchanged via connectors. Early standards bodies such as the ISO and IETF took a network-oriented approach in developing the layered protocol stacks that underlie the majority of computer communication we know today – the OSI Reference Model and the TCP/IP Model. Both are utilized through interfaces onto software services that implement the protocols.

During the 1990s these network-oriented models were supplemented with several service-oriented model organizations such as the OMG, DCE, W3C, GGF and FIPA. A service-oriented model is essentially a communication protocol stack with multiple sub-layer application protocols instead of a single layer application protocol. The FIPA model is described shortly, but first we give some additional context.

The FIPA-ACL is grounded in speech act theory which states that messages represent actions, or communicative acts – also known as speech acts or performatives. A simple example would be 'My name is John' which, when issued, informs the recipient with an item of information. The FIPA-ACL set of 22 communicative acts was based on the ARCOL proposal of France Télécom where every act is described using both a narrative form and a formal semantics based on modal logic (Garson, 1984) that specifies the effects of sending the message on the mental attitudes of the sender and receiver agents. This form of logic is consistent with the BDI – or Belief, Desires, Intention reasoning model (Rao and Georgeff, 1995).

Some of the most commonly used acts are inform, request, agree, not understood, and refuse. These capture the essence of most forms of basic communication and are described further in

Section 2.2.3. It is stated in the FIPA standards that to be fully compliant an agent must be able to receive any FIPA-ACL communicative act message and at the very least respond with a *not-understood* message if the received message cannot be processed.

Based on these communicative acts, FIPA has defined a set of interaction protocols, each consisting of a sequence of communicative acts to coordinate multi-message actions, such as the contract net for establishing agreements and several types of auctions. A selection of these protocols is highlighted in Section 2.2.3.

Within the structure of each message itself, FIPA-ACL does not mandate the use of any particular language for expressing content, although specifications do exist for several representations including FIPA-SL, FIPA-KIF and FIPA-RDF. Only FIPA-SL has been promoted to standard status.

#### 2.2.2.2 FIPA Sub-Layers

As previously mentioned, the FIPA communication stack can be separated into several sub-layers within the application layer of the classical OSI or TCP/IP stack. These are detailed as follows:

- *Sub-layer 1 (Transport)*: In the FIPA-ACL layered protocol model, the lowest application sub-layer protocol is the transport protocol. FIPA has defined message transport protocols for IIOP (IIOP, 1999), WAP (WAP) and HTTP (HTTP).
- *Sub-layer 2 (Encoding)*: Rather than send simple byte-encoded messages, FIPA defines several message representations for using higher-level data structures including XML, String and Bit-Efficient. The latter is intended for use when communicating over low-bandwidth connections.
- *Sub-layer 3 (Messaging)*: In FIPA, message structure is specified independent of the particular encoding to encourage flexibility. The important aspect at this level is that key parameters are necessary in addition to the payload or content to be exchanged, e.g. the sender and receiver, the message type (communicative act), time-outs for replies. An example of FIPA-ACL message structure is given in Section 2.2.3.
- *Sub-layer 4 (Ontology)*: The individual terms contained in the payload or content of a FIPA message can be explicitly referenced to an application-specific conceptual model or ontology. Although FIPA intrinsically allows for the use of ontologies when expressing message content, it does not specify any particular representation for ontologies or provide any domain-specific ontologies. It is, though, possible to reference Web-based ontologies if required.
- *Sub-layer 5 (Content expression)*: The actual content of FIPA messages can be of any form, but FIPA has defined guidelines for the use of general logical formulas and predicates, and algebraic operations for combining and selecting concepts. The language most often used for expressing content is FIPA-SL, with examples of logic formulae including: *not*, *or*, *implies*, *equiv*, etc., and examples of algebraic operators including *any* and *all*.
- *Sub-layer 6 (Communicative act)*: The simple classification of a message in terms of the action, or performative, that it implies. Examples include *inform*, *request* and *agree*.
- *Sub-layer 7 (Interaction protocol or IP)*: Typically messages are rarely exchanged in isolation but rather form part of some interaction sequence. FIPA defines several interaction protocols specifying typical message exchange sequences such as *request* (described in Section 2.2.3), which describes one party making a request of another which in turn should agree or refuse to comply.

#### 2.2.2.3 Agent Management

In addition to communication, the second fundamental aspect of agent systems addressed by the early FIPA specifications is agent management: a normative framework within which FIPA-compliant agents can exist, operate and be managed. It establishes the logical reference model for the creation, registration, location, communication, migration and operation of agents. The agent management reference model consists of the components depicted in Figure 2.5.

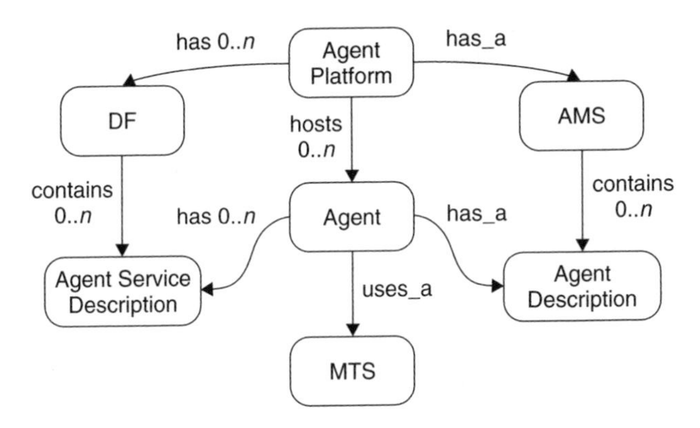


**Figure 2.5** Depiction of the agent management ontology

***Agent Platform (AP)***: This provides the physical infrastructure in which agents are deployed. The AP consists of the machines, operating systems, FIPA agent management components (described below), the agents themselves and any additional support software. The specific internal design of an AP is left to the developers of an agent system and is not a subject of FIPA standardization beyond the components discussed here. As a single AP may be spread across multiple computers, resident agents do not have to be co-located on the same host.

***Agent***: An agent is a computational process that inhabits an AP and typically offers one or more computational services that can be published as a service description. The particular design of these services, otherwise known as capabilities, is not the concern of FIPA, which only mandates the structure and encoding of messages used to exchange information between agents (and other third-party technologies if FIPA compliant). An agent must have at least one owner and must support at least one notion of identity which can be described using the FIPA Agent Identifier (AID) that labels an agent so that it may be distinguished unambiguously. An agent may be registered at a number of transport addresses at which it can be contacted.

***Directory Facilitator (DF)***: The DF is an optional component of an AP providing yellow pages services to other agents. It maintains an accurate, complete and timely list of agents and must provide the most current information about agents in its directory on a non-discriminatory basis to all authorized agents. An AP may support any number of DFs which may register with one another to form federations.

Every agent that wishes to publicize its services to other agents should find an appropriate DF and request the *registration* of its agent description. There is no intended future commitment or obligation on the part of the registering agent implied in the act of registering. Agents can subsequently request the *deregistration* of a description at which time there is no longer a commitment on behalf of the DF to broker information relating to that agent. At any time, and for any reason, an agent may request the DF to *modify* its agent description. In addition, an agent may issue a *search* request to a DF to discover descriptions matching supplied search criteria. The DF does not guarantee the validity of the information provided in response to a search request. However, the DF may restrict access to information in its directory and will verify all access permissions for agents which attempt to inform it of agent state changes.

***Agent Management System (AMS)***: The AMS is a mandatory component of an AP and is responsible for managing the operation of an AP, such as the creation and deletion of agents, and overseeing the migration of agents to and from the AP. Each agent must register with an AMS in order to obtain an AID which is then retained by the AMS as a directory of all agents present within the AP and their current state (e.g. active, suspended or waiting). Agent descriptions can be later modified

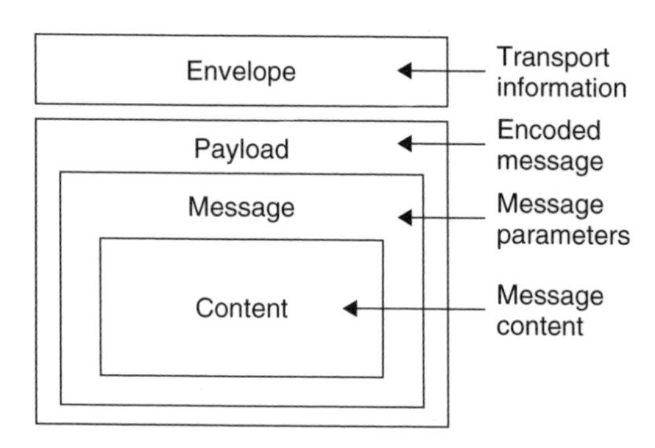


**Figure 2.6** FIPA message structure

under restriction of authorization by the AMS. The life of an agent with an AP terminates with its *deregistration* from the AMS. After deregistration, the AID of that agent can be removed by the directory and can be made available to other agents who should request it. Agent descriptions can also be *searched* for within the AMS, and the AMS is the custodian of the AP description that can be retrieved by requesting the action *get-description*.

The AMS can request that an agent performs a specific management function, such as to terminate its execution, and has the authority to enforce the operation if the request is ignored. Only a single AMS can exist in each AP and if the AP spans multiple machines, the AMS is the authority across all those machines.

***Message Transport Service (MTS)***: The MTS is a service provided by an AP to transport FIPA-ACL messages between agents on any given AP and between agents on different APs. Messages are providing a transport envelope that comprises the set of parameters detailing, for example, to whom the message is to be sent. The general structure of a FIPA-compliant message is depicted in Figure 2.6.

#### 2.2.2.4 Abstract Architecture

In addition to agent communication and agent management, between 2000 and 2002 an abstract agent architecture was created and standardized as a means to circumvent the impact on platform implementations of incremental revision to core specifications. This was achieved by abstracting the key aspects of the most critical mechanisms such as message transport and directory services into a unified specification. The overall goal of the approach is to permit the creation of systems that seamlessly integrate within their specific computing environment while interoperating with agent systems residing in separate environments.

Agent systems built according to the earlier FIPA'97 and '98 specifications can interoperate with agent systems built according to the Abstract Architecture through transport gateways with some limitations. The FIPA2000 architecture is a closer match and allows full interoperability via gateways. Because the Abstract Architecture permits the creation of multiple concrete realizations, it also provides mechanisms to permit them to interoperate including gateway transformations for both transport and encodings.

The Architecture specifies the ACL message structure, message transport, agent directory services and service directory services as mandatory. Other aspects are considered optional.

### *2.2.3 A SELECTION OF KEY FIPA SPECIFICATIONS*

To date FIPA has produced 25 standard specifications, with a further 14 remaining at the experimental stage, and 3 at the preliminary stage. The complete list of specifications is provided at the end of the bibliography. In this section we highlight a selection of the key standard specifications in a logical, rather than numerical, order.

#### 2.2.3.1 FIPA Abstract Architecture Specification (SC00001)

As discussed previously, the FIPA Abstract Architecture exists to provide a common, unchanging point of reference for FIPA-compliant implementations that captures the most critical and salient features of agent systems while providing some isolation from small iterative changes that may impact on the underlying FIPA2000 specification set.

The Architecture defines at an abstract level how two agents can locate and communicate with each other by registering themselves and exchanging messages. Agents communicate by exchanging messages which represent speech acts and which are encoded in an agent communication language. Services provide support services for agents and include the standard services of agent directory services, message transport services and service directory services. The Architecture is explicitly neutral about how services are presented, stating that they may be implemented either as agents or as software that is accessed via method invocation using programming. An agent providing a service is more constrained in its behaviour than a general-purpose agent. In particular, these agents are required to preserve the semantics of the service. This implies that these agents do not have the degree of autonomy normally attributed to agents. They may not arbitrarily refuse to provide the service.

Some of the most important items specified by the FIPA Abstract Architecture are:

- *Agent messages* are the fundamental form of communication between agents. The structure of a message is a set of key values written in FIPA-ACL. The content of the message is expressed in a content language, such as FIPA-SL or FIPA-KIF, and content expressions can be grounded by referenced ontologies. Messages can recursively contain other messages within their content and must contain key parameters such as the sender and receiver names. Prior to transmission messages must be encoded into a payload and a message transport envelope for the particular protocol in use.
- A *message transport service* is defined as the means to send and receive transport messages between agents. This is considered a mandatory element of FIPA agent systems.
- An *agent directory service* is defined as a shared information repository in which agents may publish their agent directory entries and in which they may search for agent directory entries of interest. This is considered a mandatory element of FIPA agent systems.
- A *service directory service* is defined as a shared repository in which agents and services can discover services. Services include, for example, message transport services, agent directory services, gateway services and message buffering. A service directory service can also be used to store the service descriptions of application-oriented services, such as commercial and business-oriented services. This is considered a mandatory element of FIPA agent systems.

#### 2.2.3.2 FIPA-ACL Message Structure Specification (SC00061)

A FIPA-ACL message contains a set of one or more message parameters. Precisely which parameters are needed for effective agent communication will vary according to the situation; the only parameter that is mandatory in all ACL messages is the performative, although it is expected that most ACL messages will also contain sender, receiver and content parameters. The FIPA-ACL message parameters are shown in Table 2.1 without regard to specific encodings. FIPA defines three specific encodings: String (EBNF notation), XML and Bit-Efficient.

User-defined message parameters may also be included by preceding the parameter name with the string ‘X-’.

**Table 2.1** ACL message parameters

| Parameter | Description |
|---|---|
| `Performative` | Type of the communicative act of the message |
| `sender` | Identity of the sender of the message |
| `receiver` | Identity of the intended recipients of the message |
| `reply-to` | Which agent to direct subsequent messages to within a conversation thread |
| `content` | Content of the message |
| `language` | Language in which the content parameter is expressed |
| `encoding` | Specific encoding of the message content |
| `ontology` | Reference to an ontology to give meaning to symbols in the message content |
| `protocol` | Interaction protocol used to structure a conversation |
| `conversation-id` | Unique identity of a conversation thread |
| `reply-with` | An expression to be used by a responding agent to identify the message |
| `in-reply-to` | Reference to an earlier action to which the message is a reply |
| `reply-by` | A time/date indicating by when a reply should be received |

This is a simple example of a FIPA-ACL message with a *request* performative:

```
(request
    :sender (agent-identifier :name alice@mydomain.com)
    :receiver (agent-identifier :name bob@yourdomain.com)
    :ontology travel-assistant
    :language FIPA-SL
    :protocol fipa-request
    :content
      ""((action
          (agent-identifier :name bob@yourdomain.com)
           (book-hotel :arrival 15/10/2006
                       :departure 05/07/2002 ... )
      ))""
)
```

#### 2.2.3.3 FIPA-ACL Communicative Act Library Specification (SC00037)

The FIPA-ACL defines communication in terms of a function or action, called the communicative act or CA, performed by the act of communicating. This standard provides a library of all the CAs specified by FIPA. The set of FIPA standardized CAs are listed in Table 2.2.

In general CAs are based on speech act theory (Searle, 1969) which defines the functions of simply specified actions. These functions are detailed in the FIPA CA Library specification (FIPA37), but examples include *interrogatives* which query for information, *exercitives* which ask for an action to be performed, *referentials* which share assertions about the world environment, *phatics* which establish, prolong or interrupt communication, *paralinguistics* which relate a message to other messages, and *expressives* which express attitudes, intentions or beliefs.

A message can perform several functions at the same time. For example the FIPA CA *Agree* is described as the action of agreeing to perform some action possibly in the future. This is phatic in terms of agreeing to proceed and is paralinguistic in terms of referring to another FIPA CA–hence, *Agree* is principally classified as a phatic function. All FIPA CAs inherently support the expressive function as a secondary function as they are defined in a modal logic form that

**Table 2.2** The FIPA communicative acts

| FIPA communicative act | Description |
|---|---|
| `Accept Proposal` | The action of accepting a previously submitted proposal to perform an action |
| `Agree` | The action of agreeing to perform some action, possibly in the future |
| `Cancel` | The action of one agent informing another agent that the first agent no longer has the intention that the second agent performs some action |
| `Call for Proposal` | The action of calling for proposals to perform a given action |
| `Confirm` | The sender informs the receiver that a given proposition is true, where the receiver is known to be uncertain about the proposition |
| `Disconfirm` | The sender informs the receiver that a given proposition is false, where the receiver is known to believe, or believe it likely that, the proposition is true |
| `Failure` | The action of telling another agent that an action was attempted but the attempt failed |
| `Inform` | The sender informs the receiver that a given proposition is true |
| `Inform If` | A macro action for the agent of the action to inform the recipient whether or not a proposition is true |
| `Inform Ref` | A macro action allowing the sender to inform the receiver of some object believed by the sender to correspond to a specific descriptor, for example a name |
| `Not Understood` | The sender of the act (for example, $i$) informs the receiver (for example, $j$) that it perceived that $j$ performed some action, but that $i$ did not understand what $j$ just did. A particular common case is that $i$ tells $j$ that $i$ did not understand the message that $j$ has just sent to $i$ |
| `Propagate` | The sender intends that the receiver treat the embedded message as sent directly to the receiver, and wants the receiver to identify the agents denoted by the given descriptor and send the received `propagate` message to them |
| `Propose` | The action of submitting a proposal to perform a certain action, given certain preconditions |
| `Proxy` | The sender wants the receiver to select target agents denoted by a given description and to send an embedded message to them |
| `Query If` | The action of asking another agent whether or not a given proposition is true |
| `Query Ref` | The action of asking another agent for the object referred to by a referential expression |
| `Refuse` | The action of refusing to perform a given action, and explaining the reason for the refusal |
| `Reject Proposal` | The action of rejecting a proposal to perform some action during a negotiation |
| `Request` | The sender requests the receiver to perform some action<br>One important class of uses of the request act is to request the receiver to perform another communicative act |
| `Request When` | The sender wants the receiver to perform some action when some given proposition becomes true |
| `Request Whenever` | The sender wants the receiver to perform some action as soon as some proposition becomes true and thereafter each time the proposition becomes true again |
| `Subscribe` | The act of requesting a persistent intention to notify the sender of the value of a reference, and to notify again whenever the object identified by the reference changes |

expresses attitudes, intentions and beliefs. Additionally, all FIPA CAs can refer to concepts from an explicit conceptualization defined by, for example, an ontology.

#### 2.2.3.4 FIPA-SL Content Language Specification (SC00008)

The FIPA Semantic Language (SL) is used to define the intentional semantics for the FIPA CAs as a logic of mental attitudes and actions, formalized in a first-order modal language with identity. SL is defined in terms of a string expression grammar, defined to be a sub-grammar of the more general s-expression syntax. Content expressions are defined in terms of *action expressions* or *propositions*. These in turn are represented as *well-formed formulas* (wff) consisting of terms (constant, set, sequence, functional term, action expression) and constants (numerical constants, string, datetime). A well-formed formula is constructed from an atomic formula by applying construction operators or logical connective operators. Examples of these include *negation*, *conjunction*, *disjunction*, *implication*, *equivalence*, *universal quantifier*, *belief*, *intention*, *done*, *iota*, *any* and *all*. Details of all these and others are provided in the specification document.

FIPA defines three subsets of SL (SL0, SL1 and SL2), differing in terms of which operators are supported. FIPA SL1 extends the minimal representational form of FIPA SL0 by adding Boolean connectives to represent propositional expressions, such as *not*, *and*, *or*. FIPA SL2 extends SL1 by adding construction, logic modal operators and the action operator feasible. SL2 allows first-order predicate and modal logic but is restricted to ensure that it must be decidable. Different CAs require the use of different SL subsets, e.g. *queries* requires the use of a referential operator to assign a value for the results as defined in SL2, whereas *requests* requires use of the action operator 'done' defined in SL0.

A FIPA-SL content expression may be used as the content of an ACL message. There are three cases:

1. A proposition, which may be assigned a truth value in a given context. A proposition is used in the *inform* CA and other CAs derived from it.
2. An action, which can be performed. An action may be a single action or a composite action built using the sequencing and alternative operators. An action is used as a content expression when the act is *request* and other CAs derived from it.
3. An identifying reference expression, which identifies an object in the domain. This is the referential operator and is used in the *inform-ref* macro act and other CAs derived from it.

There follows a simple example depicting an interaction between agents A and B that makes use of the *iota* operator, where agent A is supposed to have the following knowledge base KB= {P(A), Q(1, A), Q(1, B)}. The iota operator introduces a scope for the given expression (which denotes a term), in which the given identifier, which would otherwise be free, is defined. The expression `(iota x (P x))` may be read as 'the `x` such that `P` [is true] of `x`'. The iota operator is a constructor for terms which denote objects in the domain of discourse.

```
(query-ref
  :sender (agent-identifier :name B)
  :receiver (set (agent-identifier :name A))
  :content
    "((iota ?x (p ?x)))"
  :language fipa-sl
  :reply-with query1)

(inform
  :sender (agent-identifier :name A)
```

```
:receiver (set (agent-identifier :name B)
:content
  " ((= (iota ?x (p ?x)) alpha)) "
:language fipa-sl
:in-reply-to query1)
```

The only object that satisfies proposition P(x) is `alpha`; therefore, the query-ref message is replied by the inform message as shown.

#### 2.2.3.5 FIPA Request Interaction Protocol Specification (SC00026)

Depicted by the sequence diagram in Figure 2.7, the FIPA Request Interaction Protocol (IP) allows one agent, the Initiator, to *request* another, the Participant, to perform an action. The Participant processes the request and makes a decision whether to *accept* or *refuse* the request.

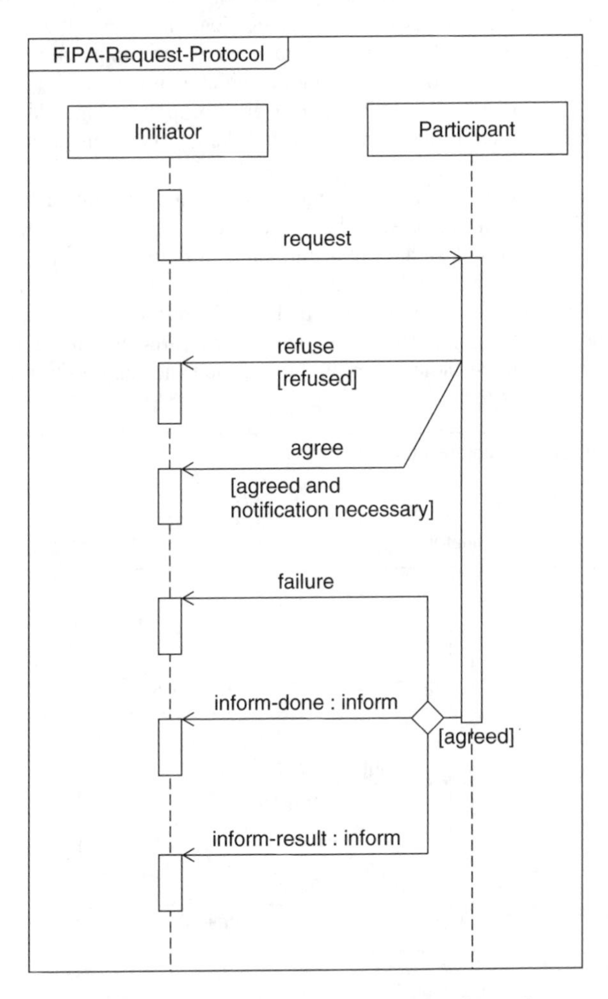


**Figure 2.7** The FIPA Request Interaction Protocol

If conditions indicate that an explicit agreement is required (that is, "notification necessary" is true), then the Participant communicates an *agree*. This agreement may be optional depending on circumstances, for example, if the requested action is very quick and can happen before a time specified in the *reply-by* parameter. Once the request has been agreed upon, then the Participant must communicate either:

- A *failure* if it fails in its attempt to fill the request,
- An *inform-done* if it successfully completes the request and only wishes to indicate that it is done, or
- An *inform-result* if it wishes to indicate both that it is done and notify the interaction Initiator of the results.

Any interaction using this interaction protocol is identified by a globally unique, non-null *conversation-id* parameter, assigned by the Initiator and set in the ACL message structure. The agents involved in the interaction must tag all of its ACL messages with this conversation identifier. This enables each agent to manage its communication strategies and activities; for example, it allows an agent to identify individual conversations and to reason across historical records of conversations.

At *any* point in the IP, the receiver of a communication can inform the sender that it did not understand what was communicated. This is accomplished by returning a *not-understood* message. Figure 2.7 does not depict a not-understood communication as it can occur at any point in the IP. The communication of a not-understood within an interaction protocol may terminate the entire IP and termination of the interaction may imply that any commitments made during the interaction are null and void.

In addition, at any point in the IP, the Initiator may cancel the interaction by initiating the Cancel Meta-Protocol shown in Figure 2.8. The conversation-id parameter of the cancel interaction is identical to the conversation-id parameter of the interaction that the Initiator intends to cancel. The semantics of cancel should roughly be interpreted as meaning that the Initiator is no longer interested in continuing the interaction and that it should be terminated in a manner acceptable to both the Initiator and the Participant. The Participant either informs the Initiator that the interaction is done using an inform-done or indicates the failure of the cancellation using a failure.

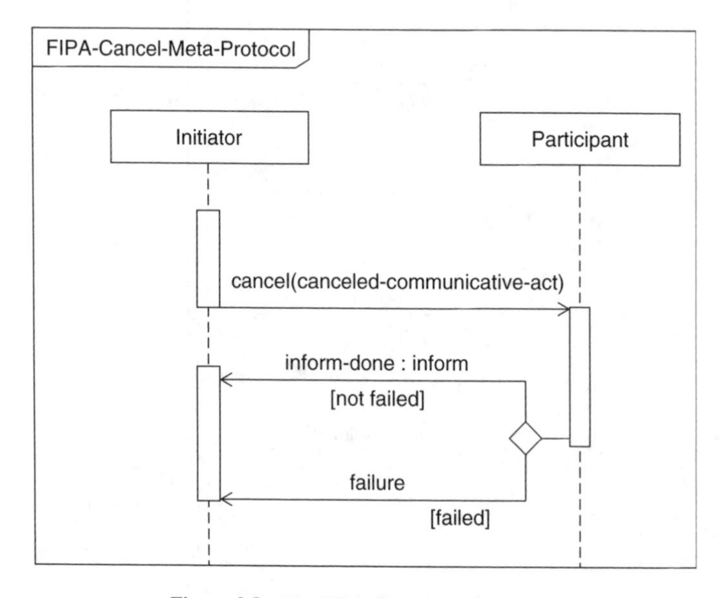


**Figure 2.8** The FIPA Cancel Meta-Protocol

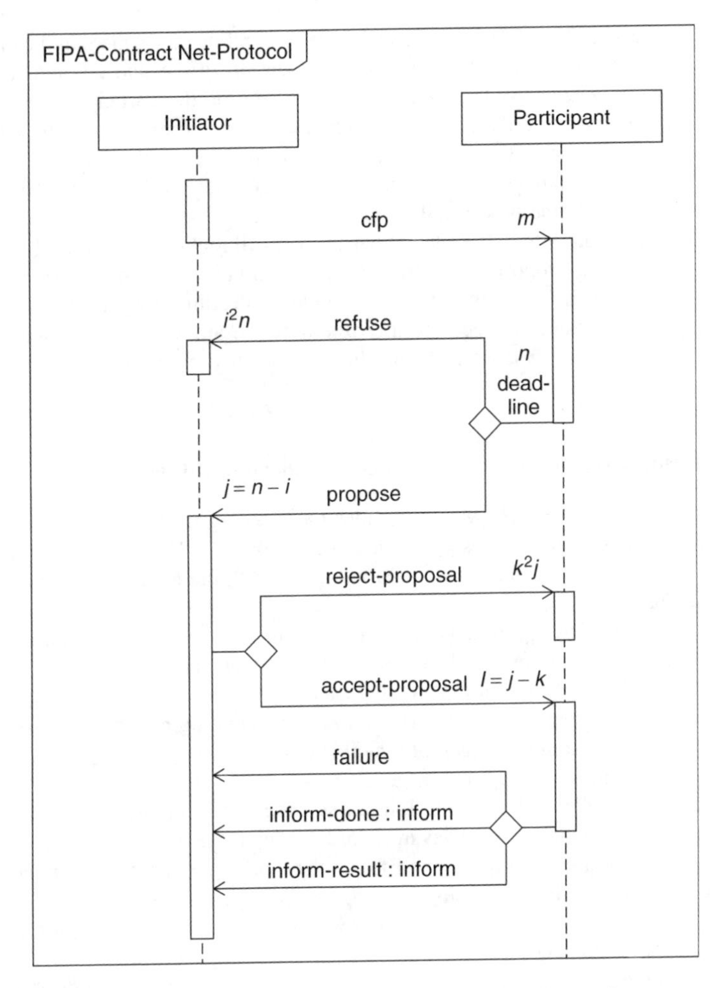


**Figure 2.9** The FIPA Contract Net Interaction Protocol

#### 2.2.3.6 FIPA Contract Net Interaction Protocol Specification (SC00026)

As an example of a slightly more complex interaction protocol, the FIPA Contract Net Interaction Protocol (IP) describes the case of one agent (the Initiator) that wishes to have some task performed by one or more other agents (the Participants) and further wishes to optimize a function that characterizes the task. This characteristic is commonly expressed as cost, but could also be soonest time to completion, fair distribution of tasks, etc. For a given task, any number of the Participants may respond with a proposal; the rest must refuse. Negotiations then continue with the Participants that proposed. The IP is depicted in Figure 2.9.

The Initiator solicits $m$ proposals from other agents by issuing a *call for proposals* (cfp) CA which specifies the task and any conditions the Initiator places upon the execution of the task. Participants receiving the call for proposals are viewed as potential contractors and are able to generate $n$ responses. Of these, $j$ are proposals to perform the task, specified as *propose* CAs.

The Participant's proposal includes the preconditions that the Participant is setting out for the task, which may be the price, time when the task will be done, etc. Alternatively, the $i = n - j$ Participants may *refuse* to propose. Once the deadline passes, the Initiator evaluates the received

$j$ proposals and selects agents to perform the task; one, several or no agents may be chosen. The $l$ agents of the selected proposal(s) will be sent an *accept-proposal* CA and the remaining $k$ agents will receive a *reject-proposal* CA. The proposals are binding on the Participant, so that once the Initiator accepts the proposal the Participant acquires a commitment to perform the task. Once the Participant has completed the task, it sends a completion message to the Initiator in the form of an inform-done or a more explanatory version in the form of an inform-result. However, if the Participant fails to complete the task, a failure message is sent.

This IP requires the Initiator to know when it has received all replies. In the case that a Participant fails to reply with either a propose or a refuse CA, the Initiator may potentially be left waiting indefinitely. To guard against this, the cfp CA includes a deadline by which replies should be received by the Initiator. Proposals received after the deadline are automatically rejected with the given reason that the proposal was late. The deadline is specified by the reply-by parameter in the ACL message.

#### 2.2.3.7 FIPA Agent Message Transport Service Specification (SC00067)

A FIPA Message Transport Service (MTS) provides the mechanism for the transfer of FIPA-ACL messages between agents using a Message Transport Protocol (MTP). The agents involved may be local to a single AP or on different APs. On any given AP, the MTS is provided by an Agent Communication Channel (ACC).

A message is made up of two parts: a message envelope for expressing transport information and a message payload comprising the encoded ACL message of the agent communication. Any MTP may use a different internal representation to describe a message envelope, but must express the same terms, represent the same semantics and perform the corresponding actions. A message envelope comprises a collection of parameters each of which is a name/value pair with the mandatory fields of `to`, `from`, `date` and `acl-representation`.

An ACC provides an MTS and is responsible for sending and receiving messages on an AP. An ACC must transfer the messages it receives in accordance with the transport instructions contained in the message envelope and is only required to read the message envelope; it is not required to parse the message payload. In performing message transfer tasks, the ACC may be required to obtain information from the AMS or DF on its own AP. Some implementations of ACCs may provide some form of buffering capability to help agents manage their messages.

Each ACC handling a message may add new information to the message envelope, but it may never overwrite existing information. ACCs can also add new parameters to a message envelope which override existing parameters that have the same parameter name; the mechanism for disambiguating message envelope entries is specified by each concrete message envelope syntax. The message forwarding behaviour of an ACC is determined by the instructions for message delivery that are expressed in the message envelope, the parameters of which are described in Table 2.3.

The recipients of a message are specified in the `to` parameter of a message envelope and take the form of AIDs. Depending upon the presence of `intended-receiver` parameter, the ACC forwards the message in one of the following ways:

- If an ACC receives a message envelope without an `intended-receiver` parameter, then it generates a new `intended-receiver` parameter from the `to` parameter (possibly containing multiple AIDs). It may also generate multiple copies of the message with different `intended-receiver` parameters if multiple receivers are specified. In all cases, the ACC is required to process all entries in the `to` field parameter and not to add and not to remove any AID that was contained in the original message. The `intended-receiver` parameters form a delivery path showing the route that a message has taken.
- If an ACC receives a message envelope with an `intended-receiver` parameter, this is used for delivery of this instance of the message and the `to` parameter is ignored.

**Table 2.3** FIPA message envelope parameters

| Parameter | Description |
|---|---|
| `to` | The names of the primary recipients of the message |
| `from` | The name of the agent that sent the message |
| `comments` | Text comments |
| `acl-representation` | The syntax representation of the message payload |
| `payload-length` | The length in bytes of the message payload |
| `payload-encoding` | The language encoding of the message payload |
| `date` | The creation date and time of the message envelope |
| `intended-receiver` | The name of the agents to whom this instance of a message is to be delivered |
| `received` | A stamp indicating the receipt of a message by an ACC |
| `transport-behaviour` | The transport requirements of the message |

- If an ACC receives a message envelope with more than one `intended-receiver` parameter, the most recent is used.

Before forwarding the message, the ACC adds a completed `received` parameter to the message envelope. Once an ACC has forwarded a message it no longer needs to keep any record of the existence of that message.

The AID given in the `to` or `intended-receiver` parameter (in the case of both parameters being present, the information in the `intended-receiver` parameter is used) of a message envelope may contain multiple transport addresses for a single receiving agent. The ACC uses the following method to try to deliver the message:

- Try to deliver the message to the first transport address in the `addresses` parameter; the first is chosen to reflect the fact that the transport address list in an AID is ordered by preference.
- If this fails, because the agent or AP was not available or because the ACC does not support the appropriate message transport protocol, etc., then the ACC creates a new `intended-receiver` parameter containing the AID with the failed transport address removed. The ACC then attempts to send the message to the next transport address in AID in the intended receiver list (now the first in the newly created `intended-receiver` parameter).
- If delivery is still unsuccessful when all transport addresses have been tried (or the AID contained no transport addresses), the ACC may try to resolve the AID using the name resolution services listed in the `resolvers` parameter of the AID. Again, the name resolution services should be tried in the order of their appearance.

Finally, if all previous message delivery attempts have failed, then an appropriate error message for the final failure is passed back to the sending agent.

An ACC uses the following rules in delivering messages to multiple intended receivers:

- If an ACC receives a message envelope with no `intended-receiver` parameter and a `to` parameter containing more than one AID, it may or may not split these up to form separate messages. Each message would contain a subset of the agents named in the `to` and `intended-receiver` parameters.
- If an ACC receives a message envelope with an `intended-receiver` parameter containing more than one AID, it may or may not split these up to form separate messages.
- If an ACC splits a message as described above, then it is enforced not to add and not to remove any AID that was contained in the original message.

The resulting messages are handled as in the single receiver case.

An agent has three options when sending a message to another agent resident on a remote AP:

1. Agent A sends the message to its local ACC using a proprietary or standard interface. The ACC then takes care of sending the message to the correct remote ACC using a suitable MTP. The remote ACC will eventually deliver the message.
2. Agent A sends the message directly to the ACC on the remote AP on which Agent B resides. This remote ACC then delivers the message to B. To use this method, Agent A must support access to one of the remote ACC's MTP interfaces.
3. Agent A sends the message directly to Agent B, by using a direct communication mechanism. The message transfer, addressing, buffering of messages and any error messages must be handled by the sending and receiving agents. This communication mode is not covered by FIPA.

Finally, the transport description forms part of an AP and is expressed in `fipa-sl0`. The following transport description is for an AP which supports IIOP and HTTP-based transports:

```
(ap-description

:name myAPDescription
:ap-services
 (set
  (ap-service
   :name myIIOPMTP
   :type fipa.mts.mtp.iiop.std
   :addresses
    (sequence
      corbaloc:iiop:agents.fipa.org:10100/acc
      IOR:00000000002233
      corbaname::agents.fipa.org:10000/nameserver#acc))
  (ap-service
   :name myHTTPMTP
   :type fipa.mts.mtp.http.std
   :addresses
    (sequence
      http://agents.fipa.org:8080/acc))) )
```

### *2.2.4 THE RELEVANCE OF FIPA TO JADE*

We recall that FIPA is based on the tenet that only the external behaviour of system components should be specified, leaving internal architecture and implementation details to the developers of individual platforms. This ensures seamless interoperation between fully compliant platforms. JADE complies with this notion by ensuring complete compatibility with the primary FIPA2000 specifications (communication, management and architecture) that provide the normative framework within which FIPA agents can exist, operate and communicate, while adopting a unique, proprietary internal architecture and implementation of key services and agents.

JADE is of course only one of several agent platforms, applications and collaborative projects that have been, and are continuing to be, compliant with the FIPA standards. This compliance has been tested on several occasions through such events as the FIPA interoperability tests conducted in 1999 and 2001 and the large-scale Agentcities project (Agentcities) that integrated several globally distributed FIPA-compliant platforms. With the continued growth of inter-organization application domains such as supply networks and inter-provider telecommunications, there is an increased likelihood that open interoperation will become an important factor.

In terms of its coverage of the FIPA standards, JADE implements the complete Agent Management specification including the key services of AMS, DF, MTS and ACC. Of course through use and experience these services have been extended with additional features, but the core compliance to FIPA has been maintained. JADE also implements the complete FIPA Agent Communication stack, ranging from the availability of FIPA-ACL for message structure, FIPA-SL for message content expression, plus support for many of the FIPA interaction and transport protocols.

An example of where JADE exceeds, but does not break, the FIPA standards, is the JADE transport mechanism which supports all specified operations, but is also capable of adapting a connection type by seamlessly selecting the best available protocol according to the particular usage situation.

Some of the ancillary aspects of the FIPA standards, such as some of the more esoteric interaction protocols and non-standardized work such as the ontology service, have yet to be developed for JADE even if programmers can find all the tools and abstractions needed to implement them. However, there are also many JADE components that go far beyond the FIPA specifications. For example, JADE provides a distributed, fault tolerant, container architecture, internal service architecture, persistent message delivery, semantic framework, security mechanisms, agent mobility support, web-service interaction, graphical interfaces, and much more. Many of these are described in the main body of this book, as is the critical aspect of agent-oriented systems that FIPA does not address – the internal architecture of agents themselves.

Due to its open source status, strong support from industry and a broad user community, JADE is now generally considered to be the leading FIPA-compliant open source agent framework.

# 3

# The JADE Platform

This chapter provides a basic overview of the JADE platform and the main components constituting its distributed architecture. Furthermore, it describes how to launch the platform with the various command-line options and how to experiment with the main graphical tools.

## 3.1 BRIEF HISTORY

The first software developments, that eventually became the JADE platform, were started by Telecom Italia (formerly CSELT) in late 1998, motivated by the need to validate the early FIPA specifications.

As is often the case, at the beginning we did not fully expect to achieve the goal of developing a platform. But, thanks to the partial financial support of the European Commission (FACTS project, ACTS AC317) and to the goodwill and capability of the team (at that time composed of Fabio Bellifemine, Agostino Poggi and Giovanni Rimassa), at a certain point it was decided to move beyond a means of simply validating the FIPA specifications towards developing a fully fledged middleware platform. The vision was to provide services to application developers and that were readily accessible and usable by both seasoned developers and newcomers with little or no knowledge of the FIPA specifications. Emphasis was placed on the simplicity and usability of the software APIs.

JADE went open source in 2000 and was distributed by Telecom Italia under the LGPL (Library Gnu Public Licence) licence. This licence assures all the basic rights to facilitate the usage of the software included in commercial products: the right to make copies of the software and to distribute those copies, the right to have access to the source code, and the right to change the code and make improvements to it. Unlike the GPL, the LGPL licence does not put any restriction on software that uses JADE, and it allows proprietary software to be merged with software covered by the LGPL. On the other hand, the licence also requires that any derivative work of JADE, or any work based on it, should be returned to the community under the same licence. Refer to (GPL-FAQ) and to the text of the licence for a comprehensive analysis of legal issues and implications.

JADE has a website, http://jade.tilab.com, from where the software, documentation, example code, and a wealth of information about usages of JADE are available. The project welcomes the participation of the open source community with a variety of means to become involved and contribute to the project; they are all detailed on the website, for example:

- Emailing jade-contrib@avalon.tilab.com with a public description of your industrial use case for JADE, research project or academic course that uses JADE, or any other public workshop or event that might be of interest to the JADE community.

*Developing Multi-Agent Systems with JADE* Fabio Bellifemine, Giovanni Caire, Dominic Greenwood

- Participating in the discussions on the JADE mailing lists by replying and giving support to less experienced users. Two mailing lists are available: jade-develop@avalon.tilab.com for discussing problems and ideas concerning the usage and the development of JADE, and jade-news@avalon.tilab.com used by the project administrators to notify the community about new software releases and JADE-related events.
- Providing bug reports and when possible, bug fixes.
- Contributing new add-ons and software modules for use by the community.

In order to better facilitate industrial involvement, in May 2003 Telecom Italia Lab and Motorola Inc. defined a collaboration agreement and formed the JADE Governing Board, a not-for-profit organization of companies committed to contributing to the development and promotion of JADE. The Board was formed as a contractual consortium with well-defined rules governing the rights and obligations toward generated IPR. The Board is open with members able to join and leave according to their needs. At the time of writing, Telecom Italia, Motorola, France Telecom R&D, Whitestein Technologies AG and Profactor GmbH have all become members of the Board.

When JADE was first made public by Telecom Italia, it was used almost exclusively by the FIPA community but as its feature set grew far beyond the FIPA specifications, so did its usage by a globally distributed developer community. It is interesting to note that JADE contributed to widespread diffusion of the FIPA specifications by providing a set of software abstractions and tools that hid the specifications themselves; programmers could essentially implement according to the specifications without the need to study them. We consider this as one of the main strengths of JADE with respect to FIPA.

One of the primary extensions of the JADE core was provided by LEAP [LEAP] (IST 1999-10211), a project partly financed by the European Commission that contributed significantly between 2000 and 2002 toward porting JADE to the Java Micro Edition and Wireless Network environment. This work was in large part led by Giovanni Caire. Today, the availability of a JADE run-time for J2ME-CLDC and -CDC platforms, and its use in addressing the problems and challenges posed by mobile telecommunications, is considered as one of JADE's leading distinguishing features.

## 3.2 JADE AND THE AGENTS PARADIGM

JADE is a software platform that provides basic middleware-layer functionalities which are independent of the specific application and which simplify the realization of distributed applications that exploit the software agent abstraction (Wooldridge and Jennings, 1995). A significant merit of JADE is that it implements this abstraction over a well-known object-oriented language, Java, providing a simple and friendly API. The following simple design choices were influenced by the agent abstraction.

***An Agent is Autonomous and Proactive***: An agent cannot provide call-backs or its own object reference to other agents in order to mitigate any chance of other entities coopting control of its services. An agent must have its own thread of execution, using it to control its life cycle and decide autonomously when to perform which actions.

***Agents Can Say 'No', and They are Loosely Coupled***: Message-based asynchronous communication is the basic form of communication between agents in JADE; an agent wishing to communicate must send a message to an identified destination (or set of destinations). There is no temporal dependency between the sender and receivers: a receiver might not be available when the sender issues the message. There is also no need to obtain the object reference of receiver agents but just name identities that the message transport system is able to resolve into proper transport addresses. It is even possible that a precise receiver identity be unknown to the sender, which instead may define a receiver set using an intentional grouping (e.g. all the agents that provide the 'Book-Selling' service) or mediated by a proxy agent (e.g. propagate this message to all agents in the domain 'selling.book.it').

Furthermore, this form of communication enables the receiver to select which messages to process and which to discard, as well as to define its own processing priority (e.g. read first all messages coming from the domain "book.it"). It also enables the sender to control its thread of execution and thus not be blocked until the receiver processes the message. Finally, it also provides an interesting advantage when implementing multi-cast communication as an atomic operation rather than as $N$ consecutive method calls (i.e. one send-type operation with a list of multiple message receivers instead of a method call for each remote object you wish to communicate with).

***The System is Peer-to-Peer***: Each agent is identified by a globally unique name (the AgentIdentifier, or AID, as defined by FIPA). It can join and leave a host platform at any time and can discover other agents through both white-page and yellow-page services (provided in JADE by AMS and the DF agents as defined also by the FIPA specifications). An agent can initiate communication with any other agent at any time it wishes and can equally be the object of an incoming communication at any time.

On the basis of these design choices, JADE was implemented to provide programmers with the following ready-to-use and easy-to-customize core functionalities.

- A *fully distributed* system inhabited by agents, each running as a separate thread, potentially on different remote machines, and capable of transparently communicating with one another, i.e. the platform provides a unique location-independent API that abstracts the underlying communication infrastructure.
- Full *compliance with the FIPA specifications*. The platform successfully participated in all FIPA interoperability events and was used as the middleware for many platforms in the Agentcities network (Agentcities). A great facilitator of this was active contribution by the JADE team to the FIPA standardization process.
- *Efficient transport of asynchronous messages* via a location-transparent API. The platform selects the best available means of communication and, when possible, avoids marshalling/unmarshalling java objects. When crossing platform boundaries, messages are automatically transformed from JADE's own internal Java representation into proper FIPA-compliant syntaxes, encodings and transport protocols.
- Implementations of both *white pages and yellow pages*. Federated systems can be implemented to represent domains and sub-domains as a graph of federated directories.
- A *simple, yet effective, agent life-cycle management*. When agents are created they are automatically assigned a globally unique identifier and a transport address which are used to register with their platform's white page service. Simple APIs and graphical tools are also provided to both locally and remotely manage agent life cycles, i.e. create, suspend, resume, freeze, thaw, migrate, clone and kill.
- Support for *agent mobility*. Both agent code and, under certain restrictions, agent state can migrate between processes and machines. Agent migration is made transparent to communicating agents that can continue to interact even during the migration process.
- A *subscription mechanism* for agents, and even external applications, that wish to subscribe with a platform to be notified of all platform events, including life-cycle-related events and message exchange events.
- A set of *graphical tools* to support programmers when debugging and monitoring. These are particularly important and complex in multi-threaded, multi-process, multi-machine systems such as a typical JADE application. As described in Section 3.7, conversations can be sniffed and emulated, and agent execution can be controlled remotely and introspected, including remote step-by-step debugging of agent execution.
- Support for *ontologies and content languages*. Ontology checking and content encoding is performed automatically by the platform with programmers able to select preferred content languages and ontologies (e.g. XML and RDF-based). Programmers can also implement new content languages to fulfil specific application requirements.

- A library of *interaction protocols* which model typical patterns of communication oriented toward the achievement of one or more goal. Application-independent skeletons are available as a set of Java classes that can be customized with application-specific code. Interaction Protocols can also be represented and implemented as a set of concurrent finite state machines.
- Integration with various *Web-based technologies* including JSP, servlets, applets and Web service technology. The platform can also be easily configured to cross firewalls and use NAT systems.
- Support for *J2ME platform and the wireless environment.* The JADE run-time is available for the J2ME-CDC and -CLDC platforms through a uniform set of APIs covering both J2ME and J2SE environments.
- An *in-process interface* for launching/controlling a platform and its distributed components from an external application.
- An *extensible kernel* designed to allow programmers to extend platform functionality through the addition of kernel-level distributed services. This mechanism is inspired by the aspect-oriented programming approach where different aspects can be woven into application code and coordinated at kernel level. To maintain compatibility with the J2ME environment which does not intrinsically support aspect-oriented code, JADE uses a special composition filter approach described in Chapter 7.

## 3.3 JADE ARCHITECTURE

Figure 3.1 shows the main architectural elements of a JADE platform. A JADE platform is composed of agent containers that can be distributed over the network. Agents live in containers which are the Java process that provides the JADE run-time and all the services needed for hosting and executing agents. There is a special container, called the *main container*, which represents the bootstrap point of a platform: it is the first container to be launched and all other containers must join to a main container by registering with it. The UML diagram in Figure 3.2 schematizes the relationships between the main architectural elements of JADE.

The programmer identifies containers by simply using a logical name; by default the main container is named 'Main Container' while the others are named 'Container-1', 'Container-2', etc. Command-line options are available to override default names.

As a bootstrap point, the main container has the following special responsibilities:

- Managing the container table (CT), which is the registry of the object references and transport addresses of all container nodes composing the platform;

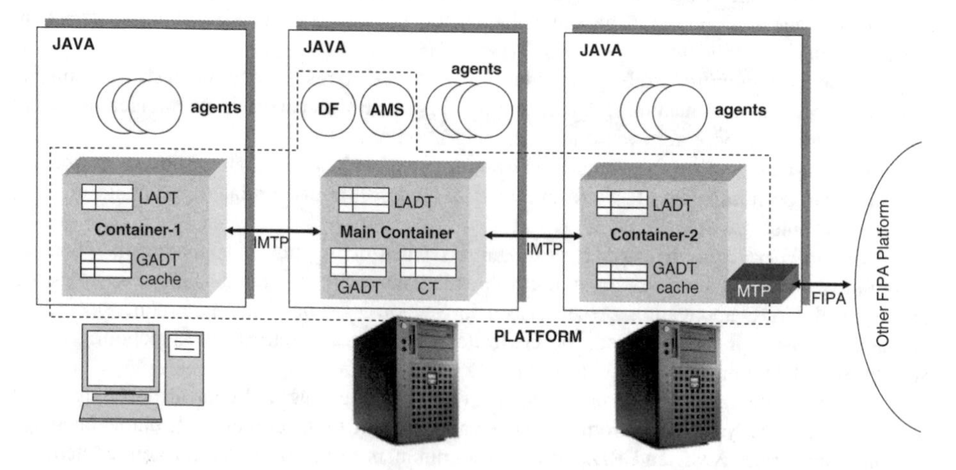


**Figure 3.1** Relationship between the main architectural elements

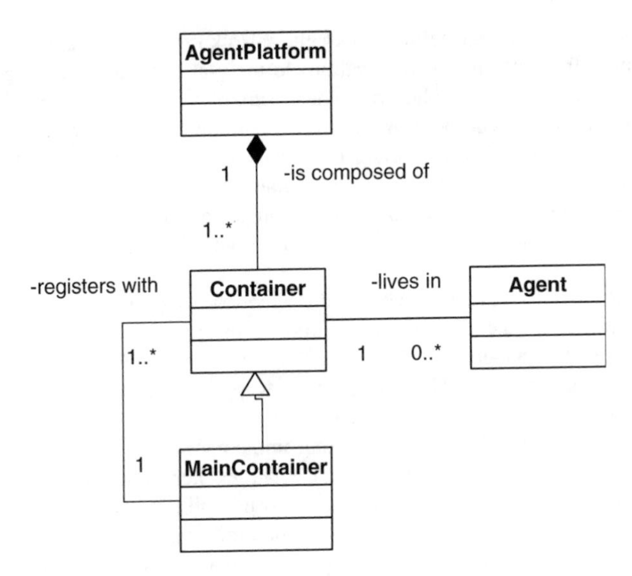


**Figure 3.2** Relationship between the main architectural elements

- Managing the global agent descriptor table (GADT), which is the registry of all agents present in the platform, including their current status and location;
- Hosting the AMS and the DF, the two special agents that provide the agent management and white page service, and the default yellow page service of the platform, respectively.

A common query is whether the main-container is a system bottleneck. In fact this is not the case as JADE provides a cache of the GADT that each container manages locally. Platform operations, in general, do not involve the main-container, but instead just the local cache and the two containers hosting the agents which are the subject and the object of the operation (e.g. sender and receiver of the message). When a container must discover where the recipient of a message lives, it first searches its LADT (local agent descriptor table) and then, only if the search fails, is the main-container contacted in order to obtain the proper remote reference which, consequently, is cached locally for future usages. Because the system is dynamic (agents can migrate, terminate, or new agents can appear), occasionally the system may use a stale cached value resulting in an invalid address. In this case, the container receives an exception and is forced to refresh its cache against the main-container. The cache replacement policy is LRU (least recently used), which was designed to optimize long conversations rather than sporadic, single message exchange conversations which are actually fairly uncommon in multi-agent applications.

However, although the main-container is not a bottleneck, it does remain a single point of failure within the platform. To manage this, Section 9.1 describes how to exploit the Main Replication Service to ensure the JADE platform remains fully operational even in the event of a main-container failure. With this service the administrator can control the level of fault tolerance of the platform, the level of scalability of the platform, and somewhat uniquely, the *level of distribution* of the platform. A control layer composed of several distributed instances of the main-container can be configured to implement a distributed bootstrapping system (i.e. several bootstrap points can be passed to each container) and a distributed control system. In the extreme case, each container can be instructed to join the Main Replication Service and act as part of the control layer.

Agent identity is contained within an Agent Identifier (AID), composed of a set of slots that comply with the structure and semantics defined by FIPA. The most basic elements of the AID are the agent name and its addresses. The name of an agent is a globally unique identifier that JADE constructs by concatenating a user-defined nickname (also known as a local name as it is sufficient for disambiguating intra-platform communication) to the platform name. The agent addresses are transport addresses inherited by the platform, where each platform address corresponds to an MTP (Message Transport Protocol) end point where FIPA-compliant messages can be sent and received. Agent programmers are also allowed to add their own transport addresses to the AID when, for any application-specific purpose, they wish to implement their own agent private MTP.

When the main-container is launched, two special agents are automatically instantiated and started by JADE, whose roles are defined by the FIPA Agent Management standard:

1. The Agent Management System (AMS) is the agent that supervises the entire platform. It is the contact point for all agents that need to interact in order to access the white pages of the platform as well as to manage their life cycle. Every agent is required to register with the AMS (automatically carried out by JADE at agent start-up) in order to obtain a valid AID. More details about the AMS are provided in Section 5.5.
2. The Directory Facilitator (DF) is the agent that implements the yellow pages service, used by any agent wishing to register its services or search for other available services. The JADE DF also accepts subscriptions from agents that wish to be notified whenever a service registration or modification is made that matches some specified criteria. Multiple DFs can be started concurrently in order to distribute the yellow pages service across several domains. These DFs can be federated, if required, by establishing cross-registrations with one another which allow the propagation of agent requests across the entire federation. Details relating to this are provided in Section 4.4.

## 3.4 COMPILING THE SOFTWARE AND LAUNCHING THE PLATFORM

All the software relating to JADE can be downloaded from the JADE website at http://jade.tilab.com. JADE-related software is divided into two sections: the main distribution and the add-ons. The add-ons in particular include self-contained modules that implement specific extended features such as codecs for given languages. In many cases these have not been developed by the JADE team directly, but by members of the open source community who decided to return their achievements to the community itself.

The main distribution is composed of five primary archive files with the following content:

- *jadeBin.zip* contains only the pre-compiled JADE Java archive (jar) files in a ready to use state.
- *jadeDoc.zip* contains the documentation, including the Administrator and Programmer guides. This documentation is also available online from the website.
- *jadeExamples.zip* contains the source code of various examples.
- *jadeSrc.zip* contains all the sources of JADE.
- *jadeAll.zip* contains all of the four files listed above.

If the above zip files are downloaded and unzipped, the directory structure should be as shown in Figure 3.3 (only the most relevant files and directories are actually shown).

Some of the important files/folders include:

- License, the open source licence that regulates all use of the software.
- The file `jade/doc/index.html` is a good starting point for beginners containing links to a variety of thematic tutorials, the Programmer and Administrator guide, javadoc documentation of all the sources, plus several other support documents.
- The `jade/lib` folder contains all the *.jar files which must be included in the Java CLASSPATH in order to run JADE. It includes the `lib/commons-codec` subdirectory where an external Base64 codec is distributed that should also be included in the Java CLASSPATH.

```
jade/
     |---License
     |---classes
     |---demo
     |---doc/
     |       |---index.html
     |---lib/
     |       |---http.jar
     |       |---iiop.jar
     |       |---jade.jar
     |       |---jadeTools.jar
     |       |---commons-codec/
     |       |                  |---commons-codec-1.3.jar
     |---src/
     |       |---demo
     |       |---examples
     |       |---FIPA
     |       |---jade
```

**Figure 3.3** JADE directory structure

- The jade/src directory contains four subdirectories. Demo contains the sources of a simple demo. Examples contains several useful source code examples of various agent fragments. FIPA contains the sources of a FIPA-defined module. Jade contains all the sources of JADE itself.

The JADE sources can be compiled using the ant tool (ANT). The most important ant targets are the following:

- jade – to compile the sources and create the. class files under the classes subdirectory;
- lib – to generate the Java archive jar files under the lib subdirectory;
- doc – to generate the javadoc documentation files under the doc subdirectory;
- examples – to compile all the examples.

Experienced users might find it useful to directly access the source code repository for which read-only access is available to the JADE community. The repository is maintained and kept up to date by an administrator; instructions on how to access it are available on the JADE website.

The lib directory contains the five archive files containing the classes needed by JADE:

- jade.jar contains all the JADE packages except add-ons, MTPs and graphical tools;
- jadeTools.jar contains all the graphical tools;
- http.jar contains the HTTP-based MTP which is also the default MTP launched at the platform start-up;
- iiop.jar contains the IIOP-based MTP. This is not often used, but is the subject of a couple of examples later in the book and it implements the FIPA IIOP MTP specs (FIPA75);
- commons-codec\commons-codec-1.3.jar contains the Base64 codec used by JADE.

The classes directory contains the class files of the examples. Note that to reduce the size of the distribution files, the examples are distributed as source code and must therefore be compiled prior to use with the command ant examples.

To launch the platform, the user must first set their local Java CLASSPATH, i.e. the set of directories and Java archive files where the Java Virtual Machine will look for bytecode (i.e. the. class and. jar files). Issues relating to the CLASSPATH remain the topic of many questions received from JADE beginners, with the most typical concerning ClassNotFound exceptions caused by incorrect configuration of the CLASSPATH parameter.

Also, remember that in order to load an agent on the platform, its class file must also be reachable via the CLASSPATH.

If, for example, JADE has been downloaded into C:\jade on a Windows platform, the CLASSPATH can typically be set using the following command-line sequence:

```
prompt> set JADE_HOME=c:\jade

prompt> set CLASSPATH=%JADE_HOME%\lib\jade.jar; %JADE_HOME%
   \lib\jadeTools.jar;  %JADE_HOME%\lib\http.jar; %JADE_HOME%
   \lib\iiop.jar; %JADE_HOME%
   \lib\commons-codec\commons-codec-1.3.jar;%JADE_HOME%\classes
```

The main-container can now be launched with the JADE GUI using the command:

```
prompt> java jade.Boot -gui
```

The result of this should be as shown in Figure 3.4.

The first part of this output is the disclaimer printed out each time a JADE run-time is started. Following that, all the standard JADE platform services are initialized, which implement the various functionalities provided by the container. These are detailed later in Chapter 7. Since this instance of the JADE run-time is a main container, an HTTP MTP is started by default and its local address printed. Finally, a notification indicates that a container called 'main container' is ready; the JADE platform is now ready for use.

As mentioned, the command-line option -gui has the effect of launching the primary JADE graphical interface, shown in Figure 3.5. This GUI is actually provided by a JADE system agent called the Remote Monitoring Agent (RMA) and allows a platform administrator to manipulate and monitor the running platform.

It should be noted that use of the RMA GUI, and all other graphical tools, can negatively impact system performance. This is one reason why the -gui option is provided. If performance is a

```
prompt> java jade.Boot -gui
29-dic-2005 16.41.11 jade.core.Runtime beginContainer
INFO: ----------------------------------
    This is JADE snapshot - revision 5752 of 2005/07/15 14:22:11
    downloaded in Open Source, under LGPL restrictions,
    at http://jade.tilab.com/
----------------------------------------
29-dic-2005 16.41.14 jade.core.BaseService init
INFO: Service jade.core.management.AgentManagement initialized
29-dic-2005 16.41.15 jade.core.BaseService init
INFO: Service jade.core.messaging.Messaging initialized
29-dic-2005 16.41.15 jade.core.BaseService init
INFO: Service jade.core.mobility.AgentMobility initialized
29-dic-2005 16.41.15 jade.core.BaseService init
INFO: Service jade.core.event.Notification initialized
29-dic-2005 16.41.16 jade.core.messaging.MessagingService boot
INFO: MTP addresses:
http://anduril:7778/acc
29-dic-2005 16.41.19 jade.core.AgentContainerImpl joinPlatform
INFO: ----------------------------------
Agent container Main-Container@JADE-IMTP://NBNT2004130496 is ready.
```



**Figure 3.4** Standard output snapshot at the JADE start-up

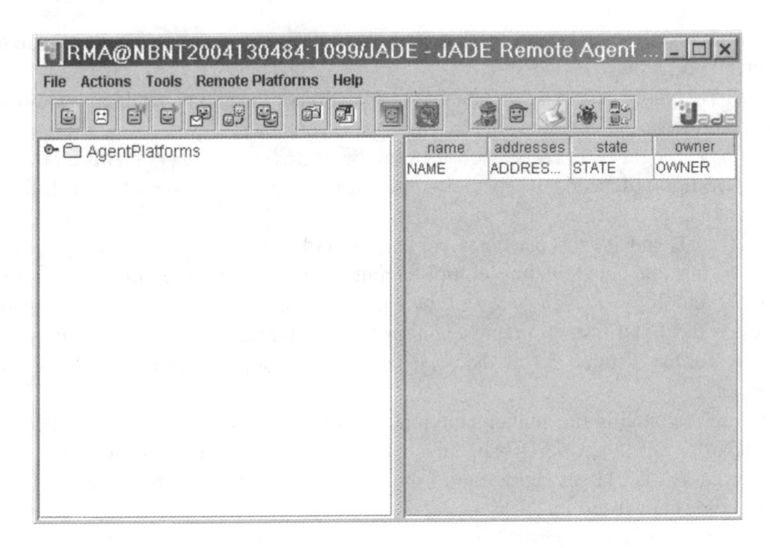


**Figure 3.5** GUI of the JADE RMA

concern, it is suggested not to use the RMA GUI at deployment time, rather to limit its use to system monitoring as required.

Now that the main-container has been initialized, any number of other containers can be launched on the various hosts composing the platform. If, for example, the host name of the machine where the main container was launched is 'anduril', the following command will launch a non-main-container on the current host and attach it to the main-container running on the host specified by `-host`.

```
prompt> java jade.Boot -container -host anduril
```

Appendix A details the syntax of the JADE command line and provides the complete list of available options.

## 3.5 JADE PACKAGES

The JADE platform sources are organized into a hierarchy of Java packages and sub-packages, where each package, in principle, contains the set of classes and interfaces that implement a specific functionality. The main packages are:

- `jade.core` implements the kernel of JADE, the distributed run-time environment that supports the entire platform and its tools. It includes the fundamental `jade.core.Agent` class as well as all the basic run-time classes needed to implement agent containers. It also includes a set of sub-packages each implementing a specific kernel-level service (see Chapter 7). These are:
  - `jade.core.event` that implements the distributed event notification service. This allows subscribers to be notified of system events generated by the various distributed components of a platform;
  - `jade.core.management` that implements the distributed agent life-cycle management service;
  - `jade.core.messaging` that implements the message distribution service;
  - `jade.core.mobility` that implements the agent mobility and cloning service, including the transfer of both the state and the code of an agent;

- `jade.core.nodeMonitoring` that enables containers to monitor each other and discover unreachable or dead containers;
- `jade.core.replication` that enables replication of the main container as a failover option in the event of serious faults in the original main container;
- `jade.core.behaviours` is a sub-package of `jade.core` that contains a hierarchy of the core application-independent behaviours. A JADE behaviour represents a task that an agent can carry out, as described in Section 4.2.

- `jade.content` and its sub-packages contain the collection of classes that support programmers in creating and manipulating complex content expressions according to a given content language and ontology. This includes all the coded mechanisms necessary for automatic conversion between the JADE internal representation format and the FIPA-compliant message content transmission format. Section 5.1 includes a detailed description of the functionalities implemented in this package.
- `jade.domain` contains the implementation of the AMS and DF agents, as specified by the FIPA standards, plus their JADE-specific extensions that will be described later. Each sub-package contains the classes representing the various entities of a predefined JADE ontology. These ontologies are listed in Table 3.1.
- `jade.gui` contains some general-purpose Java components and icons that can be useful to build Swing-based GUIs for JADE agents. The package provides several ready-to-use graphical components for displaying typical JADE abstractions, e.g. the AID, the ACLMessage, and the AgentDescription.
- `jade.imtp` contains the JADE IMTP (Internal Message Transport Protocol) implementations. In particular, the sub-package `jade.imtp.rmi` is the default IMTP of JADE which is based upon Java RMI (RMI).
- `jade.lang.acl` contains support for the FIPA Agent Communication Language (ACL) including the ACLMessage class, the parser, the encoder, and a helper class for representing templates of ACL messages.
- `jade.mtp` contains the set of Java interfaces that should be implemented by a JADE MTP. It also contains two sub-packages with an implementation based upon the HTTP protocol (which is the default implementation) and one based upon the IIOP protocol.

**Table 3.1** Predefined JADE ontologies

| Ontology | Package | Description |
|---|---|---|
| `FIPA-Agent-Management` | `jade.domain.FIPAAgent Management` | Entities, exceptions and actions needed to interact with the AMS and the DF, according to the FIPA specs |
| `JADE-Agent-Management` | `jade.domain.JADEAgent Management` | JADE extensions to the FIPA-agent-management ontology |
| `JADE-Introspection` | `jade.domain. introspection` | JADE extensions related to the monitoring of platform events |
| `JADE-Mobility` | `jade.domain.mobility` | JADE extensions related to agent mobility |
| `JADE-Persistence` | `jade.domain.persistence` | JADE extensions related to agent persistence |
| `DFApplet-Management` | `jade.domain.DFGUI Management` | Ontology used by the DF GUI to interact with the DF. It allows multiple GUIs of the same DF, including GUIs implemented as applets |

- `jade.proto` contains the implementations of some general-purpose interaction protocols, including some of those specified by FIPA. Section 5.4 provides a description of several interaction protocols.
- `jade.tools` contains the implementation of all the JADE graphical tools, which are described in Section 3.7.
- `jade.util` contains several miscellaneous utility classes.
- `jade.wrapper` together with the `jade.core.Profile` and `jade.core.Runtime` classes provides support for the JADE in-process interface that allows external Java applications to use JADE as a library, as described in Section 5.6.
- `FIPA` is a package that includes the IDL (Interface Definition Language) module specified by FIPA for the IIOP-based MTP.

## 3.6 MESSAGE TRANSPORT SERVICE

According to the FIPA specifications, a Message Transport Service (MTS) is one of the three most important services that an agent platform is required to provide (the other two being the Agent Management Service and the Directory Facilitator). An MTS manages all message exchange within and between platforms.

### *3.6.1 MESSAGE TRANSPORT PROTOCOLS*

To promote interoperability between different (i.e. with non-JADE) platforms, JADE implements all the standard Message Transport Protocols (MTPs) defined by FIPA, where each MTP includes the definition of a transport protocol and a standard encoding of the message envelope.

As was shown in Figure 3.4, one of the messages provided to the standard output when launching a main container reports something similar to the following:

```
INFO: MTP addresses:
http://anduril:7778/acc
```

By default, JADE always starts an HTTP-based MTP with the initialization of a main container; no MTP is activated on normal containers. In effect this creates a server socket on the main container host and listens for incoming connections over HTTP at the URL specified in the second line shown above. Whenever an incoming connection is established and a valid message received over that connection, the MTP routes the message to its final destination which, in general, is one of the agents located within the distributed platform. Internally, the platform uses a proprietary transport protocol called IMTP (Internal Message Transport Protocol) which is described in the next section. JADE performs message routing for both incoming and outgoing messages using a single-hop routing table that requires direct IP visibility among containers.

Using command-line options, any number of MTPs can be activated on any JADE container, including MTPs implementing different transport protocols. MTPs can also be 'plugged in' and instantiated at run-time by exploiting the RMA GUI console. This allows the platform administrator the freedom to manipulate topology by, for example, isolating a host with an open connection with an external network in order to improve security. When a new MTP is activated on a container, the entire JADE platform gains a new transport address, a new endpoint where messages can be received. This address is also added to the following data structures:

- The platform profile, obtainable from the AMS by using the action `get-description`.
- All the `ams-agent-description` objects contained within the AMS repository, obtainable through a search operation.
- The local Agent Identifier (AID) that any agent on any container can obtain with the `getAID()` method of the `Agent` class.

**Table 3.2** JADE MTPs available in the public domain

| Transport protocol | Message encoding | Developer |
|---|---|---|
| HTTP and HTTPS | XML | Universitat Autonoma of Barcelona (UAB), Spain |
| IIOP (Sun implementation) | CORBA IDL | JADE team |
| IIOP (ORBacus implementation) | CORBA IDL | Giovanni Rimassa, Università di Parma, Italy |
| JMS | Java data structure | Edward Curry, University of Galway |
| Jabber XMPP | Java data structure | Universitat Politècnica of Valencia, Spain |

The public Java interface `jade.mtp.MTP` allows the development of MTPs customized for specific application requirements and network environments. The `MTP` interface models a bidirectional channel that can both send and receive ACL messages by extending the `jade.mtp.OutChannel` and `jade.mtp.InChannel` interfaces that represent one-way channels. The companion `jade.mtp.TransportAddress` interface is a simple representation for an URL providing access to fields such as protocol, host, port and file part. At this time the MTPs listed in Table 3.2 are available in the public domain, each deployed as a separate jar file.

While the HTTP and IIOP MTPs are included with the main JADE distribution, all others must be downloaded as add-ons available from the JADE website. As already stated, HTTP is the default MTP that is launched with a main container. HTTP was selected as the default MTP as the specific implementation provided by UAB offers a number of advantages:

- The local port numbers for incoming and outgoing connections can be selected for stricter firewall configuration using the parameters `jade_mtp_http_port` and `jade_mtp_http_outPort`.
- A proxy can be configured and used for all outgoing connections using the parameters `jade_mtp_http_proxyHost` and `jade_mtp_http_proxyPort`.
- Performance can be fine-tuned through the use of persistent connections: instead of performing TCP handshaking for every message, connections can be cached and reused when messages are frequently exchanged between two different platforms. This is controlled with the parameters `jade_mtp_http_numKeepAlive` and `jade_mtp_http_timeout`.
- HTTPS can be used to establish secure and authenticated channels between platforms. In order to use HTTPS, a transport address must simply begin with **`https`**. It should be noted of course that, although HTTPS improves security, it incurs a consequent loss in performance; a rough estimation indicates that the HTTPS MTP is around 15% slower than the standard HTTP MTP.

The various command-line parameters for configuring the HTTP and HTTPS MTPs are listed in Appendix A.

The `-mtps <className>` command-line option allows specification of the preferred MTP, when a new container is launched, by specifying the fully qualified name of the Java class that implements the MTP interface. If the MTP supports activation on specific server addresses, then the address URL can be provided in parentheses immediately following the class name. If multiple MTPs are to be activated, they can be provided in a semicolon-delimited list.

For example, the following command-line launches a new container with two MTPs: an HTTPS MTP listening at the URL `https://anduril:8080/acc` and an IIOP MTP listening at the default address, where `anduril` is the host name of the container:

```
prompt> java jade.Boot -container -host anduril -mtps
    jade.mtp.http.MessageTransportProtocol(https://anduril:8080/acc);
    jade.mtp.iiop.MessageTransportProtocol
```

Alternatively, the JADE RMA console allows more flexible management of MTPs by allowing their activation and deactivation while the platform is running. A right click on an agent container tree node in the left panel of the RMA GUI produces a pop-up context menu which includes the two items *Install a new MTP* and *Uninstall an MTP*. Choosing the former causes a dialog to appear where the user can select the container to install the new MTP, the fully qualified name of the class implementing the protocol, and (if it is supported by the chosen protocol) the preferred listening transport address. If *Uninstall an MTP* is selected, a dialog appears where the user can select an MTP from a list of those currently installed and remove it from the platform.

Some applications might not need any communication outside the local platform. In such cases, the command-line option `-nomtp` will inhibit the creation of the default HTTP MTP on the main container. This of course will isolate the platform from all external communication with remote platforms. It is worth noting here that a *remote container* only implies a container that is not hosted on the same host as a main container, but which is within the same platform; in other words, a remote container is not to be considered part of a remote platform. Containers in the same platform always communicate with the JADE IMTP (see Section 3.6.2) due to the intrinsic performance gains.

### *3.6.2 IMTP*

The JADE IMTP (Internal Message Transport Protocol) is exclusively used for exchanging messages between agents living in different containers of the same platform. It is considerably different from inter-platform MTPs, such as HTTP. First, as it is used for internal platform communication only, it does not need to be compatible with any FIPA standards; it can be proprietary and thus designed to enhance platform performance. The JADE IMTP is in fact used not only to transport messages, but also to transport internal commands needed to manage the distributed platform, as well as to monitor the status of remote containers. For instance, it is used to transport a command to order shut down of a container, as well as for monitoring when a container is killed or becomes otherwise unreachable.

JADE was designed to allow selection of the IMTP at platform launch time. For the time being, two main IMTP implementations are available. One is based on Java RMI and is the default option. The second is based on a proprietary protocol using TCP sockets that circumvents the absence of support for Java RMI in the J2ME environment; this is started by default when initiating the JADE LEAP platform as will be described in Chapter 8. Both implementations provide several configuration options to allow fine-tuning of the IMTP for specific network and device characteristics.

#### 3.6.2.1 RMI-IMTP

The RMI-IMTP is implemented by the package `jade.imtp.rmi`. When the main container boots up, it searches for an RMI registry [RMI] on the local host and binds its object reference; if none can be found, it creates a new registry. When a non-main container boots up, it locates the RMI registry on the specified main container host and looks up the object reference of the main container. It then invokes the remote method `addNode()` of the main container to join the platform and registers its own reference with the main container.

Agent messages and system control information exchanged between containers is implemented through a Command pattern where the requestor node (i.e. a container) creates a proper `Command` object and passes this object, with an execution requests, to the executor node.

The following two command-line arguments are available for the RMI-IMTP:

```
-host <hostName>
```

which specifies the host that is running the main container to register with; its default value is localhost. This option should also be used when launching the main container to override the value of `localhost`, e.g. to include the full domain of the host with `-host anduril.cselt.it` when `localhost` would have returned only ‘anduril’.

```
-port <portNumber>
```

which specifies the port number where the RMI registry created by the main container should accept look-up and bind requests. Its default value is 1099.

## 3.7 ADMIN AND DEBUGGING TOOLS

Multi-agent applications are, in general, quite complex. They are often distributed across several hosts; they are composed of perhaps hundreds of multi-threaded processes (i.e. containers with several agents, each agent owning its own thread); they are dynamic in that agents can appear, disappear and migrate. These facets alone imply difficulties in management and especially debugging. To mitigate this somewhat, JADE has an event notification service which forms the basis of the JADE RMA management console and a set of graphical tools which are provided to help in the management and debugging phase. All these tools are packaged into the *jadeTools.jar* archive file.

This section describes the various tools provided with the JADE distribution, the Event Notification Service (ENS) and the JADE Tools model. Additional information is provided to assist users with creating their own tools.

To help explain the platform tools, the following code illustrates a simple `HelloWorldAgent` that implements the following cyclic behaviour: each time a message is received, it prints the message onto the standard output and replies to the sender with a 'Hello!' message. The various code sections will be described in later chapters; for the time being we will simply launch and interact with the agent.

```
import jade.core.Agent;
import jade.core.behaviours.CyclicBehaviour;
import jade.lang.acl.ACLMessage;

public class HelloWorldAgent extends Agent {

  public void setup() {

    System.out.println("Hello. My name is "+getLocalName());

    addBehaviour(new CyclicBehaviour() {
      public void action() {
        ACLMessage msgRx = receive();
        if (msgRx != null) {
          System.out.println(msgRx);
          ACLMessage msgTx = msgRx.createReply();
          msgTx.setContent("Hello!");
          send(msgTx);
        } else {
          block();
        }
      }
    });
  }
}
```

Having compiled the agent, we can launch a new container with our agent inside it by using the following command line where `Pet` is the agent local name and `HelloWorldAgent` is the fully qualified Java class name (in this case we did not use a package name). The two parameters are separated by a colon.

```
prompt> java jade.Boot -container -host anduril Pet:HelloWorldAgent
```

### *3.7.1 THE PLATFORM MANAGEMENT CONSOLE*

The JADE RMA (Remote Monitoring Agent) is a system tool that implements a graphical platform management console. The tool is implemented by the class `jade.tools.rma.rma` but it is more usually launched directly from the command line using the option `-gui`. It provides a visual interface to monitor and administer a distributed JADE platform composed of one or several hosts and container nodes. It includes a 'Tools' menu through which other tools can be launched. Several RMAs can be launched on the same platform if a different agent name is assigned to each instance.

At start-up the RMA agent subscribes with the AMS to be notified of all platform-level events; Figure 3.5 shows its graphical user interface. The left panel provides a view of the platform topology represented as a tree of containers whose leaves are agents. This panel is implemented by the class `jade.gui.AgentTree` and is reused by most other tools. In particular, there are three types of nodes: agent platform, container and agent. For each node, the life cycle of the represented entity can be controlled via a pop-up menu that appears by right-clicking on the relative node.

If an agent is selected, the pop-up menu allows the agent to be suspended, resumed, killed, cloned, saved, frozen or migrated to another container. It also allows the composition and sending of a custom, ad-hoc message.

If a container is selected, the pop-up menu allows the creation of a new agent, loading of an existing agent, installation or removal of an MTP, saving/loading of the container including all its agents and termination of the container. Using the RMA launched by the example in the previous section, we can try creating a new `HelloWorldAgent` that we call Bill as shown in Figure 3.6.

If a platform is selected, the pop-up menu allows the platform profile to be viewed, i.e. the data structure, called AP for Agent Platform, which describes a FIPA-compliant platform and lists the platform name and its available services. The menu also allows management of platform MTPs, i.e. installation and removal of MTPs on/from specified containers.

Note that the root of the tree is called 'Agent Platforms' in the plural. This is indicative of the fact that the RMA can be used to control a set of platforms, provided they are all FIPA-compliant. Of course, the degree of control is limited when interacting with a remote platform as only those management messages and actions defined by FIPA can be used, instead of those available through the JADE IMTP within any single JADE platform. It is, for example, possible to view the AP Description of a remote platform and its list of active agents. However, as the container abstraction is not specified by FIPA, the tree view of a remote platform cannot be directly represented. In order to communicate with a remote platform, the identifier of its AMS must be provided (i.e. the AMS AID), which must include its name and at least one valid transport address. This can be simply tried as shown in Figure 3.7, by cheating the RMA agent and requesting it to communicate with your local AMS as if it were a remote platform: after the pop-up menu, when requested to

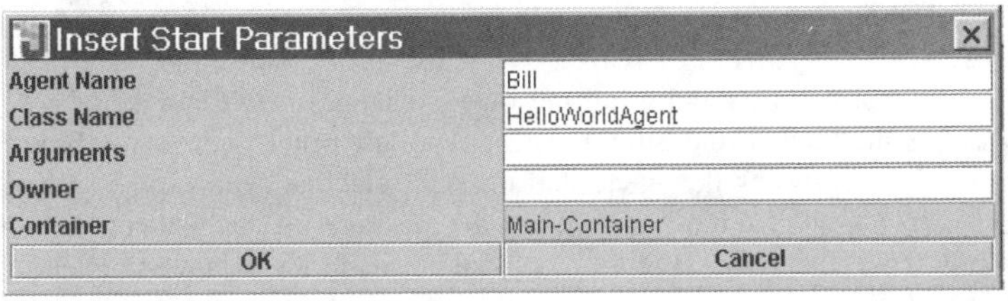

**Figure 3.6** GUI to launch a new agent

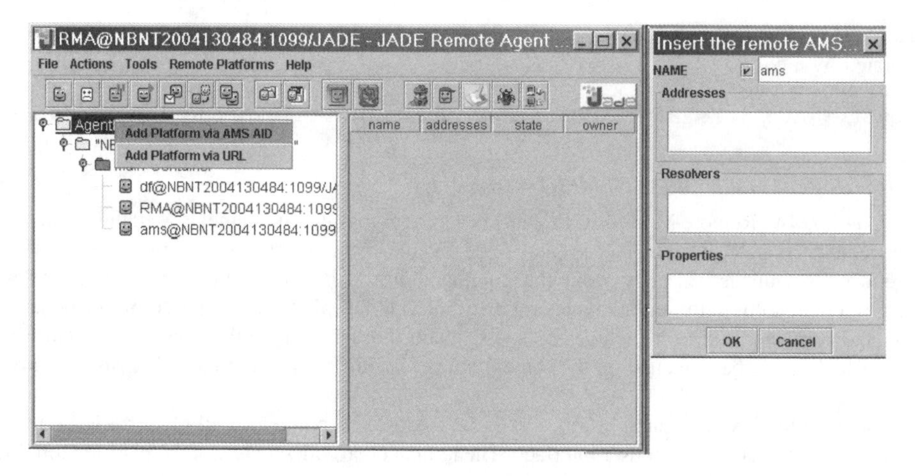


**Figure 3.7** Communicating with a remote platform

insert the AID of the AMS, type 'ams' and check the box to indicate that it is not a GUID. A second platform should now appear within the tree of platforms that, incidentally, has the same AP Description and the same list of agents as your own platform.

### *3.7.2 THE DUMMYAGENT*

The DummyAgent is a very simple tool that is useful for sending stimuli, in the form of custom ACL messages, to test the behaviour of another agent. It is implemented by the class `jade.tools.DummyAgent.DummyAgent`. Its sole capability is to send and receive custom messages that can be composed using a simple GUI and loaded/saved from/into a file. When an application agent has been launched, a DummyAgent can be used to stimulate it by sending user-specified messages and analysing its reaction in terms of received messages. It is a simple but effective tool which is typically used extensively during application development. Figure 3.8 shows the DummyAgent GUI with the right panel dedicated to showing the list of sent and received messages. The left panel is used to compose custom messages. Both of these panels and other components are reusable and provided as separate Java classes; in particular `jade.gui.AclGui` and `jade.gui.AIDGui` are useful classes for composing/visualizing an ACLMessage and an AID.

Many instances of the DummyAgent can be launched as and where required, both from the Tool menu of the RMA and from the command line, as in the following example:

```
prompt> java jade.Boot myDummy:jade.tools.DummyAgent.DummyAgent
```

### *3.7.3 THE SNIFFER AGENT*

While all the other tools are for the most part used for debugging a single agent, this tool is extensively used for debugging, or simply documenting conversations between agents. It is implemented by the class `jade.tools.sniffer.Sniffer`. The 'sniffer' subscribes to a platform AMS to be notified of all platform events and of all message exchanges between a set of specified agents.

Figure 3.9 shows the GUI of the Sniffer Agent. The left panel is the same browser as that of the RMA, but used for browsing the agent platform and selecting agents to be sniffed. The canvas on the right provides a graphical representation of the messages exchanged between sniffed agents, where each arrow represents a message and each colour identifies a conversation.

When the user decides to sniff an agent or a group of agents, every message directed to, or coming from, that agent/group is tracked and displayed in the sniffer GUI. The user can select and

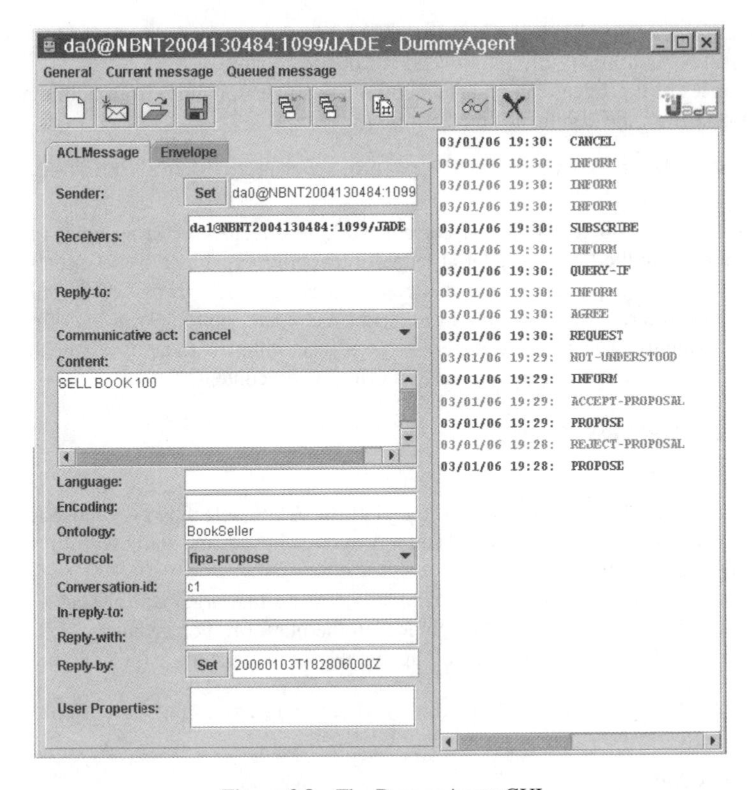


**Figure 3.8** The DummyAgent GUI

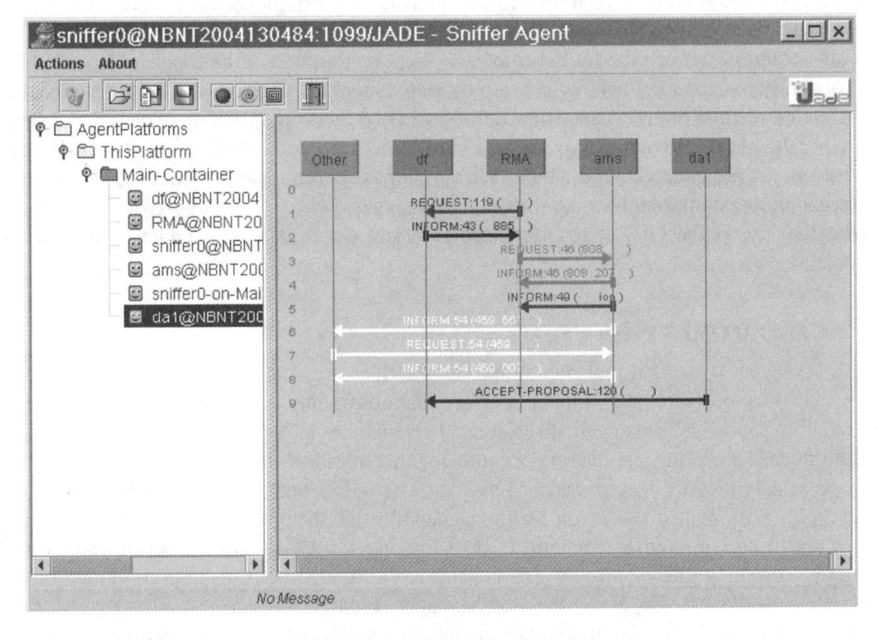


**Figure 3.9** The Sniffer Agent GUI

view the details of every individual message, save the message to disk as a text file or serialize an entire conversation as a binary file (e.g. useful for documentation).

Many instances of this agent can be launched on any single container from both the Tools menu of the RMA and from the command line as follows:

```
prompt> java jade.Boot mySniffer:jade.tools.sniffer.Sniffer
```

If a file named 'sniffer.inf' exists in the current working directory or in a parent of the working directory, it is read by the Sniffer Agent at launch time and treated as a list of agents to sniff and, optionally, a filter on the performatives of the ACL messages. The format of this file is a simple sequence of lines, where each line contains an agent name and, optionally, a list of performatives. The wildcard symbols '*' and '?' can also be used according to their usual regular expression meaning. For instance, a file 'sniffer.inf' with the following content:

```
ams inform propose
d*
```

tells the sniffer to sniff the following agents: the platform AMS (selecting only those ACL messages whose performative is inform or propose), and any agent whose name starts with 'd' (with no filter on the message performatives).

A list of agent names can also be passed as a command-line argument at start-up treated as a list of agents to be sniffed as soon as they appear in the platform. For instance, with the following command-line a sniffer is started to sniff all agents whose name starts with a 'd':

```
prompt> java jade.Boot mySniffer:jade.tools.sniffer.Sniffer(d*)
```

### *3.7.4 THE INTROSPECTOR AGENT*

While the Sniffer Agent is useful to sniff, monitor and debug conversations between agents, the Introspector Agent should be used to debug the behaviour of a single agent. This tool in fact allows an agent's life cycle, and its queues of sent and received messages, to be monitored and controlled. It also allows the queue of scheduled behaviours (see Section 4.2) to be monitored, including the useful capability of executing behaviours step by step. Note that a behaviour step is an execution of the `action()` method of an instance of a Behaviour class and is not directly related to the stepwise execution of Java code. In summary, this tool allows the introspection of an agent's execution, i.e. which behaviours are executed and which behaviours pass into the sleeping queue, and monitoring of its reactions to external stimuli, i.e. incoming messages.

Figure 3.10 shows the GUI of the Introspector Agent when, in this example, introspecting the DF agent.

### *3.7.5 THE LOG MANAGER AGENT*

The Log Manager Agent is a tool that simplifies the dynamic and distributed management of the logging facility by providing a graphical interface that allows the logging levels of each component of the JADE platform to be changed at run-time. This includes all those components being executed at remote nodes, including application-specific logging messages. The log manager exploits the underlying capabilities of the `java.util.logging` APIs on which JADE logging is based. Each class uses its own instance of `Logger` named after the fully qualified class name. Each `Logger` object can be configured with its own logging level and its own set of Handlers. This configuration can be either static – by specifying a `java.util.logging` configuration file at launch time – or dynamic, by using the Log Manager Agent.

For instance, the following command line launches a JADE container and specifies a configuration file to initialize the logging system of the JVM:

```
prompt> java -Djava.util.logging.config.file=logging.properties
    jade.Boot -container
```

The logging configuration file is in standard `java.util.Properties` format. Refer to the javadoc documentation of the java class `java.util.logging.LogManager` for detailed information on configuration information and format of this file. In the example above, if the content of the file *logging.properties* is as follows:

```
handlers = java.util.logging.ConsoleHandler
.level = OFF
jade.core.messaging.level = FINEST
```

the launched JADE container will record only the log messages of the Messaging sub system, i.e. from all the classes in the package `jade.core.messaging`.

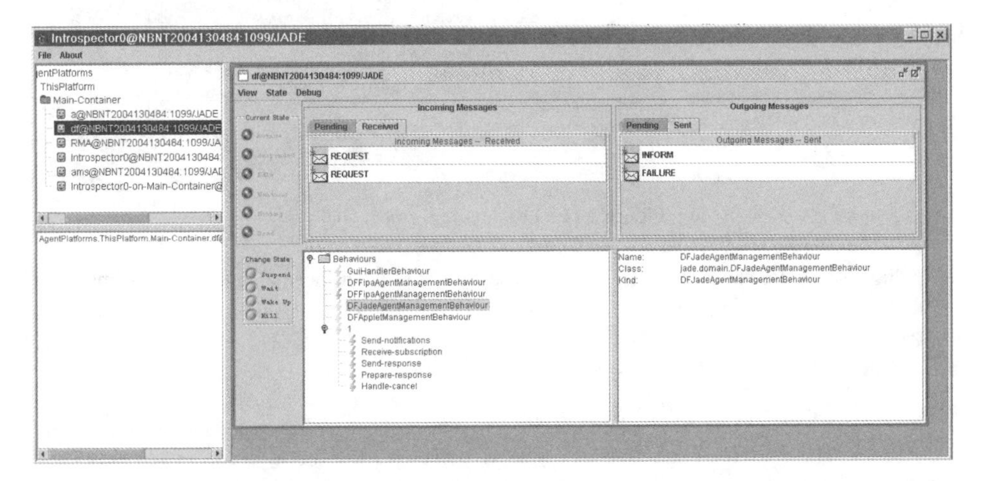


**Figure 3.10** Introspector Agent GUI

An administrator wishing to modify the logging level of a specific class at run-time (e.g. to debug a reported problem without rebooting the platform), can use the Log Manager Agent. As shown in Figure 3.11, the right panel of its GUI lists all the existing `Logger` objects with their current value of logging level and logging handlers. With this console, these values may be modified dynamically at run-time. By using the RMA console, an instance of a Log Manager Agent can be remotely launched on any container of the platform.

Of course, if application developers use the `java.util.logging` system, the Log Manager Agent can also be used to manage application-specific Loggers.

### *3.7.6 THE EVENT NOTIFICATION SERVICE AND THE JADE TOOL MODEL*

The Event Notification Service (ENS) is a platform-level service (see Chapter 7) that manages the distributed notification of all the events generated by each node of the platform. The service is called `jade.core.event.Notification`, is implemented in the `jade.core.event` package and launched by default in every container. Each time an event is generated by a container (e.g. agent born, message sent), it is intercepted by the ENS and routed to all agents that have previously subscribed to be notified of such types of event. If no agent is subscribed, the ENS has a negligible performance overhead. In fact, the only platform nodes that incur a performance penalty, due to event notification, are the container where the subscribed agent lives and the container that generates

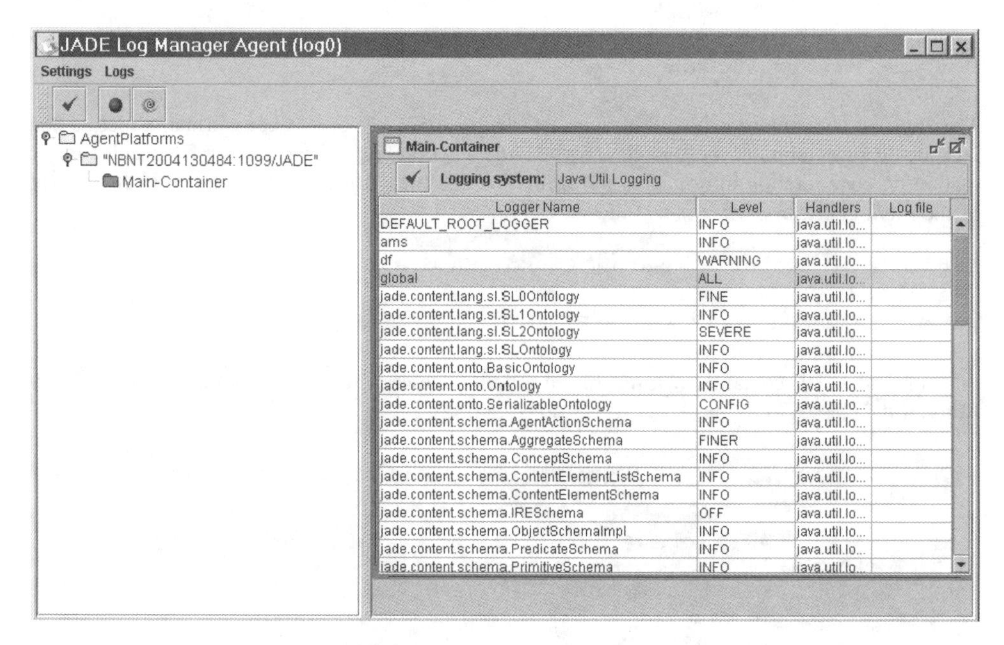


**Figure 3.11** Log Manager Agent GUI

the notified event. Because all tool agents can be activated when needed, even at run-time during the platform operation, improved performance can be achieved by starting them only if and when necessary.

There are four major types of events:

1. Life-cycle related events, also called *platform-type* events because they always involve the main container. These events relate to agent life-cycle changes (i.e. born, dead, moved, suspended, resumed, frozen, thawed) and to container life-cycle changes (i.e. added, removed).
2. *MTP-type* events generated by the platform when an MTP is (de)activated and when a message is sent/received by/from an MTP, i.e. when there is some inter-platform communication.
3. *Message-passing-type* events generated when an ACL message is sent, received, routed or posted into the agent message queue. These are the events that the Sniffer is typically used to monitor.
4. *Agent-internal-type* events which relate to changes in the state of agent behaviours. These are the events that the Introspector is typically used to monitor.

Agents interact with the ENS by exchanging ACL messages with the AMS, the ultimate authority of the platform. More detailed platform type and MTP type of events are fired on the main container with the AMS able to directly notify interested agents. Agents can subscribe to the AMS to be notified about events of these types by using the `AMSSubscriber` behaviour class included in the `jade.domain.introspection` package. The usage of this class is described in detail in Section 5.5.2.

Message-passing-type and agent-internal-type of events on the other hand are only fired on the container where the agent that generates them lives. Transferring them to the main container in fact would sensibly decrease the overall performances of the platform. As a consequence in order to be notified about message-passing and agent-internal events related to a given target agent, an observer agent must request the AMS to explicitly sniff and debug the target agents respectively. This can be done by means of the SniffOn and DebugOn actions of the JADEManagementOntology. Section 5.5.1 describes in detail how to request platform management operations to the AMS. As a result of the SniffOn action a proper ToolNotifier auxiliary agent is created in the container where the target agent lives. This notifier listens to local message-passing events and forwards them to the observer agent. Similarly as a result of the DebugOn action a ToolNotifier is created to listen to agent-internal events generated by the target agent and to forward them to the observer agent. The JADE event notification system is depicted in Figure 3.12.

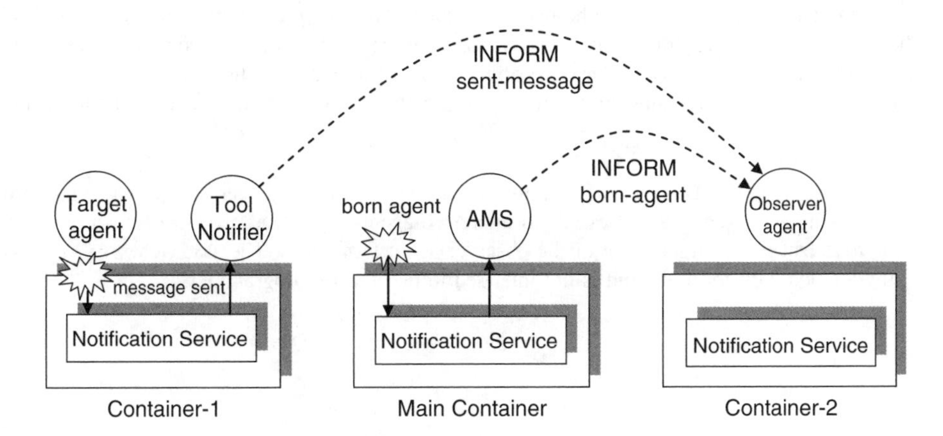


**Figure 3.12** The JADE event notification system

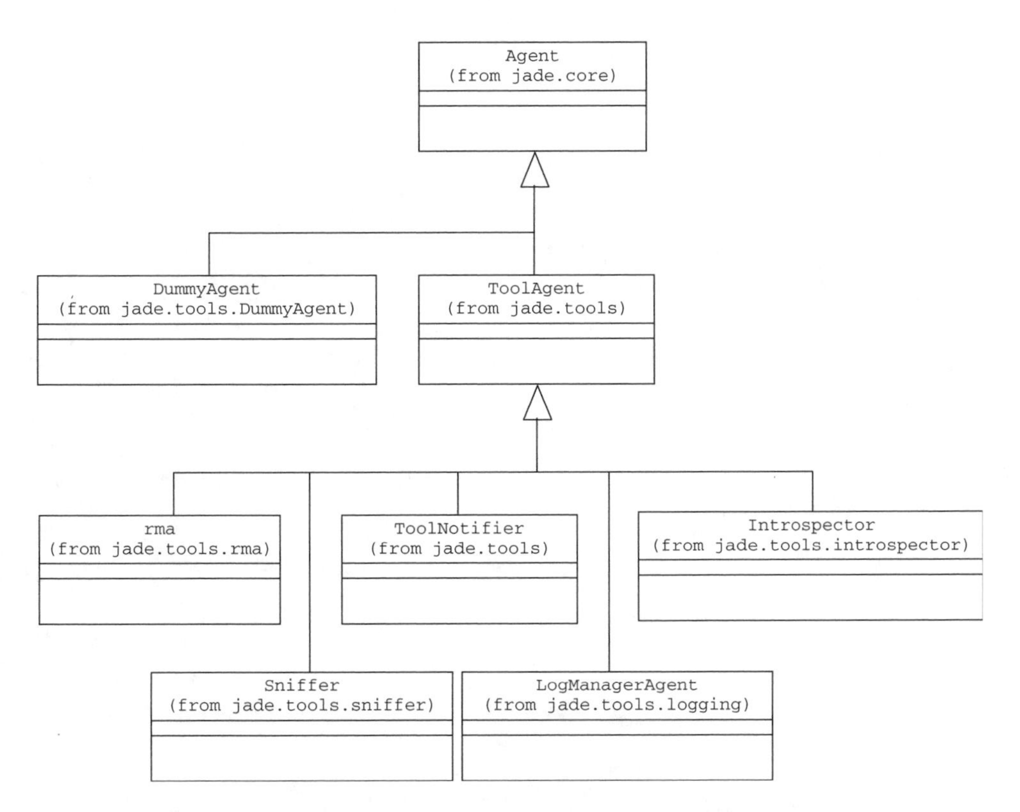


**Figure 3.13** Class diagram of the JADE tools

All JADE tools, except the DummyAgent, inherit from the `jade.tools.ToolAgent` class that provides the ability to receive event notifications in a uniform way. Figure 3.13 is a UML class diagram of the JADE tools.

Each tool is implemented as an Agent that extends the basic `jade.core.Agent` class. This enables some important features and simplifications:

1. The life cycle of a JADE tool can be managed exactly as any other agent of the platform.
2. The messaging capability of the basic Agent class can be used to allow interaction between the tool and the AMS, in particular for subscribing to platform event notifications.
3. Several instances of the same tool can coexist on the same platform and even in the same container.

The class `jade.tools.ToolNotifier` implementing the auxiliary agents required to forward message-passing and agent-internal events to interested agents is itself a `ToolAgent`. In this way it is able to detect if the target agent or the observer agent terminates. The `ToolNotifier` class is tightly coupled with the ENS and is not intended to be used by programmers.

# 4

# Programming with JADE – Basic Features

In Chapter 3 we gave an overview of the JADE platform, outlining its high-level architecture and presenting its functionality. In this section we start describing how to develop multi-agent systems with JADE, focusing on the basic features the framework provides. These consist of creating agents, making agents execute tasks, making agents talk together, and publishing and discovering 'services' in the yellow pages catalogue. By means of these features, that together involve fewer than 10 classes of the JADE library, it is already possible to implement distributed applications with a certain degree of complexity. In Chapter 5, we will describe additional features that target specific problems. JADE, however, adopts a 'pay as you go' philosophy, implying that developers need not be concerned with advanced features until they have the need or desire to do so.

As mentioned in Chapter 3, JADE is a full Java tool and therefore creating a JADE-based multi-agent system only implies creating Java classes without any significant expertise in Java programming.

In order to illustrate the steps required to develop applications with JADE, this section introduces a simple case study, which will be used throughout the book, where an agent-based system enables users to trade second-hand books. In this book-trading system there will be two types of agent: sellers and buyers. Each buyer agent takes as input some books to buy and tries to find agents selling them at an acceptable price. Similarly each seller agent takes as input some books to sell and tries to do so at the highest possible price. Both buyers and sellers implement some simple strategies and carry out negotiations to achieve the best result for the users they represent. Both buyers and sellers may appear and disappear dynamically in the system. All issues related to purchased book delivery and payment are considered out of scope, and this is not taken into account.

## 4.1 CREATING AGENTS

Creating a JADE agent is as simple as defining a class that extends the `jade.core.Agent` class and implementing the `setup()` method as exemplified in the code below.

```
import jade.core.Agent;

public class HelloWorldAgent extends Agent {
  protected void setup() {
```

*Developing Multi-Agent Systems with JADE* Fabio Bellifemine, Giovanni Caire, Dominic Greenwood

```
      // Printout a welcome message
      System.out.println("Hello World. I'm an agent!");
   }
}
```

More appropriately a class, such as the `HelloWorldAgent` class shown above, represents a type of agent exactly as a normal Java class represents a type of object. Several instances of the `HelloWorldAgent` class can be launched at run-time. Unlike normal Java objects, which are handled by their references, an agent is always instantiated by the JADE run-time and its reference is never disclosed outside the agent itself (unless of course the agent does that explicitly). Agents never interact through method calls but rather by exchanging asynchronous messages, as will be described in Section 4.3.

The `setup()` method is intended to include agent initializations. The actual job an agent has to perform is typically carried out within 'behaviours' as will be described in Section 4.2. Examples of typical operations that an agent performs in its `setup()` method are: showing a GUI, opening a connection to a database, registering the services it provides in the yellow pages catalogue (see Section 4.4) and starting the initial behaviours. It is good practice not to define any constructor in an agent class and to perform all initializations inside the `setup()` method. This is because at construction time the agent is not yet linked to the underlying JADE run-time and thus some of the methods inherited from the `Agent` class may not work properly.

### *4.1.1 AGENT IDENTIFIERS*

Consistent with the FIPA specifications, each agent instance is identified by an 'agent identifier'. In JADE an agent identifier is represented as an instance of the `jade.core.AID` class. The `getAID()` method of the `Agent` class allows retrieval of the local agent identifier. An `AID` object includes a globally unique name (GUID) plus a number of addresses. The name in JADE has the form *<local-name>*@*<platform-name>* such that an agent called *Peter* living on a platform called *foo-platform* will have *Peter@foo-platform* as its globally unique name. The addresses included in the AID are the addresses of the platform the agent inhabits, as described in Section 3.6. These addresses are only used when an agent needs to communicate with another agent living on a different compliant FIPA platform.

The `AID` class provides methods to retrieve the local name (`getLocalName()`), the GUID (`getName()`) and the addresses (`getAllAddresses()`). We can therefore enrich the welcome message of our `HelloWorldAgent` as follows:

```
protected void setup() {
   // Printout a welcome message
   System.out.println("Hello World. I'm an agent!");
   System.out.println("My local-name is "+getAID().getLocalName());
   System.out.println("My GUID is "+getAID().getName());
   System.out.println("My addresses are:");
   Iterator it = getAID().getAllAddresses();
   while (it.hasNext()) {
      System.out.println("- "+it.next());
   }
}
```

The local name of an agent is assigned at start-up time by the creator and must be unique within the platform. If an agent with the same local name already exists in the platform, the JADE run-time prevents the creation of the new agent. Knowing the local name of an agent, its AID can be obtained as follows:

```
String localname = "Peter";
AID id = new AID(localname, AID.ISLOCALNAME);
```

The platform name is automatically appended to the GUID of the newly created AID by the JADE run-time. Similarly, knowing the GUID of an agent, its AID can be obtained as follows:

```
String guid = "Peter@foo-platform";
AID id = new AID(guid, AID.ISGUID);
```

### *4.1.2 AGENT INITIALIZATION*

The `HelloWorldAgent` class described previously can be compiled, as with normal Java classes, by typing:

```
javac  -classpath <JADE-classes> HelloWorldAgent.java
```

Of course the JADE libraries must be in the Classpath for the compilation to succeed. At that point, in order to execute a Hello-World agent, i.e. an instance of the `HelloWorldAgent` class, the JADE run-time must be started and a local name for the agent to execute must be chosen:

```
java  -classpath <JADE-classes>;. jade.Boot Peter:HelloWorldAgent
```

This command starts the JADE run-time and tells it to launch an agent whose local name is *Peter* and whose class is `HelloWorldAgent`. Again both the JADE libraries and the `HelloWorldAgent` class must be in the Classpath. As a result of the typed command, just after the JADE initialization messages described in Section 3.4, the following printouts produced by the Hello-World agent should appear.

```
Hello World. I'm an agent!
My local-name is Peter
My GUID is Peter@anduril:1099/JADE
My addresses are:
- http://anduril:7778/acc
```

The local name of the agent is *Peter* as we specified in the command line. Since we did not specify any platform name, JADE created one by default using the local host and port of the main container: *anduril:1099/JADE*. Therefore the GUID of the agent is *Peter@anduril:1099/JADE*. It is important to note that, though this GUID may look like an address, it is NOT an address. In order to assign a name to a platform the `-name` option must be specified when launching the main container as follows:

```
java  -classpath <JADE-classes> jade.Boot -name foo-platform
   Peter:HelloWorldAgent
```

If we restart JADE using this option, the GUID of agent Peter will become '*Peter@foo-platform*'. Other than the `-name` parameter there are several configuration options that can be specified when starting the JADE run-time such as `-gui` which is used to activate the JADE administration GUI described in Section 3.7.1. The most important options are described throughout this book with a consolidated list available in Appendix A.

Finally, we observe that the AID of agent Peter includes only one address as there is only one MTP active in the platform at present.

There are other ways to launch agents such as by means of the administration GUI as described in Section 3.7.1 or via code that issues a request to the AMS (see Section 5.5), or by using the in-process interface (see Section 5.6).

### *4.1.3 AGENT TERMINATION*

After printing the welcome message, even if it does not have anything else to do, our Hello-World agent is still alive. In order to make it terminate its `doDelete()` method must be called. Similar to the `setup()` method that is invoked to initialize an agent, the `takeDown()` method is invoked just before an agent terminates in order to perform various clean-up operations.

### *4.1.4 PASSING ARGUMENTS TO AN AGENT*

Agents can take start-up arguments which are retrieved, as an array of `Object`, by means of the `getArguments()` method of the `Agent` class. When launching an agent on the command-line as shown in Section 3.7, start-up arguments can be specified in parentheses and separated by spaces[1] as exemplified below:

```
java  -cp <...> jade.Boot -name foo-platform
   Peter:HelloWorldAgent(arg1 arg2 arg3);
```

Of course in the command-line case, only `String` arguments can be specified, but when launching an agent directly from code, generic `Object` arguments can be passed. If we modify the `setup()` method of our `HelloWorldAgent` class as follows:

```
protected void setup() {
  // Printout a welcome message
  System.out.println("Hello World. I'm an agent!");
  System.out.println("My local-name is "+getAID().getLocalName());
  System.out.println("My GUID is "+getAID().getName());
  System.out.println("My addresses are:");
  Iterator it = getAID().getAllAddresses();
  while (it.hasNext()) {
    System.out.println("- "+it.next());
  }
  System.out.println("My arguments are:");
  Object[] args = getArguments();
  if (args != null) {
    for (int i = 0; i < args.length; ++i) {
      System.out.println("- "+args[i]);
    }
  }
}
```

then, when executing the same command-line invocation, the following output is received:

```
java -cp <...> jade.Boot -name foo-platform
   Peter:HelloWorldAgent(arg1 arg2 arg3)
...
Hello World. I'm an agent!
```

[1] When using the LEAP version of JADE for wireless devices (see Chapter 8), arguments are separated by commas (',') instead of spaces.

```
My local-name is Peter
My GUID is Peter@anduril:1099/JADE
My addresses are:
- http://anduril:7778/acc
My arguments are:
- arg1
- arg2
- arg3
```

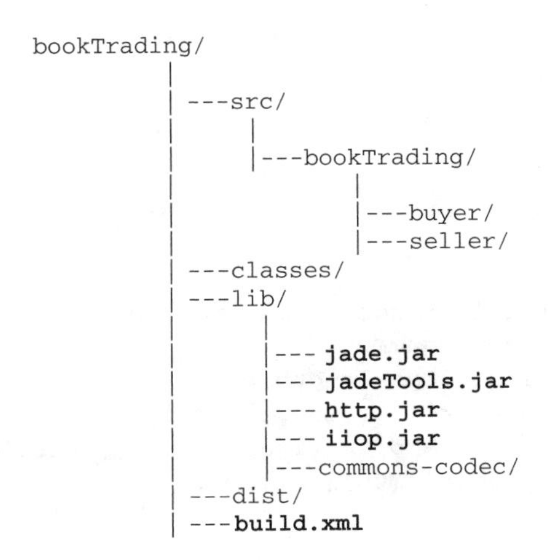


**Figure 4.1** Book-trading project directory structure

### *4.1.5 SETTING UP THE BOOK-TRADING PROJECT*

At this point we have all the elements needed to start setting up the book-trading project. To compile and create the jar files that users will need to access the system, we adopt the [ANT] build tool from the Apache Software Foundation available for download at http://ant.apache.org. We organize our project as depicted in Figure 4.1.

In the *src* directory we will keep all sources. We start by adding a root package called `bookTrading` and two sub-packages for the buyer agent sources and for the seller agent sources, respectively. The *classes* directory is where we will put the .class files produced by the compilation. In the *lib* directory we copy the JADE jar files and the book-trading jar file that includes all book-trading related classes. Finally in the *dist* directory we will store the. zip file to be distributed to users. In addition, we create an ANT build.xml file to tell ANT how to build the book-trading project.

Having established the project environment we can start writing the skeleton of the `BookBuyerAgent` class by implementing buyer agents. The `BookSellerAgent` is of course very similar, but at this early stage we will assume that seller agents are fixed, their names known a priori, and passed to each buyer agent as arguments. When we describe the yellow pages service in Section 4.4, we will remove this limitation.

```
package bookTrading.buyer;

import jade.core.Agent;
import jade.core.AID;

import java.util.Vector;
```

```
import java.util.Date;

public class BookBuyerAgent extends Agent {
  // The list of known seller agents
  private Vector sellerAgents = new Vector();

  // The GUI to interact with the user
  private BookBuyerGui myGui;

  /**
   * Agent initializations
   */
  protected void setup() {
    // Printout a welcome message
    System.out.println("Buyer-agent "+getAID().getName()+" is
       ready.");

    // Get names of seller agents as arguments
    Object[] args = getArguments();
    if (args != null && args.length > 0) {
      for (int i = 0; i < args.length; ++i) {
        AID seller = new AID((String) args[i], AID.ISLOCALNAME);
        sellerAgents.addElement(seller);
      }
    }

    // Show the GUI to interact with the user
    myGui = new BookBuyerGuiImpl();
    myGui.setAgent(this);
    myGui.show();
  }

  /**
   * Agent clean-up
   */
  protected void takeDown() {
    // Dispose the GUI if it is there
    if (myGui != null) {
      myGui.dispose();
    }

    // Printout a dismissal message
    System.out.println("Buyer-agent "+getAID().getName()+"
       terminated.");
  }

  /**
   * This method is called by the GUI when the user inserts a new
   * book to buy
   * @param title The title of the book to buy
   * @param maxPrice The maximum acceptable price to buy the book
   * @param deadline The deadline by which to buy the book
```

```
   */
  public void purchase(String title, int maxPrice, Date deadline) {
    // To be implemented
  }
}
```

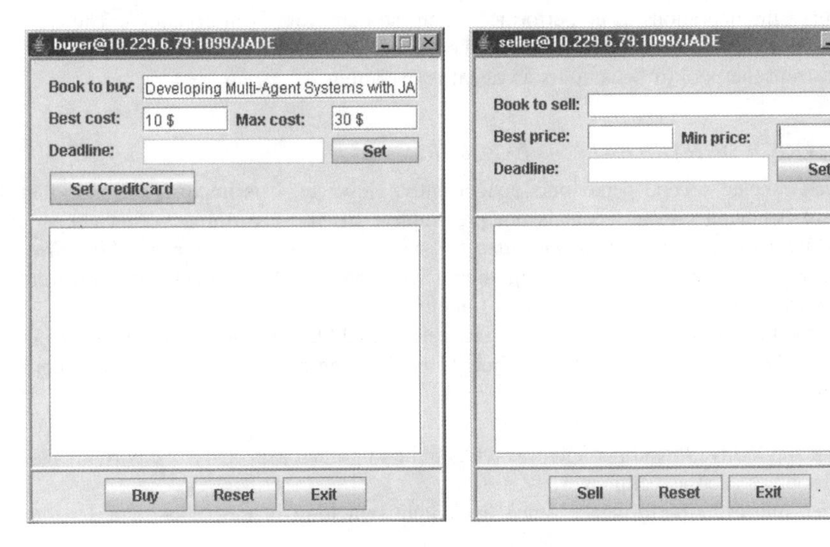


**Figure 4.2** Buyer and seller GUIs

Both buyer and seller agents should have a GUI to interact with their users. For instance, the GUI of a buyer agent would allow a user to specify the titles of books he wants to buy and other information (such as maximum price and a deadline) required by the buyer agent to carry out negotiations. Figure 4.2 shows the GUI of buyer and seller agents. Since the development of these GUIs is outside the scope of this book, we just focus on the interface provided by each GUI and assume the implementation classes to already be available.[2] The definition of the `BookBuyerGui` interface is as follows (again, it would be very similar for the seller agents):

```
package bookTrading.buyer;

public interface BookBuyerGui {
  void setAgent(BookBuyerAgent a);
  void show();
  void hide();
  void notifyUser(String message);
}
```

When the user fills in the title of a book and other relevant information and presses the 'Purchase' button, the GUI invokes the `purchase()` method of the underlying buyer agent.

## 4.2 AGENT TASKS

As mentioned in Section 4.1, the actual job, or jobs, an agent has to do is carried out within 'behaviours'. A behaviour represents a task that an agent can carry out and is implemented as an

[2] The complete source code of the book-trading case study (including GUI sources) is available in the JADE website among the examples.

object of a class that extends `jade.core.behaviours.Behaviour`. To make an agent execute the task implemented by a behaviour object, the behaviour must be added to the agent by means of the `addBehaviour()` method of the `Agent` class. Behaviours can be added at any time when an agent starts up (in the `setup()` method) or from within other behaviours.

Each class extending `Behaviour` must implement two abstract methods. The `action()` method defines the operations to be performed when the behaviour is in execution. The `done()` method returns a `boolean` value to indicate whether or not a behaviour has completed and is to be removed from the pool of behaviours an agent is executing.

### *4.2.1 BEHAVIOUR SCHEDULING AND EXECUTION*

An agent can execute several behaviours concurrently. However, it is important to note that the scheduling of behaviours in an agent is not pre-emptive (as for Java threads), but cooperative. *This means that when a behaviour is scheduled for execution its* `action()` *method is called and runs until it returns.* Therefore it is the programmer who defines when an agent switches from the execution of one behaviour to the execution of another.

This approach often creates difficulties for inexperienced JADE developers and must always be kept in mind when writing JADE agents. Though requiring an additional effort, this model does have several advantages:

- It allows a single Java thread per agent which is quite important especially in environments with limited resources such as cellphones.
- It provides improved performance since behaviour switching is far faster than Java thread switching.
- It eliminates all synchronization issues between concurrent behaviours accessing the same resources since all behaviours are executed by the same Java thread. This also results in a performance enhancement.
- When a behaviour switch occurs, the status of an agent does not include any stack information, implying that it is possible to take a 'snapshot' of it. This allows the implementation of some important advanced features, such as saving the status of an agent in a persistent storage for later resumption (agent persistency), or transferring the agent to another container for remote execution (agent mobility). Advanced features will be addressed in detail later on in this book.

The path of execution of the agent thread[3] is depicted in Figure 4.3.

It is important to note that a behaviour such as that shown below will prevent any other behaviour from being executed because its `action()` method will never return.

```
public class OverbearingBehaviour extends Behaviour {
  public void action() {
    while (true) {
      // do something
    }
  }

  public boolean done() {
    return true;
  }
}
```

[3] In JADE there is a single Java thread per agent. Since JADE agents are written in Java, however, programmers may start new Java threads at any time as they require. In this case note should be taken of the fact that the advantages discussed in this section are no longer valid.

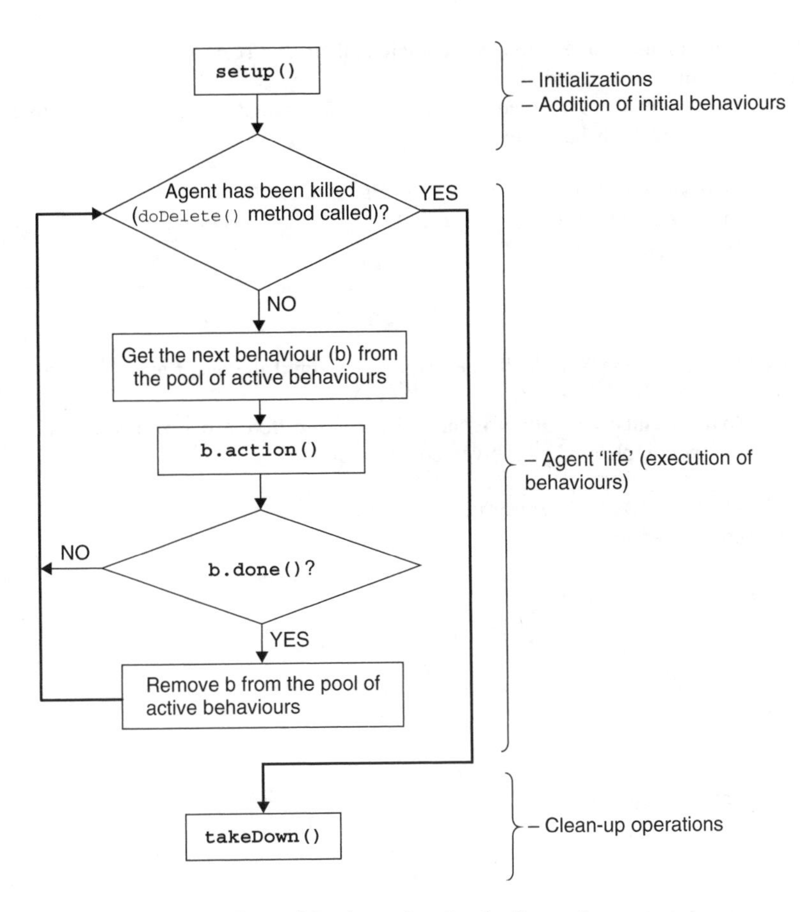


**Figure 4.3** Agent thread path of execution

When there are no behaviours available for execution the agent's thread goes to sleep in order not to consume CPU time. The thread is woken up again once a behaviour becomes available for execution again.

### *4.2.2 ONE-SHOT BEHAVIOURS, CYCLIC BEHAVIOURS AND GENERIC BEHAVIOURS*

The three primary behaviour types available with JADE are as follows:

1. 'One-shot' behaviours are designed to complete in one execution phase; their `action()` method is thus executed only once. The `jade.core.behaviours.OneShotBehaviour` class already implements the `done()` method by returning `true` and can be conveniently extended to implement new one-shot behaviours.

```
public class MyOneShotBehaviour extends OneShotBehaviour {
  public void action() {
    // perform operation X
  }
}
```

   In this example, operation X is performed only once.

2. 'Cyclic' behaviours are designed to never complete; their `action()` method executes the same operations each time it is called. The `jade.core.behaviours.CyclicBehaviour` class already implements the `done()` method by returning `false` and can be conveniently extended to implement new cyclic behaviours.

```
public class MyCyclicBehaviour extends CyclicBehaviour {
  public void action() {
    // perform operation Y
  }
}
```

   In this example, operation Y is performed repetitively until the agent executing the behaviour terminates.

3. Generic behaviours embed a status trigger and execute different operations depending on the status value. They complete when a given condition is met.

```
public class ThreeStepBehaviour extends Behaviour {
  private int step = 0;

  public void action() {
    switch (step) {
    case 0:
      // perform operation X
      step++;
      break;
    case 1:
      // perform operation Y
      step++;
      break;
    case 2:
      // perform operation Z
      step++;
      break;
    }
  }

  public boolean done() {
    return step == 3;
  }
}
```

   In this example, the `step` member variable implements the status of the behaviour. Operations X, Y and Z are performed sequentially after which the behaviour completes.

JADE also provides the possibility of composing behaviours together to create complex behaviours. This feature, that is particularly convenient when implementing complex tasks, is described in Section 4.3.5.

### *4.2.3 MORE ABOUT BEHAVIOURS*

All behaviours inherit the `onStart()` and `onEnd()` methods from the `Behaviour` class. These methods are executed only once just before the first call to the `action()` method and just after the done() method returns true. They are intended to perform task-specific initialization and termination operations. Unlike the action() and done() methods that are declared abstract, they have a default empty implementation allowing developers to implement them only if they so desire.

A behaviour can be aborted at any time by calling the removeBehaviour() method of the Agent class. A call to this method removes the referred behaviour from the pool of behaviours currently executed by the agent. Consequently if a behaviour is aborted using the removeBehaviour() method, its onEnd() method is not called.

Every behaviour has a member variable called myAgent that points to the agent that is executing the behaviour. This provides an easy way to access an agent's resources from within the behaviour.

Finally it is important to remember that once a Behaviour object has been executed, if it has to be executed a second time, it is necessary to invoke its reset() method first. Not doing this may lead to unexpected results.

### *4.2.4 SCHEDULING OPERATIONS*

JADE provides two ready-made classes (in the jade.core.behaviours package) which can be implemented to produce behaviours that execute at selected points in time.

1. The WakerBehaviour has action() and done() methods pre-implemented to execute the onWake() abstract method after a given timeout (specified in the constructor) expires. After the execution of the onWake() method the behaviour completes.

```
public class MyAgent extends Agent {
  protected void setup() {
    System.out.println("Adding waker behaviour");
    addBehaviour(new WakerBehaviour(this, 10000) {
      protected void onWake() {
        // perform operation X
      }
    } );
  }
}
```

   In this example, operation X is performed 10 seconds after the 'Adding waker behaviour' text is printed.

2. The TickerBehaviour has action() and done() methods pre-implemented to execute the onTick() abstract method repetitively, waiting a given period (specified in the constructor) after each execution. A TickerBehaviour never completes unless it is explicitly removed or its stop() method is called.

```
public class MyAgent extends Agent {
  protected void setup() {
    addBehaviour(new TickerBehaviour(this, 10000) {
      protected void onTick() {
        // perform operation Y
      }
    } );
  }
}
```

   In this example, operation Y is performed periodically every 10 seconds.

### *4.2.5 BEHAVIOURS REQUIRED IN THE BOOK TRADING EXAMPLE*

Having described the basic types of behaviour, let us move on to analyse which behaviours are required by the book-buyer and book-seller agents in our book-trading example.

#### 4.2.5.1 Book-Buyer Agent Behaviours

When a buyer agent is requested to buy a book, the simplest approach it can adopt is to periodically perform the task to ask all known seller agents if they have the target book available for sale, and if so, to provide an offer. Depending on this and on the range of price specified by its user, the buyer agent can then ask the seller that provided the best offer to sell the book. We can implement this functionality by using a `TickerBehaviour` that, at each tick, adds another behaviour that talks with the seller agents. This `TickerBehaviour` is added to the agent in the `purchase()` method described in Section 4.1.5 and terminates when the book is successfully purchased or the deadline specified by the user is reached. The `purchase()` method will now look like this:

```
public void purchase(String title, int maxPrice, Date deadline) {
  addBehaviour(new PurchaseManager(this, title, maxPrice,
     deadline));
}
```

The `PurchaseManager` behaviour must now be implemented as an inner class of the `BookBuyerAgent`. Implementing behaviour classes as inner classes of the agent that will execute them is typically[4] good practice as it allows behaviours to directly access agent resources such as the member variable `myGui` of the `BookBuyerAgent` class. The `PurchaseManager` behaviour uses a very simple strategy to optimize the price of the book to buy: at each negotiation the acceptable price is increased linearly until it reaches the maximum acceptable price when the deadline is about to expire. Our new behaviour is coded as follows:

```
private class PurchaseManager extends TickerBehaviour {
  private String title;
  private int maxPrice;
  private long deadline, initTime, deltaT;

  private PurchaseManager(Agent a, String t, int mp, Date d) {
    super(a, 60000); // tick every minute
    title = t;
    maxPrice = mp;
    deadline  = d.getTime();
    initTime = System.currentTimeMillis();
    deltaT = deadline -  initTime;
  }

  public void onTick() {
    long currentTime = System.currentTimeMillis();
    if (currentTime > deadline) {
      // Deadline expired
      myGui.notifyUser("Cannot buy book "+title);
      stop();
    }
```

[4] This is not the case for general-purpose behaviours that can be executed by several types of agent.

```
    else {
      // Compute the currently acceptable price and start a
         negotiation
      long elapsedTime = currentTime - initTime;
      int acceptablePrice = maxPrice * (elapsedTime / deltaT);
      myAgent.addBehaviour(new BookNegotiator(title,
         acceptablePrice, this));
    }
  }
}
```

Next the buyer agent requires a `BookNegotiator` behaviour which will negotiate with seller agents. The description of this is deferred until Section 4.3.5 when we discuss agent communication, but for the moment we will just point out that, besides the title of the book to negotiate and the acceptable price, the `BookNegotiator` behaviour takes a pointer to the `PurchaseManager` behaviour that started it. This is because, if the negotiation is successful, the `BookNegotiator` must stop the `PurchaseManager` to avoid purchasing additional copies of the same book.

Since a user may ask his agent to buy several books concurrently, at any given point in time a buyer agent may have several `PurchaseManager` and several `BookNegotiator` behaviours running in parallel.

#### 4.2.5.2 Book-Seller Agent Behaviours

The users of seller agents must provide them with the titles of books to sell together with an initial price, a minimum price, and a deadline. Sellers adopt a linearly decreasing pricing strategy. Therefore, for each book available for sale the seller agents require a `TickerBehaviour` (`PriceManager`) behaviour that simply decreases the current price linearly until the deadline arrives. The behaviour must also remove the book from the catalogue when the deadline expires. The catalogue of books available for sale is implemented as a table that maps the title of a book onto the related `PriceManager` behaviour. Moreover, a seller agent needs two `CyclicBehaviour` (`CallForOfferServer` and `PurchaseOrderServer`) behaviours that serve incoming requests from buyers to obtain an offer for a given book and to purchase a given book respectively. These two behaviours will be shown in Section 4.3 when we will describe agent communication.

Based on this, the code framework for the seller agent is as follows:

```
import jade.core.Agent;
import jade.core.behaviours.*;

import java.util.*;

public class BookSellerAgent extends Agent {
  // The catalogue of books available for sale
  private Map catalogue = new HashMap();
  // The GUI to interact with the user
  private BookSellerGui myGui;

  /**
   * Agent initializations
   */
  protected void setup() {
    // Create and show the GUI
    myGui = new BookSellerGuiImpl();
```

```
    myGui.setAgent(this);
    myGui.show();

    // Add the behaviour serving calls for price from buyer agents
    addBehaviour(new CallForOfferServer());

    // Add the behaviour serving purchase requests from buyer agents
    addBehaviour(new PurchaseOrderServer());
  }

  /**
   * Agent clean-up
   */
  protected void takeDown() {
    // Dispose the GUI if it is there
    if (myGui != null) {
      myGui.dispose();
    }

    // Printout a dismissal message
    System.out.println("Seller-agent "+getAID().getName()+"
       terminating.");
  }

  /**
   * This method is called by the GUI when the user inserts a new
   * book for sale
   * @param title The title of the book for sale
   * @param initialPrice The initial price
   * @param minPrice The minimum price
   * @param deadline The deadline by which to sell the book
   */
  public void putForSale(String title, int initPrice, int minPrice,
     Date deadline) {
    addBehaviour(new PriceManager(this, title, initPrice, minPrice,
       deadline);
  }
}
```

The `PriceManager` behaviour is now implemented as an inner class of the `BookSellerAgent` class. Note that we add the book for sale to the catalogue in the `onStart()` method of the `PriceManager` and not directly in the `putForSale()` method as would be more intuitive. This is because the `putForSale()` method is executed by the GUI thread; it is considered good practice to always access agent resources only via methods executed by the agent thread. This approach ensures that there will be no threading or synchronization problems.

```
private class  PriceManager extends TickerBehaviour {
  private String title;
  private int minPrice, currentPrice, deltaP;
  private long initTime, deadline, deltaT;

  private PriceManager(String t, int ip, int mp, Date d) {
```

```
      title = t;
      initPrice = ip;
      currentPrice = initPrice;
      deltaP = initPrice - mp;
      deadline = d.getTime();
      initTime = System.currentTimeMillis();
    }

    public void onStart() {
      // Insert the book in the catalogue of books available for sale
      catalogue.put(title, this);
      super.onStart();
    }

    public void onTick() {
      long currentTime = System.currentTimeMillis();
      if (currentTime > deadline) {
        // Deadline expired
        myGui.notifyUser("Cannot sell book "+title);
        catalogue.remove(title);
        stop();
      }
      else {
        // Compute the current price
        long elapsedTime = currentTime - initTime;
        currentPrice = initPrice - deltaP * (elapsedTime / deltaT);
      }
    }

    public int getCurrentPrice() {
      return currentPrice;
    }
  }
```

## 4.3 AGENT COMMUNICATION

Agent communication is probably the most fundamental feature of JADE and is implemented in accordance with the FIPA specifications described in Section 2.2.

The communication paradigm is based on *asynchronous message passing*. Thus, each agent has a 'mailbox' (the agent message queue) where the JADE run-time posts messages sent by other agents. Whenever a message is posted in the mailbox message queue the receiving agent is notified. However, when, or if, the agent picks up the message from the queue for processing is a design choice of the agent programmer. This process is depicted in Figure 4.4.

The particular format of messages in JADE is compliant with that defined by the FIPA-ACL message structure described in Section 2.2.3. Each message includes the following fields:

- The *sender* of the message.
- The list of *receivers*.
- The communicative act (also called the '*performative*') indicating what the sender intends to achieve by sending the message. For instance, if the performative is REQUEST, the sender wants the receiver to perform an action, if it is INFORM the sender wants the receiver to be aware of a fact, if it is a PROPOSE or a CFP (Call for Proposals), the sender wants to enter into a negotiation.

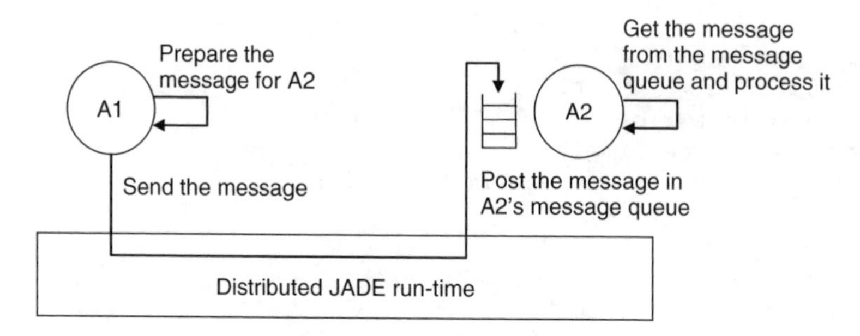


**Figure 4.4** The JADE asynchronous message passing paradigm

- The *content* containing the actual information to be exchanged by the message (e.g., the action to be performed in a REQUEST message, or the fact that the sender wants to disclose in an INFORM message, etc.).
- The content *language* indicating the syntax used to express the content. Both the sender and the receiver must be able to encode and parse expressions compliant with this syntax for the communication to be effective.
- The *ontology* indicating the vocabulary of the symbols used in the content. Both the sender and the receiver must ascribe the same meaning to these symbols for the communication to be effective.
- Some additional fields used to control several concurrent conversations and to specify timeouts for receiving a reply such as *conversation-id, reply-with, in-reply-to* and *reply-by*.

A message in JADE is implemented as an object of the `jade.lang.acl.ACLMessage` class that provides `get` and `set` methods for accessing all fields specified by the ACL format. All performatives defined in the FIPA specification are mapped as constants in the `ACLMessage` class.

### *4.3.1 SENDING MESSAGES*

Sending a message to another agent is as simple as filling out the fields of an `ACLMessage` object and then calling the `send()` method of the `Agent` class. The code below creates a message to inform an agent whose nickname is *Peter* that *today it's raining*:

```
ACLMessage msg = new ACLMessage(ACLMessage.INFORM);
msg.addReceiver(new AID("Peter", AID.ISLOCALNAME));
msg.setLanguage("English");
msg.setOntology("Weather-forecast-ontology");
msg.setContent("Today it's raining");
send(msg);
```

ACL performatives defined by FIPA have well-defined formal semantics that can be exploited to make an agent automatically take proper decisions when a message is received. This advanced feature is not used in our book-trading case study, but will be described later in Chapter 11. Instead we will select the performatives to use in the messages exchanged between buyer and seller agents on the basis of their intuitive meaning. In particular we can conveniently use the CFP (call for proposal) performative for messages that buyer agents send to seller agents to request an offer for a book. The `PROPOSE` performative can be used for messages carrying seller offers and the `ACCEPT_PROPOSAL` performative for messages carrying offer acceptances, i.e. purchase orders. Finally the `REFUSE` performative will be used for messages sent by seller agents when the requested book is not in their catalogue.

In order to keep things as simple as possible, we will put the title of the book to buy into the content of CFP messages sent by buyer agents. Similarly, the content of `PROPOSE` messages carrying seller agents' offers will be the price of the book. Here is how a CFP message can be created and sent by a buyer agent:

```
// Message carrying a call for offer
ACLMessage cfp = new ACLMessage(ACLMessage.CFP);
for (int i = 0; i < sellerAgents.lenght; ++i) {
  // Send this message to all seller agents
  cfp.addReceiver(sellerAgents[i]);
}
cfp.setContent(targetBookTitle);
myAgent.send(cfp);
```

### *4.3.2 RECEIVING MESSAGES*

As previously mentioned, the JADE run-time automatically posts messages into a receiver's private message queue as soon as they arrive. An agent can pick up messages from its message queue by means of the `receive()` method. This method returns the first message in the message queue (thus causing it to be removed), or `null` if the message queue is empty, and immediately returns.

```
ACLMessage msg = receive();
if (msg != null) {
  // Process the message
}
```

### *4.3.3 BLOCKING A BEHAVIOUR WAITING FOR A MESSAGE*

Programmers typically need to implement behaviours that process messages received from other agents. This is the case for the `CallForOfferServer` and `PurchaseOrderServer` behaviours, introduced in Section 4.2.5.2, where we need to serve messages from buyer agents. Such behaviours must be continuously running (cyclic behaviours) and, at each execution of their `action()` method, must check if a message has been received and process it. In our case the two behaviours are very similar. Here we show the `CallForOfferServer` behaviour; the `PurchaseOrderServer` source is available from the online JADE example library.

```
/**
   Inner class CallForOfferServer.
   This is the behaviour used by Book-seller agents to serve
      incoming call for offer from buyer agents.
   If the indicated book is in the local catalogue, the seller agent
      replies with a PROPOSE message specifying the price. Otherwise
      a REFUSE message is sent back.
 */
private class CallForOfferServer extends CyclicBehaviour {
  public void action() {
    ACLMessage msg = myAgent.receive();
    if (msg != null) {
      // Message received. Process it
      String title = msg.getContent();
      ACLMessage reply = msg.createReply();

      PriceManager pm = (PriceManager) catalogue.get(title);
```

```
      if (pm != null) {
        // The requested book is available for sale. Reply with the
           price
        reply.setPerformative(ACLMessage.PROPOSE);
        reply.setContent(String.valueOf(pm.getCurrentPrice()));
      }
      else {
        // The requested book is NOT available for sale.
        reply.setPerformative(ACLMessage.REFUSE);
      }
      myAgent.send(reply);
    }
  }
}  // End of inner class CallForOfferServer
```

As usual we implement the `CallForOfferServer` behaviour as an inner class of the `BookSellerAgent` class. This simplifies things as we can directly access the catalogue of books for sale. Of course this approach is only recommended, not mandatory.

The `createReply()` method of the `ACLMessage` class automatically creates a new `ACLMessage`, automatically setting the receivers and any necessary fields for controlling the conversation (e.g. `conversation-id, reply-with, in-reply-to`).

With reference to Figure 4.3, however, we may notice that as soon as we add the above behaviour, the agent's thread starts a continuous loop that is extremely CPU intensive. On the other hand, we would like the `action()` method of the `CallForOfferServer` behaviour to be executed only when a new message is received. In order to effect this we must use the `block()` method of the `Behaviour` class, which, in spite of what the method name suggests, is not a blocking call, but just marks the behaviour as 'blocked' so that the agent no longer schedules it for execution. When a new message is inserted into the agent's message queue all blocked behaviours become available for execution again so that they have a chance to process the received message. The `action()` method must therefore be modified as follows:

```
public void action() {
  ACLMessage msg = myAgent.receive();
  if (msg != null) {
    // Message received. Process it
    ...
  }
  else {
    block();
  }
}
```

**The above code is the typical, and strongly recommended, pattern for receiving messages inside a behaviour**.

### *4.3.4 SELECTING MESSAGES FROM THE MESSAGE QUEUE*

Considering that both the `CallForOfferServer` and `PurchaseOrderServer` behaviours are cyclic behaviours with an `action()` method that starts with a call to `myAgent.receive()`, you may have noticed a problem: how can we be sure that the `CallForOfferServer` behaviour reads from the agent's message queue only messages carrying calls for offer and the `PurchaseOrderServer` behaviour reads only messages carrying purchase orders? To solve this problem we must modify our current code by specifying 'templates' to be used when calling the `receive()` method. When a template is specified, the `receive()` method returns the first message matching it (if any) and ignores all non-matching messages. Such templates are implemented as instances of the `jade.lang.acl.MessageTemplate` class that provides a number of factory methods to create templates in a very simple and flexible way.

As mentioned in Section 4.3.1, we use the CFP performative for messages carrying calls for offer and the `ACCEPT_PROPOSAL` performative for messages carrying proposal acceptances, i.e. purchase orders. Therefore we can modify the `action()` method of the `CallForOfferServer` such that the call to `myAgent.receive()` ignores all messages except those whose performative is CFP:

```
private MessageTemplate mt =
   MessageTemplate.MatchPerformative(ACLMessage.CFP);

public void action() {
  ACLMessage msg = myAgent.receive(mt);
  if (msg != null) {
    // CFP Message received. Process it
    ...
  }
  else {
    block();
  }
}
```

### 4.3.5 COMPLEX CONVERSATIONS

The `BookNegotiator` behaviour discussed in Section 4.2.5.1 represents an example of a behaviour carrying out a 'complex' conversation. A conversation is a sequence of messages exchanged by two or more agents with well-defined causal and temporal relations. The `BookNegotiator` behaviour sends a CFP message to several agents (the known seller agents) and receives all responses. After that, if at least one `PROPOSE` reply is received, it must send a further `ACCEPT_PROPOSAL` message (to the seller agent that made the best proposal) and wait for a confirmation. Whenever a conversation such as this has to be carried out, it is a good practice to specify the conversation control fields in the messages exchanged within the conversation. This allows the easy creation of unambiguous templates matching the possible replies.

```
/**
   Inner class BookNegotiator.
   This is the behaviour used by Book-buyer agents to actually
      negotiate with seller agents the purchase of a book.
 */
private class BookNegotiator extends Behaviour {
  private String title;
  private int maxPrice;
  private PurchaseManager manager;

  private AID bestSeller; // The seller agent who provides the best
     offer
  private int bestPrice;  // The best offered price
  private int repliesCnt = 0; // The counter of replies from seller
     agents
```

```
  private MessageTemplate mt; // The template to receive replies
  private int step = 0;

  public BookNegotiator(String t, int p, PurchaseManager m) {
    super(null);
    title = t;
    maxPrice = p;
    manager = m;
  }

  public void action() {
    switch (step) {
    case 0:
      // Send the cfp to all sellers
      ACLMessage cfp = new ACLMessage(ACLMessage.CFP);
      for (int i = 0; i < sellerAgents.length; ++i) {
        cfp.addReceiver(sellerAgents[i]);
      }
      cfp.setContent(title);
      cfp.setConversationId("book-trade");
      cfp.setReplyWith("cfp"+System.currentTimeMillis()); // Unique
         value
      myAgent.send(cfp);
      // Prepare the template to get proposals
      mt = MessageTemplate.and(
            MessageTemplate.MatchConversationId("book-trade"),
            MessageTemplate.MatchInReplyTo(cfp.getReplyWith()));
      step = 1;
      break;
    case 1:
      // Receive all proposals/refusals from seller agents
      ACLMessage reply = myAgent.receive(mt);
      if (reply != null) {
        // Reply received
        if (reply.getPerformative() == ACLMessage.PROPOSE) {
          // This is an offer
          int price = Integer.parseInt(reply.getContent());
          if (bestSeller == null || price < bestPrice) {
            // This is the best offer at present
            bestPrice = price;
            bestSeller = reply.getSender();
          }
        }
        repliesCnt++;
        if (repliesCnt >= sellerAgents.length) {
          // We received all replies
          step = 2;
        }
      }
      else {
        block();
      }
```

```
      break;
    case 2:
      if (bestSeller != null && bestPrice <= maxPrice) {
        // Send the purchase order to the seller that provided the
           best offer
        ACLMessage order = new
           ACLMessage(ACLMessage.ACCEPT_PROPOSAL);
        order.addReceiver(bestSeller);
        order.setContent(title);
        order.setConversationId("book-trade");
        order.setReplyWith("order"+System.currentTimeMillis());
        myAgent.send(order);
        // Prepare the template to get the purchase order reply
        mt = MessageTemplate.and(
               MessageTemplate.MatchConversationId("book-trade"),
               MessageTemplate.MatchInReplyTo
                  (order.getReplyWith()));
        step = 3;
      }
      else {
        // If we received no acceptable proposals, terminate
        step = 4;
      }
      break;
    case 3:
      // Receive the purchase order reply
      reply = myAgent.receive(mt);
      if (reply != null) {
        // Purchase order reply received
        if (reply.getPerformative() == ACLMessage.INFORM) {
          // Purchase successful. We can terminate
          myGui.notifyUser("Book "+title+" successfully purchased.
             Price = "
            + bestPrice);
          manager.stop();
        }
        step = 4;
      }
      else {
        block();
      }
      break;
    }
  }

  public boolean done() {
    return step == 4;
  }
} // End of inner class BookNegotiator
```

Complex conversations are typically carried out following a well-defined interaction protocol, such as those defined by FIPA. JADE provides rich support for several of the most commonly used interaction protocols in the `jade.proto` package. The conversation we implemented above, for example, follows a 'Contract-net' protocol which could be very easily implemented by exploiting the `jade.proto.ContractNetInitiator` class. This will be described further in Section 5.4.

### *4.3.6 RECEIVING MESSAGES IN BLOCKING MODE*

Besides the `receive()` method, the `Agent` class also provides the `blockingReceive()` method that, as the name suggests, is a blocking call: it does not return until there is a message in the agent's message queue. An overloaded version that takes a `MessageTemplate` as a parameter (it does not return until there is a message matching the specified template) is also available.

It is important to stress that the `blockingReceive()` methods actually blocks the agent thread. Therefore if you call `blockingReceive()` from within a behaviour, this prevents all other behaviours from executing until the call to `blockingReceive()` returns. Taking this into account, a good programming practice is to receive messages using `blockingReceive()` in the `setup()` and `takeDown()` methods; use `receive()` in combination with `Behaviour.block()` (as shown in Section 4.3.3) within behaviours.

## 4.4 AGENT DISCOVERY: THE YELLOW PAGES SERVICE

In the code we have written so far we have assumed that there is a fixed set of seller agents (passed to each buyer agent as start-up arguments). In this chapter we describe how remove this assumption by exploiting the yellow pages service provided by the JADE platform to allow buyer agents to dynamically discover available seller agents at a given point in time.

### *4.4.1 THE DF AGENT*

A 'yellow pages' service allows agents to publish descriptions of one or more services they provide in order that other agents can easily discover and exploit them. This is depicted in Figure 4.5.

Any agent can both register (publish) services and search for (discover) services. Registrations, deregistrations, modifications and searches can be performed at any time during an agent's lifetime.

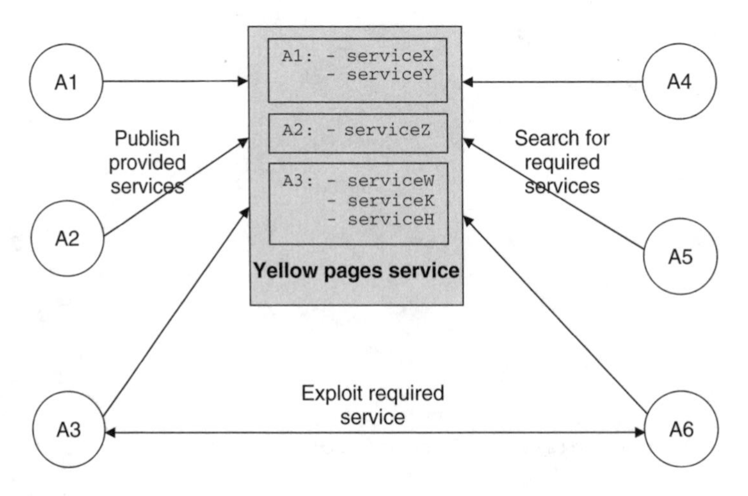


**Figure 4.5** The yellow pages service

The yellow pages service in JADE, in accordance with the FIPA Agent Management specification, is provided by a specialized agent called the DF (Directory Facilitator). Every FIPA-compliant platform should host a default DF agent (whose local name is 'df@<platform-name>'). Other DF agents can be deployed if required and several DF agents (including the default) can be federated to provide a single distributed yellow pages catalogue.

### *4.4.2 INTERACTING WITH THE DF*

As the DF is an agent, it is possible to interact with it as with any other agent by exchanging ACL messages using a proper content language (e.g. the SL0 language) and a proper ontology (e.g. the FIPA-agent-management ontology) as defined in the FIPA specifications. In order to simplify these interactions, JADE provides the `jade.domain.DFService` class with which it is possible to publish and search for services through a variety of method calls.

#### 4.4.2.1 Publishing Services

An agent wishing to publish one or more services must provide the DF with a description that includes its own AID, a list of provided services and optionally the list of languages and ontologies that other agents must use to interact with it. Each published service description must include the service type, the service name, the languages and ontologies required to use the service and a collection of service-specific properties in the form of key-value pairs. The `DFAgentDescription`, `ServiceDescription` and `Property` classes, included in the `jade.domain.FIPAAgentManagement` package, represent these abstractions.

In order to publish a service an agent must create a proper description (as an instance of the `DFAgentDescription` class) and call the `register()` static method of the `DFService` class. With reference to our book-trading example, seller agents register their ability to sell books (a service of type 'Book-selling') in their `setup()` method as follows:

```
protected void setup() {
  ...
  // Register the book-selling service in the yellow pages
  DFAgentDescription dfd = new DFAgentDescription();
  dfd.setName(getAID());
  ServiceDescription sd = new ServiceDescription();
  sd.setType("Book-selling");
  sd.setName(getLocalName()+"-Book-selling");
  dfd.addServices(sd);
  try {
    DFService.register(this, dfd);
  }
  catch (FIPAException fe) {
    fe.printStackTrace();
  }
  ...
}
```

Note that in this simple example we do not specify any language, ontology or service-specific properties.

When an agent terminates it is good practice to deregister published services:

```
protected void takeDown() {
  // Deregister from the yellow pages
```

```
    try {
      DFService.deregister(this);
    }
    catch (FIPAException fe) {
      fe.printStackTrace();
    }
    ...
  }
```

*4.4.3 SEARCHING FOR SERVICES*

An agent wishing to search for services must provide the DF with a template description. The result of the search is a list of all the descriptions that match the provided template. According to the FIPA specifications, a description matches the template if all the fields specified in the template are present in the description with the same values.

The `search()` static method of the `DFService` class can be used as exemplified in the code used by the book buyer agents to maintain an up-to-date the list of seller agents:

```
public class BookBuyerAgent extends Agent {
  // The list of known seller agents
  private Vector sellerAgents;

  protected void setup() {
    ...
    // Update the list of seller agents every minute
    addBehaviour(new TickerBehaviour(this, 60000) {
      protected void onTick() {
        // Update the list of seller agents
        DFAgentDescription template = new DFAgentDescription();
        ServiceDescription sd = new ServiceDescription();
        sd.setType("Book-selling");
        template.addServices(sd);
        try {
          DFAgentDescription[] result = DFService.search(myAgent,
             template);
          sellerAgents.clear();
          for (int i = 0; i < result.length; ++i) {
            sellerAgents.addElement(result[i].getName());
          }
        }
        catch (FIPAException fe) {
          fe.printStackTrace();
        }
      }
    } );
    ...
```

Note that the search is repeated once every minute since seller agents may dynamically appear and disappear in the system. The JADE DF also provides a subscription mechanism that allows agents to be notified as soon as other agents register or deregister given services. Exploiting this mechanism (that would be more appropriate in our case) requires initiating a FIPA-Subscribe protocol with the DF. We will therefore describe this in Section 5.4 when discussing interaction protocols.

## 4.5 AGENTS WITH A GUI

A typical problem that developers have to manage is how JADE agents should interact with their GUI (graphical user interface) and vice versa. The issue is that proper inter-thread communication programming patterns must be used between the agent thread that must be woken whenever an ACL message is received and the AWT event-dispatching thread that the AWT wakes up whenever AWT components (i.e. GUI components) fire different types of events (e.g. the user clicked a button).

Typical problems that should be avoided are reacting to an AWT event by blocking the event dispatcher thread until an ACL message is received, or updating the GUI from within the agent thread, or modifying unsynchronized variables from both threads. There follows a few recommendations of good programming practices to avoid some of these problems.

### *4.5.1 GOOD PROGRAMMING PRACTICE FOR AN AWT ACTIONLISTENER*

When an agent has a GUI, it typically needs to react to user actions, such as starting a new conversation when the user presses a button. When the AWT action event occurs, the method `actionPerformed()` is invoked by the event dispatcher thread on the `ActionListener` registered for that event source. Within the body of this method, a good programming practice is to prepare a JADE `Behaviour` object and schedule it for execution by the agent thread.

There follows a fragment of code from the RMA Agent of JADE (class `jade.tools.rma.rma`). This method is invoked when the user interacts with the GUI and selects an agent to kill. The method shows how a behaviour is instantiated, its arguments prepared, and then scheduled for execution.

```
public void actionPerformed(ActionEvent e) {
  /*omissis*/
  AgentTree.Node curNode =
                   (AgentTree.Node)panel.treeAgent.tree
                      .getSelectionPath();
  rma.killAgent(new AID(curNode.getName(), AID.ISLOCALNAME));
  /*omissis*/
}

public void killAgent(AID name) {
  KillAgent ka = new KillAgent();
  ka.setAgent(name);
  try {
    Action a = new Action();
    a.setActor(getAMS());
    a.setAction(ka);
    ACLMessage requestMsg = getRequest();
    requestMsg.setOntology(JADEManagementOntology.NAME);
    getContentManager().fillContent(requestMsg, a);
    addBehaviour(new AMSClientBehaviour("KillAgent", requestMsg));
  } catch(Exception fe) {
     fe.printStackTrace();
  }
}
```

Recall that behaviours are scheduled for execution only after the `setup()` method of the agent object has terminated. Behaviours are always executed by the agent thread. As a consequence, no synchronization between different behaviours is required.

Of course, in some cases, adding a behaviour might be an unnecessary burden when reaction to an AWT action might be simply modifying the value of a variable or preparing and sending an

ACLMessage (remember that message delivery is fully asynchronous). These are all valid activities for the AWT thread and, in general, do not cause any problem. In contrast, blocking calls (e.g. `Agent.blockingReceive()`) should never be executed within the AWT thread.

### *4.5.2 GOOD PROGRAMMING PRACTICE FOR GUI MODIFICATION BY THE AGENT THREAD*

Having said that the agent has its own thread of execution, expert Java programmers will immediately infer that updating the GUI from within this thread may lead to unexpected problems due to synchronization issues. AWT, Swing, MIDP (and most other UI frameworks) provide a proper ad hoc method that queues a `Runnable` object and causes it to be executed asynchronously on the GUI event dispatching thread:

- `java.awt.EventQueue.invokeLater()` for AWT;
- `javax.swing.SwingUtilities.invokeLater()` for Swing;
- `javax.microedition.lcdui.Display.callSerially()` for MIDP.

Therefore, the recommended programming practice is to encapsulate into a `Runnable` object all accesses to GUI objects from a JADE behaviour and, in general, from a thread which is not the `EventDispatchThread`. Then, the proper method should be used to submit this `Runnable` object to the `EventDispatchThread`.

There follows a fragment of code from the RMA Agent of JADE class (`jade.tools.rma.rma`). This method is invoked when the RMA Agent receives a message informing that a new agent has been created on a given container. The method shows how the agent thread creates an instance of a new `Runnable` object and submits it to the `EventDispatchThread`. This `Runnable` object is responsible for updating the GUI by creating a new `javax.swing.tree.TreeNode` object and adding it to the `JTree` of the given container.

```
public void addAgent(final String containerName, final AID agentID)
    {
  Runnable addIt = new Runnable() {
    public void run() {
      String agentName = agentID.getName();
      AgentTree.Node node = tree.treeAgent.createNewNode(agentName,
         1);
      /* [omissis] */
      tree.treeAgent.addAgentNode((AgentTree.AgentNode)node,
         containerName, agentName, agentAddresses, "FIPAAGENT");
    }
  };
  SwingUtilities.invokeLater(addIt);
}
```

# 5

# Programming with JADE – Advanced Features

In Chapter 4, the basic features of the JADE platform have been described. With these features it is already possible to develop distributed multi-agent systems such as the book-trading case study. However, implementing real-world applications by means of only these features is in general quite complex. Many of the problems that developers have to face in such situations can be better addressed by using the advanced features of the JADE platform that are presented in this chapter. These concern the manipulation of complex content expressions by means of ontologies and content language codecs, the possibility of building complex tasks (behaviours) by composing simple ones, the support for creating conversations ruled by the interaction protocols defined by FIPA and more.

## 5.1 ONTOLOGIES AND CONTENT LANGUAGES

As described in Section 4.3, the actual information that is transferred from the sender to the receivers of an ACL message is included in the content slot of the message. According to the FIPA specifications the value of this slot is either a string or a raw sequence of bytes. In the simple messages we have seen so far the content was just the title of a book or an integer representing the price of a book. In more realistic cases, however, agents often need to communicate more complex information. Let us get back to the book-trading case study presented in Chapter 4 and consider that there may be two or more books titled the same. In order to unequivocally identify a book, it may this be necessary to specify at least the title, the authors and, if present, the editor. When representing complex information such as a book with this additional information as a string, it is necessary to adopt a well-defined syntax so that the content of a message can be parsed by the receiver to extract each specific piece of information (the title, the authors and the editor in our case). According to FIPA terminology this syntax is known as a *content language*. FIPA does not mandate a specific content language but defines and recommends the SL language to be used when communicating with the AMS and DF. An example of book-related information encoded with the SL language is as follows:

```
(Book :title "Programming Multi Agent Systems with JADE" :authors
   (sequence "F. Bellifemine" "G. Caire" "D. Greenwood") :editor
   Wiley)
```

*Developing Multi-Agent Systems with JADE* Fabio Bellifemine, Giovanni Caire, Dominic Greenwood

When receiving such a string an agent must be able to parse the SL syntax in order to actually understand the information it represents. Additionally, it must have some shared understanding with the sender of the book concept and the symbols ('book', 'title', 'authors' and 'editor') used to express this structure. This set of concepts and the symbols used to express them are known as an *ontology*. Unlike content languages that are typically domain independent, ontologies are typically specific to a given domain. Thus, the concept of book, for example, is relevant in our book-trading domain while it would be most likely meaningless in a telecommunications management application.

Clearly, while String representation of complex information is suitable for embedding the information inside an ACL message, it is rather inconvenient when an agent has to process it. With reference to the above example, each time an agent needs to extract the editor of a given book it needs to parse the entire content expression. However, as JADE agents are Java-based, content information can conveniently be represented using Java objects. For example, representing the selected parameters relating to the 'Programming Multi Agent Systems with JADE' book as an instance (a Java object) of an application-specific class would look something like this:

```
public class Book {
  String title;
  List authors;
  String editor;

  public String getTitle() {return title;}
  public void setTitle(String t) {title = t;}
  public List getAuthors() {return authors;}
  public void setAuthors(List l) {authors = l;}
  public String getEditor() {return editor;}
  public void setEditor(String s) {editor = s;}
  ...
}
```

setting `title = "Programming Multi Agent Systems with JADE"` and so on.

It is clear, however, that, even though this eases information handling inside an agent, each time a message is exchanged:

1. The sender needs to convert its internal representation into the corresponding ACL content expression representation, and the receiver needs to perform the opposite conversion.
2. The receiver should perform a number of semantic checks to verify that the received information complies with the rules (for instance that the editor of a book is actually a string) of the ontology shared by the communicating agents.

The support for content languages and ontologies provided by JADE is designed to automatically perform all the above conversion and check operations, as depicted in Figure 5.1. This allows developers to manipulate information within their agents as Java objects without the need for any additional marshalling or unmarshalling work.

One query here is why is Java serialization not enough? Well of course Java serialization is a very simple and powerful means to convert Java objects into sequences of bytes. In fact, as JADE agents are essentially just pieces of Java code, developers are free to use serialization to insert Java objects into the content slot of ACL messages. However, Java serialization does have some disadvantages. Firstly, it is only applicable in a Java environment. If a JADE agent has to communicate with another agent living on a remote FIPA-compliant platform other than JADE, there is absolutely no guarantee that the receiver can understand a message whose content slot was encoded using Java serialization. Secondly, Java serialization produces a non-human-readable format. In many cases being able to read the content slot of a message is very helpful when investigating problems, such

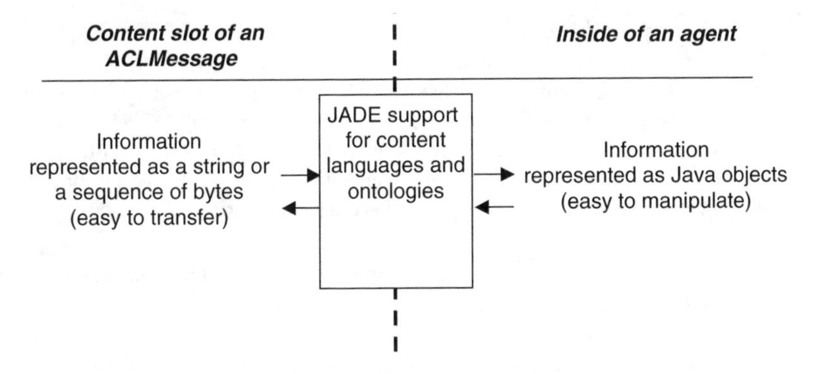


**Figure 5.1** The JADE support for content languages and ontologies

as when using the Sniffer agent described in Section 3.7.3. Thirdly, an agent receiving a message has no means of determining the kind of object it will obtain when decoding the content slot – any serializable object could be received in principle.

One technology we may consider as a readily available means to solve this problem is XML. As a cross-platform, language-independent representation which is also human-readable, XML or DTDs would allow the relatively simplistic definition of structures containing the concepts that agents share when communicating and acting in a multi-agent system. As with Java serialization, XML is also an option JADE programmers may employ, and is actually adopted by members of the JADE open source community to help deal with complex content expressions: JAXB (JSR222), for instance, is commonly used to encode Java beans back and forth into XML. Again, however, using XML requires developers to have some additional skills with respect to pure Java programming. Moreover, XML encoding and decoding tools such as JAXB are often not entirely straightforward to understand and use. Java developers often prefer writing code rather than dealing with XML files.

The support for ontologies provided by JADE is available as an option to ease the programmer's burden in dealing with complex content. Developers are of course free to choose between the dedicated JADE classes, Java serialization, XML or other technology. The goal of this section is not to provide an in-depth explanation of the JADE content language and support for ontologies, but rather to provide usage guidelines. Readers interested in the details of these features may refer to the JADE Tutorial on Content Languages and Ontologies available on the JADE website.

### *5.1.1 MAIN ELEMENTS*

The conversion and check operations described in the previous section are carried out by a **content manager** object (i.e. an instance of the `ContentManager` class included in the `jade.content` package). Each JADE agent embeds a content manager accessible through the `getContent-Manager()` method of the `Agent` class. The `ContentManager` class provides all the methods needed to transform Java objects into strings (or sequences of bytes) and to insert them into the content slot of `ACLMessages`, and vice versa.

The content manager provides a convenient interface to access the conversion functionality but actually just delegates the conversion and check operations to an *ontology* (i.e. an instance of the `Ontology` class included in the `jade.content.onto` package) and a *content language codec* (i.e. an instance of the `Codec` interface included in the `jade.content.lang` package). More specifically, the ontology validates the information to be converted from a semantic point of view while the codec performs the translation into strings (or sequences of bytes) according to the syntactic rules of the related content language.

### *5.1.2 THE CONTENT REFERENCE MODEL*

In order for JADE to perform proper semantic checks on a given content expression it is necessary to classify all elements in the domain of discourse (i.e. elements that can appear within a valid sentence sent by an agent as the content of an ACL message), according to their generic semantic characteristics. This classification is derived from the ACL language defined by FIPA that requires the content of each ACL message to have proper semantics according to the performative of the ACL message. At the highest level we distinguish between predicates and terms:

- *Predicates* are expressions that say something about the status of the world and can be either `true` or `false`, e.g.:

  ```
  (Works-for (Person :name John) (Company :name "Telecom Italia"))
  ```

  This states that 'the person John works for the company Telecom Italia'. Predicates can be used effectively as the content of an INFORM or QUERY-IF message, for example, both of which express facts, while would make no sense if used as the content of a REQUEST message.
- *Terms* are expressions identifying entities (abstract or concrete) that 'exist' in the world and that agents may reason about. They are further classified into:
  - *Primitives* are atomic entities such as strings and integers.
  - *Concepts* are entities with a complex structure that can be defined in terms of slots, e.g.:

    ```
    (Person :name John :age 33)
    ```

  - Concepts typically make no sense if used directly as the content of an ACL message. In general they are referenced inside predicates and other concepts such as

    ```
    (Book :title "The Lord Of The Rings":author (Person :name "J.R.R.
    Tolkien"))
    ```

  - *Agent actions* are special concepts that indicate actions that can be performed by some agents, e.g.:

    ```
    (Sell (Book :title "The Lord Of The Rings") (Person :name John))
    ```

    It is useful to treat agent actions separately since, unlike 'normal' concepts, they are the meaningful contents of certain types of ACLMessage such as REQUEST. Communicative acts (i.e. ACL messages) are themselves agent actions.
  - *Aggregates* are entities that are groups of other entities, e.g.:

    ```
    (sequence (Person :name John) (Person :name Bill))
    ```

  - *Identifying referential expressions (IRE)* are expressions that identify the entity (or entities) for which a given predicate is true, e.g.:

    ```
    (all ?x (Works-for ?x (Company :name "Telecom Italia"))
    ```

    This identifies 'all the elements x for which the predicate `(Works-for x (Company :name "Telecom Italia"))` is `true`, i.e. all the people that work for the company `"Telecom`

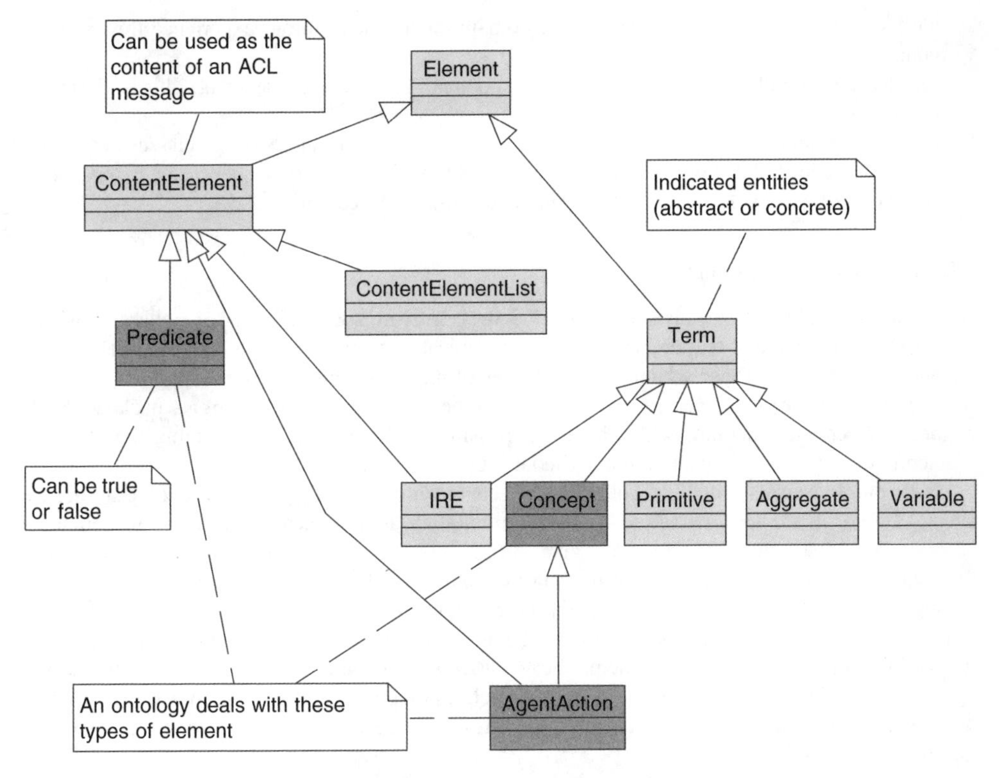


**Figure 5.2** The content reference model

`Italia"`). These expressions require variables and are most typically used in queries (e.g. as the content of a QUERY_REF message).

- *Variables* are expressions, typically used in queries, that indicate a generic entity that is not known a priori.

A fully expressive content language should be able to represent and distinguish between all the above types of element. An ontology for a given domain is a set of schemas defining the structure (essentially names and slots) of the predicates, concepts and agent actions that are pertinent to that domain.

The final content reference model, depicted in Figure 5.2, includes two more element types. The *ContentElementList* is a list of the three primary types: predicates, agent actions and IREs. The *ContentElement* is a super-type from which the primary and list types inherit.

### *5.1.3 USING JADE CONTENT LANGUAGE AND ONTOLOGY SUPPORT*

Exploiting the JADE content language and ontology support to allow agents to discourse and reason about facts and knowledge related to a given domain is achieved by the following steps:

1. Define an ontology including the schemas for the types of predicate, agent action and concept that are pertinent to the addressed domain. This is discussed in Section 5.1.3.1.
2. Develop proper Java classes for all types of predicate, agent action and concept in the ontology. This is discussed in Section 5.1.3.2.
3. Select a suitable content language among those directly supported by JADE. This is discussed in Section 5.1.3.3. New user-defined content languages can be created by implementing the codec

interface, but, since content languages are domain independent, in our experience this is never required.
4. Register the defined ontology and the selected content language with the agent. This is discussed in Section 5.1.3.4.
5. Create and handle content expression as Java objects that are instances of the classes developed in step 2 and let JADE translate these Java objects to/from strings or sequences of bytes that fit the `content` slot of `ACLMessages`. This is discussed in Section 5.1.3.5.

#### 5.1.3.1 Defining an Ontology

An ontology in JADE is an instance of the `jade.content.onto.Ontology` class to which schemas have been added that define the types of predicates, agent actions and concepts relevant to the addressed domain. These schemas are instances of the `PredicateSchema`, `AgentActionSchema` and `ConceptSchema` classes included in the `jade.content.schema` package. Each of these classes have methods with which it is possible to declare the slots defining the structure of each type of predicate, agent action and concept.

As an ontology is essentially a collection of schemas that typically does not evolve during an agent lifetime, it is good practice to declare the ontology as a singleton object and to define an ad hoc class (that extends `jade.content.onto.Ontology`) with a static method to access this singleton object. This allow the same ontology object (and all included schemas) to be shared among different agents in the same Java Virtual Machine (JVM).

In our book-trading example we can model the addressed domain by means of a simple ontology that includes one concept (`BOOK`), one predicate (`COSTS`, associating a book with a price) and one agent action (`SELL`, to be applied to a book). Each ontology in JADE normally extends a basic ontology, represented as a singleton object of the `jade.content.onto.BasicOntology` class, that includes the schemas for:

- the primitive types (`STRING, INTEGER, FLOAT, ...`),
- the aggregate type, and
- some generic (i.e. not belonging to any specific domain) predicates, agent actions and concepts among which is the AID concept identifying an agent.

In order to declare that an ontology `O1` extends an ontology `O2` (i.e. all predicates, agent actions and concepts included in `O2` are also included in `O1`), it is sufficient to pass `O2` as a parameter when `O1` is constructed.

Finally, with reference to the book-trading project structure set up in Section 4.1.5, since the ontology is shared among buyers and sellers, we place the `BookTradingOntology` class in an ad hoc package, `ontology`, just under the top-level package `bookTrading`. Taking into account all the above issues, the ontology for the book-trading domain can be defined as follows:

```
package bookTrading.ontology;

import jade.content.onto.*;
import jade.content.schema.*;

public class BookTradingOntology extends Ontology {
  // The name identifying this ontology
  public static final String ONTOLOGY_NAME =
    "Book-trading-ontology";

  // VOCABULARY
```

```
  public static final String BOOK = "Book";
  public static final String BOOK_TITLE = "title";
  public static final String BOOK_AUTHORS = "authors";
  public static final String BOOK_EDITOR = "editor";

  public static final String COSTS = "Costs";
  public static final String COSTS_ITEM = "item";
  public static final String COSTS PRICE = "price";

  public static final String SELL = "Sell";
  public static final String SELL_ITEM = "item";

  // The singleton instance of this ontology
  private static Ontology theInstance = new BookTradingOntology();

  // Retrieve the singleton Book-trading ontology instance
  public static Ontology getInstance() {
    return theInstance;
  }

  // Private constructor
  private BookTradingOntology() {
    // The Book-trading ontology extends the basic ontology
    super(ONTOLOGY_NAME, BasicOntology.getInstance());

    try {
      add(new ConceptSchema(BOOK), Book.class);
      add(new PredicateSchema(COSTS), Costs.class);
      add(new AgentActionSchema(SELL), Sell.class);

      // Structure of the schema for the Book concept
       ConceptSchema cs = (ConceptSchema) getSchema(ITEM);
      cs.add(BOOK_TITLE, (PrimitiveSchema)
          getSchema(BasicOntology.STRING));
      cs.add(BOOK_AUTHORS, (PrimitiveSchema)
          getSchema(BasicOntology.STRING), 0,
             ObjectSchema.UNLIMITED);
      cs.add(BOOK_EDITOR, (PrimitiveSchema)
          getSchema(BasicOntology.STRING), ObjectSchema.OPTIONAL);

      // Structure of the schema for the Costs predicate
      ...
      // Structure of the schema for the Sell agent action
       AgentActionSchema as = (AgentActionSchema) getSchema(SELL);
      as.add(SELL_ITEM, (ConceptSchema) getSchema(BOOK));
    }
    catch (OntologyException oe) {
       oe.printStackTrace();
    }
  }
}
```

From this code it is possible to observe that:

- Each schema added to the ontology is associated with a Java class, e.g. the schema for the `BOOK` concept is associated with the `Book.java` class. While using the defined ontology, expressions indicating books will be instances of the `Book` class. These Java classes must have a proper structure as described in Section 5.1.3.2.
- Each slot in a schema has a name and a type, i.e. values for that slot must comply with a given schema.
- A slot can be declared as `OPTIONAL` meaning that its value can be `null`. Otherwise a slot is considered `MANDATORY`. If a `null` value for a `MANDATORY` slot is encountered in the validation of a content expression, an `OntologyException` is thrown.
- A slot can have cardinality >1, i.e. values for that slot are aggregates of elements of a given type. For example, the `authors` slot in the schema for the `BOOK` concept can contain 0 or more elements of type String.

It is also possible to define specialization/extension relationships among concepts. For example, if we had to extend our system to support trading of other types of goods such as CDs, we could define a generic `ITEM` schema and add it as super-schema to both the `BOOK` schema, the `CD` schema and so on. A schema automatically inherits all slots included in its super-schemas. Super-schemas are added by means of the `addSuperSchema()` method of the `ConceptSchema` class.

#### 5.1.3.2 Developing Ontological Java Classes

As mentioned in Section 5.1.3.1, each schema included in an ontology is associated with a Java class (or interface). Clearly the structure of these classes must be coherent with the associated schemas, i.e. they must obey the following rules:

1. They must implement a proper interface:
   - If the schema is a `ConceptSchema` the class must implement (either directly or indirectly) the `Concept` interface.
   - If the schema is a `PredicateSchema` the class must implement (either directly or indirectly) the `Predicate` interface.
   - If the schema is a `AgentActionSchema` the class must implement (either directly or indirectly) the `AgentAction` interface.

   The above interfaces are simply tagged interfaces (i.e. they do not include any method) and are part of a hierarchy that follows the content reference model presented in Section 5.1.2. They are included in the `jade.content` package.
2. They must have the proper inheritance relations, i.e. if S1 is a super-schema of S2 then the class C2 associated with schema S2 must extend the class C1 associated with schema S1.
3. They must have the proper member fields and accessor methods:
   - For each slot in schema S1 with name `Nnn` and type (i.e. whose schema is) S2, the class C1 associated with schema S1 must have two accessor methods with the following signature:

     ```
     public void setNnn(C2 c);
     public C2 getNnn();
     ```

     where C2 is the class associated with schema S2. In particular, if S2 is a schema defined in the BasicOntology, then
     - if S2 is the schema for STRING → C2 is `java.lang.String`
     - if S2 is the schema for INTEGER → C2 is `int`, `long`, `java.lang.Integer` or `java.lang.Long`[1]

[1] Users can choose among these options according to their preferences.

  - if S2 is the schema for BOOLEAN → C2 is `boolean` or `java.lang.Boolean`
  - if S2 is the schema for FLOAT → C2 is `float`, `double`, `java.lang.Float` or `java.lang.Double`
  - if S2 is the schema for DATE → C2 is `java.util.Date`
  - if S2 is the schema for BYTE_SEQUENCE → C2 is `byte[]`
  - if S2 is the schema for AID → C2 is `jade.core.AID`
- For each slot in schema S1 with name `Nnn`, type S2 and cardinality $>1$, the class C1 associated with schema S1 must have two accessor methods with the following signature:

```
public void setNnn(jade.util.leap.List l);
public jade.util.leap.List getNnn();
```

To exemplify the rules that ontological classes must obey, the classes associated with the `BOOK` concept and the `COSTS` predicate in the book-trading example are as follows:

```
// Class associated to the BOOK schema
package bookTrading.ontology;

import jade.content.Concept;
import jade.util.leap.List;

public class Book implements Concept {
  private String title;
  private List authors;
  private String editor;

  public String getTitle() {
    return title;
  }
  public void setTitle(String title) {
    this.title = title;
  }
  public List getAuthors() {
    return authors;
  }
  public void setAuthors(List authors) {
    this.authors = authors;
  }
  public String getEditor() {
    return editor;
  }
  public void setEditor(String editor) {
    this.editor = editor;
  }

}


// Class associated to the COSTS schema
package bookTrading.ontology;

import jade.content.Predicate;
```

```
import jade.core.AID;

public class Costs implements Predicate
  private Book item;
  private int price;

  public Book getItem() {
    return item;
  }
  public void setItem(Book item) {
    this.item = item;
  }
  public int getPrice() {
    return price;
  }
  public void setPrice(int price) {
    this.price = price;
  }
}
```

The `jade.util.leap` package includes a set of classes and interfaces that provide the same features as the Java Collection Framework (JCF), but is designed to operate with MIDP. Using these classes allows the development of applications that are portable across all Java editions, configurations and profiles. Of course, developers who do not have to take MIDP compatibility into account may not wish to deal with these classes, instead preferring to use those provided by default in the `java.util` package. In this case, using standard `java.util` classes instead of those provided by the `jade.util.leap` package can be simply achieved by passing the `CFReflectiveIntrospector` (Collection Framework-based Reflective Introspector) to the constructor of the ontology that is being defined. This is coded as follows:

```
private BookTradingOntology() {
  super(ONTOLOGY_NAME, BasicOntology.getInstance(),
        new CFReflectiveIntrospector());
```

Actually an ontology delegates to an introspector (i.e. an instance of the `jade.content.onto.Introspector` interface) all operations associated with reading and writing fields of the ontological Java classes. By writing a proper introspector developers can therefore modify the rules that ontological Java classes must obey, as they see fit.

#### 5.1.3.3 Selecting a Content Language

The `jade.content` package directly includes codecs for two content languages: the SL language and the LEAP language. In addition, a codec for an XML-based syntax is available in the XMLCodec add-on found in the add-ons area of the JADE website. All these codecs support the content reference model described in Section 5.1.2. A codec for content language *L* is a Java object able to manage content expressions written in the *L* language. In the great majority of the cases a developer can just adopt one of the three content languages mentioned above and use the related codec without any additional effort. This section gives some hints to assist with choosing which one. If a developer wants his agents to ‘speak’ a different content language he has to create a proper codec by implementing the `jade.content.lang.Codec` interface. This is not a trivial task, however, and in our experience is rarely, if ever, required.

The **SL content language** is a human-readable string-encoded content language (i.e. a content expression in SL is a string), based on an S-Expression syntax. All content examples in this book are expressed in SL, and in general we recommend adoption of SL for open agent-based applications where agents produced by different developers and running on different platforms must communicate. SL has a number of useful operators including logical operators such as `AND`, `OR` and `NOT`, and modal operators such as `BELIEF`, `INTENTION` and `UNCERTAINTY`. The reader should refer to the specific tutorial on the JADE website for a description of how to properly use them. Moreover, an additional property of SL is that it is human-readable, which can be very helpful when debugging and testing an application.

The **LEAP content language** is a non-human-readable byte-encoded content language (i.e. a content expression in LEAP is a sequence of bytes) that has been defined specifically for JADE within the LEAP project (LEAP). It is therefore clear that only JADE agents will be able to natively 'speak' the LEAP language. There are some cases in which the LEAP language is preferable over SL:

- The `LEAPCodec` class is lighter than the `SLCodec` class, thus when there are strong memory limitations the LEAP language is preferable.
- Unlike the LEAP language, the SL language does not support sequences of bytes.

The developer should also take into account that the SL language deals particularly with agent actions. All agent actions in SL must be inserted into the ACTION construct (included in the `BasicOntology` and implemented by the `jade.content.onto.basic.Action` class) that associates the agent action to the `AID` of the agent that is intended to perform the action. Therefore the expression

```
(Sell
      (Book :title "Developing Multi Agent Systems with JADE" ...)
)
```

cannot be used directly as the content of e.g. a REQUEST message even if it corresponds to an agent action in the content reference model. In fact the SL grammar does not allow it as a first-level expression. The following expression must be used instead:

```
(action
  (agent-identifier :name seller-X)
  (Sell
    (Book :title "Developing Multi Agent Systems with JADE" ...)
  )
)
```

The **XML content language** uses an XML syntax as exemplified below:

```
<action>
  <agent-identifier>
    <name>seller-X</name>
  </agent-identifier>
  <Sell>
    <Book>
      <title>Developing Multi Agent Systems with JADE</title>
    </Book>
  </Sell>
</action>
```

This codec is particularly useful when a set of ontological entities has to be exported or imported to/from an external system.

#### 5.1.3.4 Registering Content Languages and Ontologies with an Agent

Before an agent can actually use an ontology and a content language, it must register them with its content manager. This operation is typically, but not necessarily, performed during agent set-up (i.e. in the `setup ()` method of the `Agent` class). The following code shows this registration in the case of the seller agent (the buyer agent looks the same) assuming that the SL language is selected for our book-trading example.

```
public class BookSellerAgent extends Agent {
  ...
  private Codec codec = new SLCodec();
  private Ontology ontology = BookTradingOntology.getInstance();
  ...
  protected void setup() {
    ...
    getContentManager().registerLanguage(codec);
    getContentManager().registerOntology(ontology)
    ...
  }
  ...
}
```

From now on the content manager will associate the registered `Codec` and `Ontology` objects to the strings returned by their respective `getName()` methods. Note that while it is generally good practice to have a singleton `Ontology` object, this is not the case for `Codec` objects as synchronization problems can arise during parsing operations.

#### 5.1.3.5 Creating and Manipulating Content Expressions as Java Objects

Having defined an ontology (and the classes associated with the types of predicate, agent action and concept it includes), selected a proper language, and registered them to the agent's content manager, creating and manipulating content expressions as Java objects is straightforward. For instance, the code below shows how the `CallForOfferServer` behaviour presented in Section 4.3.3 can be modified to take advantage of the JADE support for managing ontologies and content languages.

```
private class CallForOfferServer extends CyclicBehaviour {
  public void action() {
    ACLMessage msg = myAgent.receive();
    if (msg != null) {
      // Message received. Process it
      ACLMessage reply = msg.createReply();
      try {
        ContentManager cm = myAgent.getContentManager();
        Action act = (Action) cm.extractContent(msg);
        Sell sellAction = (Sell) act.getAction();
        Book book = sellAction.getItem();

        PriceManager pm = (PriceManager)
          catalogue.get(book.getTitle());
```

```
        if (pm != null) {
          // The requested book is available for sale
          reply.setPerformative(ACLMessage.PROPOSE);
          ContentElementList cel = new ContentElementList();
          cell.add(act);
          Costs costs = new Costs();
          costs.setItem(book);
          costs.setPrice(pm.getCurrentPrice());
          cel.add(costs);
          cm.fillContent(reply, cel);
        }
        else {
          // The requested book is NOT available for sale.
          reply.setPerformative(ACLMessage.REFUSE);
        }
      }
      catch (OntologyException oe) {
        oe.printStackTrace();
        reply.setPerformative(ACLMessage.NOT_UNDERSTOOD);
      }
      catch (CodecException ce) {
        ce.printStackTrace();
        reply.setPerformative(ACLMessage.NOT_UNDERSTOOD);
      }
      myAgent.send(reply);
    }
  }
} // End of inner class CallForOfferServer
```

In the `extractContent()` and `fillContent()` methods the seller agent's content manager obtains from its internal table of registered ontologies and codecs those entries corresponding to the values of the `:ontology` and `:language` slots of the message passed as a parameter and lets them perform the necessary conversion and check operations.

This is how the `:content` slot of the `PROPOSE` reply would look in the case of a call for offer, referring to the 'Developing Multi Agent System with JADE' book being received by a seller agent called seller@JADE-book-trading that is currently selling that book at 30 euros.

```
((action
   (agent-identifier :name seller@JADE-book-trading)
   (Sell
     (Book :title "Developing Multi Agent Systems with JADE"
           :authors (sequence Bellifemine Caire Greenwood)
           :editor Wiley) ) )
 (Costs
   (Book :title "Developing Multi Agent Systems with JADE"
         :authors (sequence Bellifemine Caire Greenwood)
         :editor Wiley)
   30))
```

#### 5.1.3.6 Combining Ontologies

The support for content languages and ontologies included in the `jade.content` package provides an easy way to combine ontologies thus aiding code reuse. In particular, it is possible to define a

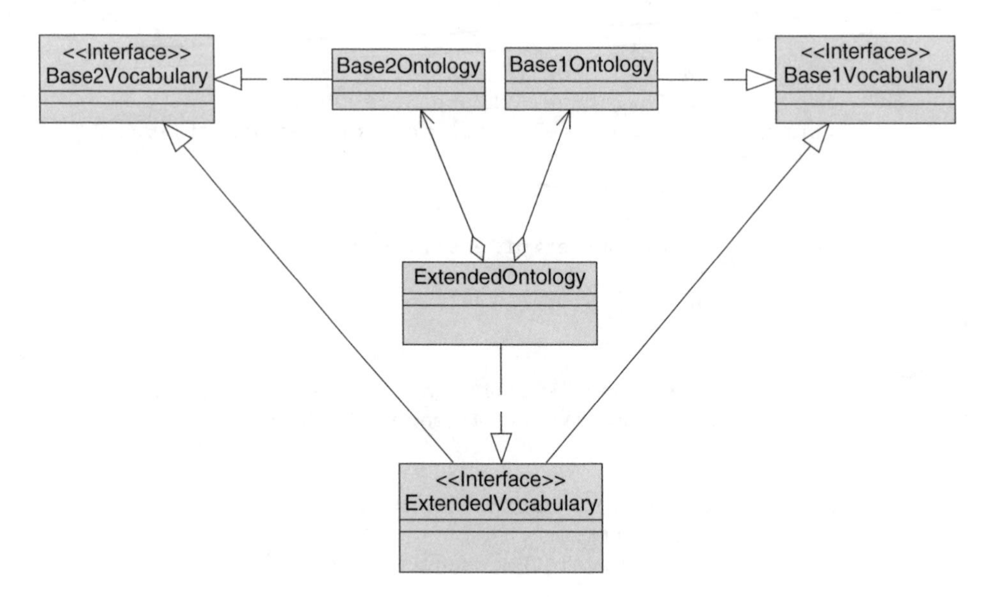


**Figure 5.3** The vocabulary interface pattern

new ontology that extends one or more (previously defined) ontologies by simply specifying the extended ontologies as parameters in the constructor used to create the new ontology.

When doing this we suggest employing the vocabulary interface pattern shown in Figure 5.3. All the symbols used for names of concepts, predicates, agent actions and their slots are grouped into an interface representing the vocabulary. In our book-trading example we could therefore have a BookTradingVocabulary interface as follows:

```
public interface BookTradingVocabulary {
  // VOCABULARY
  public static final String BOOK = "Book";
  public static final String BOOK_TITLE = "title";
  public static final String BOOK_AUTHORS = "authors";
  public static final String BOOK_EDITOR = "editor";

  public static final String COSTS = "Costs";
  public static final String COSTS_ITEM = "item";
  public static final String COSTS PRICE = "price";

  public static final String SELL = "Sell";
  public static final String SELL_ITEM = "item";
}
```

The ontology class implements the vocabulary interface.

When extending one or more ontologies the resulting ontology class includes the extended ones, while the resulting vocabulary interface inherits from the extended ones. By organizing things in this way, each vocabulary constant can be accessed as if it were defined in the resulting ontology class regardless of where it is actually defined.

### *5.1.4 USING PROTÉGÉ AND THE BEAN GENERATOR ADD-ON TO CREATE JADE ONTOLOGIES*

When dealing with large ontologies, developing the ontology definition class (i.e. the schemas) and the Java classes representing the predicates, agent actions and concepts 'by hand', as described in Sections 5.1.3.1 and 5.1.3.2, can be quite time-consuming. Thanks to a JADE add-on called the *Bean Generator*, created by C. J. van Aart from the Department of Social Science Informatics at the University of Amsterdam, it is possible to define the ontology using the [Protégé] tool and then let the Bean Generator automatically create the ontology definition class plus the predicates, agent actions and concepts classes. The Bean Generator is described in more detail in Section 13.1

## 5.2 COMPOSING BEHAVIOURS TO CREATE COMPLEX TASKS

As described in Section 4.2, tasks in JADE are implemented as instances of classes that extend `jade.core.behaviours.Behaviour` and implement the `action()` and `done()` methods. It is quite clear, however, that when dealing with complex tasks that involve several computational steps, possibly mixed with conversations with other agents and so on, it is not convenient to implement the complete task logic within the `action()` method of a single 'fat' behaviour. Let us consider, for example, the `BookNegotiator` behaviour presented in Section 4.3.5. Though it involves simply exchanging a few messages and taking one decision (whether or not to accept an offer from one of the seller agents), its `action()` method is already quite complex.

A simpler and cleaner approach to implement complex tasks in JADE is behaviour composition – creating complex tasks by composing simple behaviours. The basis for this feature is provided by the `CompositeBehaviour` class included in the `jade.core.behaviours` package. A composite behaviour (an instance of the `CompositeBehaviour` class) is itself a behaviour that embeds a number of child sub-behaviours. The `CompositeBehaviour` class already implements the `action()` method in such a way that, each time it is called, it invokes the `action()` method of one of its children. The policy used to select which child to fire at each round is delegated to the `scheduleFirst()` (first round) and `scheduleNext()` (successive rounds) methods. These methods are declared abstract and must be defined in `CompositeBehaviour` subclasses that implement the actual types of compositions.

Three ready-to-use types of composite behaviour are provided in the JADE distribution. These are `SequentialBehaviour`, `FSMBehaviour` and `ParallelBehaviour` and will be detailed in the following sections. The complete hierarchy of classes included in the `jade.core.behaviour` package is depicted in Figure 5.4. The classes in the dashed box in Figure 5.4 follow the composite pattern described in Section 5.2.5. Sub-behaviours embedded in a composite behaviour can also be composite behaviours, thus making it possible to create very complex tasks by hierarchically combining simple behaviours together.

Typically developers do not need to directly extend `CompositeBehaviour` and just use its concrete subclasses `SequentialBehaviour`, `FSMBehaviour` and `ParallelBehaviour`. In case an ad hoc sub-behaviour scheduling policy is required, the reader is referred to Section 5.2.5 where more details about the `CompositeBehaviour` class are provided.

### *5.2.1 THE SEQUENTIALBEHAVIOUR CLASS*

The `SequentialBehaviour` class implements a composite behaviour that schedules its children according to a very simple sequential policy. It starts with the first child; when this is finished (i.e. its `done()` method returned `true`) it moves to the second one and so on. When the last child is completed, the whole sequential behaviour terminates. Figure 5.5 depicts the logical flow of operations

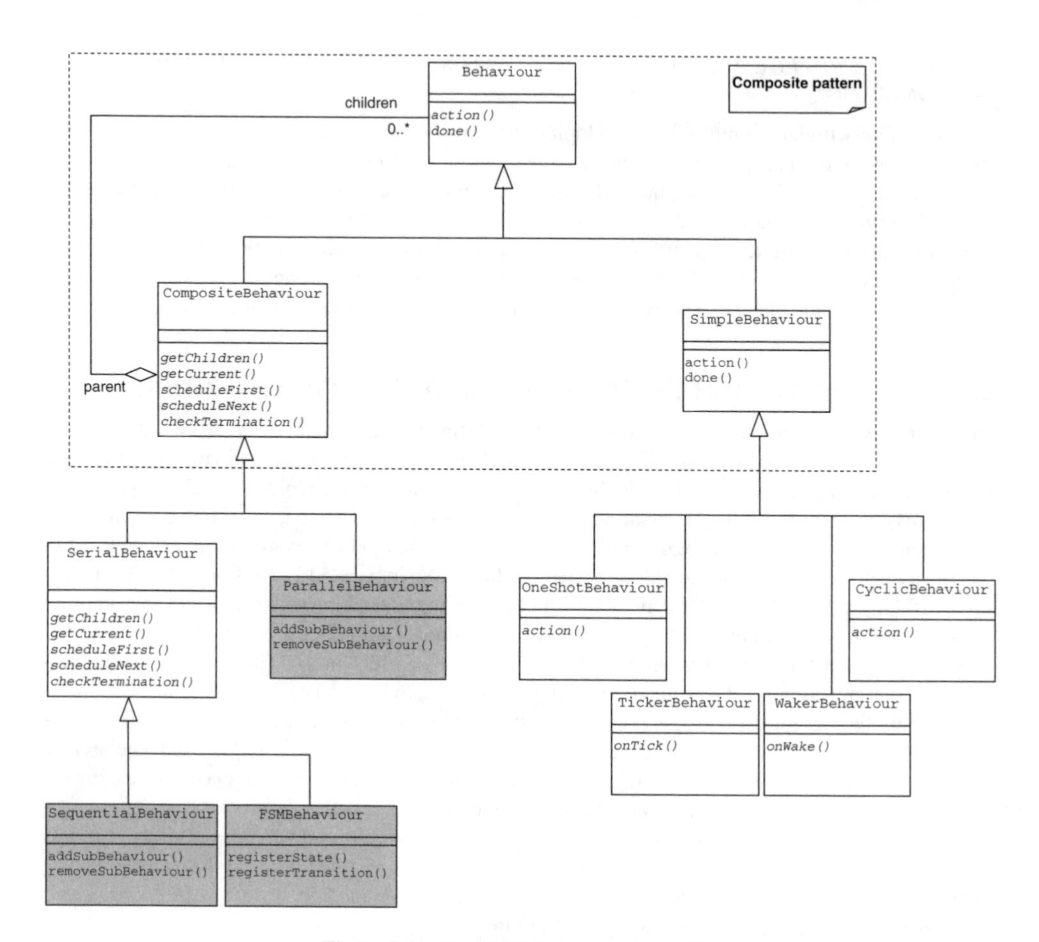


**Figure 5.4** The JADE behaviours hierarchy

that are performed each time a `SequentialBehaviour` is scheduled and its `action()` method is executed.

Sub-behaviours in a sequential behaviour are added by means of the `addSubBehaviour()` method. The order in which sub-behaviours are added determines the order in which they are scheduled by the sequential behaviour. As an example, the `ThreeStepBehaviour` class presented in Section 4.2.2 could have been conveniently implemented using the `SequentialBehaviour` class as follows:

```
SequentialBehaviour threeStepBehaviour = new
   SequentialBehaviour(anAgent);
threeStepBehaviour.addSubBehaviour(new OneShotBehaviour(anAgent) {
  public void action() {
       // perform operation X
  }
} );
threeStepBehaviour.addSubBehaviour(new OneShotBehaviour(anAgent) {
  public void action() {
       // perform operation Y
  }
```

```
} );
threeStepBehaviour.addSubBehaviour(new OneShotBehaviour(anAgent) {
  public void action() {
      // perform operation Z
  }
} );
```

Executing the `threeStepBehaviour` above is completely equivalent to executing an instance of the `ThreeStepBehaviour` class presented as an example in Section 4.2.2. The code in this example, however, is considerably more compact.

### *5.2.2 THE FSMBEHAVIOUR CLASS*

The `FSMBehaviour` class implements a composite behaviour that schedules its children according to a finite state machine (FSM) whose states correspond to the FSM behaviour children. The

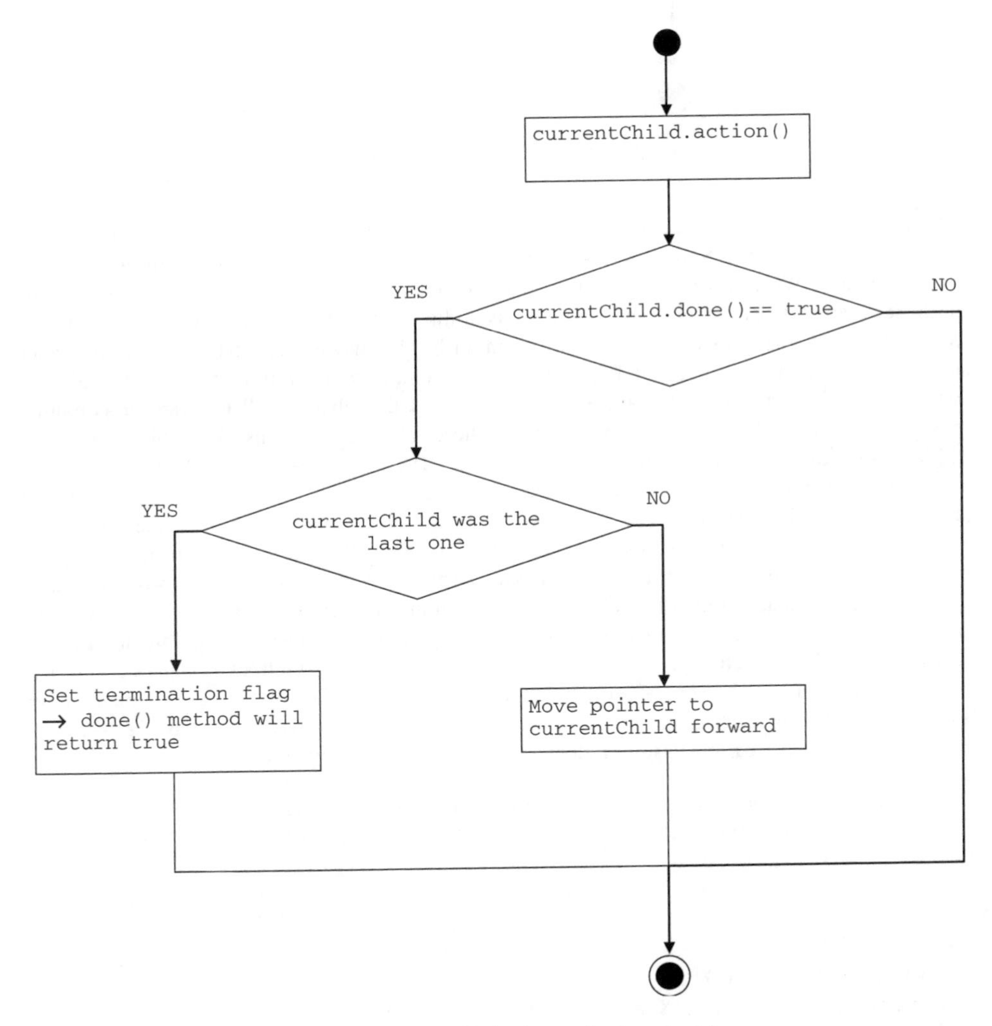


**Figure 5.5** SequentialBehaviour action() method flow

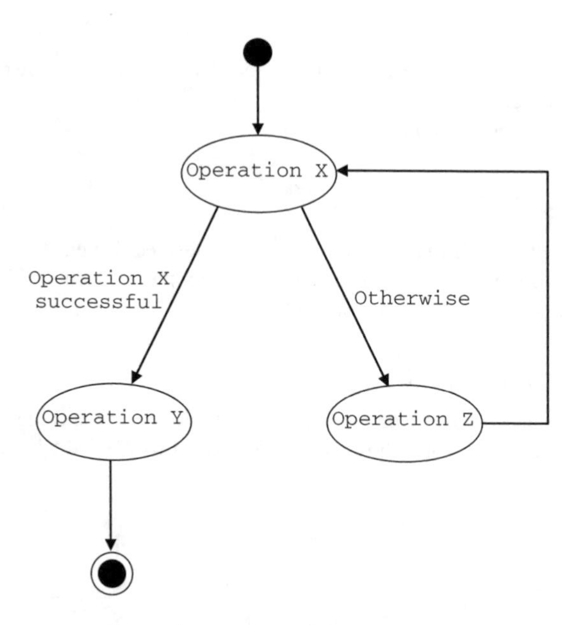


**Figure 5.6** A simple finite state machine

FSMBehaviour class provides methods to register sub-behaviours as FSM states and to register transitions between states. Similar to a sequential behaviour, an FSM behaviour keeps a pointer to the current child. When this is completed (its done() method returned true), the FSM behaviour checks its internal transition table and, on the basis of this, selects the new child to fire the next time its action() method is executed. Transitions in an FSM behaviour are marked with an integer label. When the current child of the FSM behaviour is completed, the return value of its onEnd() method is taken as an exit value and is matched against the labels of all the transitions exiting from the current child state. The first transition whose label matches this exit value is followed and its destination state becomes the new current child. The registerState() method, used to add states to an FSMBehaviour instance, accepts two arguments: a String defining the name of the state that is being registered and a Behaviour that will be executed in that state. The registerTransition() method, used to add transitions to an FSMBehaviour instance, accepts three arguments: two Strings defining the source state and the destination state of the transition and an int value defining the label marking the transition. The registerFirstState() and registerLastState() methods allow us to register the entrance state and the termination states of the FSM respectively. It should be noticed, however, that while there must be only one entrance state, there can be several termination states. The whole FSM behaviour terminates when a termination state is reached and fully executed. As an example, a task that behaves as the FSM depicted in Figure 5.6 can be easily implemented with the following code:

```
FSMBehaviour sampleFSM = new FSMBehaviour(anAgent);
sampleFSM.registerFirstState(new OneShotBehaviour(anAgent) {
  public void action() {
    // Perform operation X
  }

  public int onEnd() {
    return (operation X successful ? 1 : 0);
  }
```

```
}, "X");
sampleFSM.registerLastState(new OneShotBehaviour(anAgent) {
  public void action() {
    // Perform operation Y
  }
}, "Y");
sampleFSM.registerState(new OneShotBehaviour(anAgent) {
  public void action() {
    // Perform operation Z
  }
}, "Z");
sampleFSM.registerTransition("X", "Y", 1);
sampleFSM.registerTransition("X", "Z", 0);
sampleFSM.registerDefaultTransition("Z", "X", new String[]{"X",
   "Z"});
```

The `registerDeafultTransition()` method of the `FSMBehaviour` class allows the definition of a default transition between two states. A default transition is not marked with any label and is followed if and only if all other transitions (if any) exiting from the same state are not followed. Both the `registerTransition()` and the `registerDefaultTransition()` methods exist in an overloaded version taking a further `String[]` parameter. This parameter indicates a set of FSM states that must be reset when the registered transition is followed. This is very useful when registering 'backward transitions', i.e. transitions that lead to states that have already been visited. In fact, as explained in Section 4.2.3, any `Behaviour` object that has been executed once, must be reset by calling its `reset()` method before it can be executed again. For instance, with reference to Figure 5.6, if the transition from Z to X is followed, state X and possibly state Z will be executed one more time. Before this occurs they must be reset to avoid undesired effects.

### *5.2.3 THE PARALLELBEHAVIOURS CLASS*

The `ParallelBehaviour` class implements a composite behaviour that schedules its children in parallel. As usual, when dealing with JADE behaviours, scheduling is cooperative and not preemptive. This means that each time the `action()` method of a parallel behaviour is executed, it invokes the `action()` method of the current child and then moves the pointer forward to the next child regardless of whether the latter was completed or not. Sub-behaviours in a parallel behaviour are added by means of the `addSubBehaviour()` method. A parallel behaviour can be instructed to terminate when all of its children have completed or, alternatively, when the first child completes. The termination policy is selected at instantiation time by specifying in the constructor either the `WHEN_ALL` or the `WHEN_ANY` constants defined in the `ParallelBehaviour` class. A typical use of the `ParallelBehaviour` class with the `WHEN_ANY` termination policy is to abort a task in case it does not complete within a given timeout, as exemplified in the code below:

```
Behaviour task = new MyTask();
ParellelBehaviour pb = new ParallelBehaviour(anAgent,
    ParallelBehaviour.WHEN_ANY);
pb.addSubBehaviour(task);
pb.addSubBehaviour(new WakerBehaviour(anAgent, 60000) {
  public void onWake() {
    System.out.println("timeout expired");
  }
});
```

### *5.2.4 SHARING DATA AMONG SUB-BEHAVIOURS: THE DATASTORE*

When composing behaviours within a sequential, FSM or parallel behaviour it is typically the case that a sub-behaviour will need to access some data produced by other sub-behaviours. Of course these data cannot be passed as parameters to the sub-behaviour constructor since all children of a composite behaviour are typically constructed before the whole composite behaviour is executed. As usual when behaviours need to share data, it is possible to use member variables of the agent or of the parent composite behaviour. Let us assume, for example, that we need a sequential behaviour that at step $n$ must receive a message and at step $n + 1$ must do some processing of the received message. We could create a code similar to this:

```
public class MySequentialBehaviour extends SequentialBehaviour {
  private ACLMessage receivedMsg;

  public MySequentialBehaviour(Agent a) {
    super(a);

    // ...
    addSubBehaviour(new SimpleBehaviour(a) {
      private boolean finished = false;

      public void action() {
        receivedMsg = myAgent.receive();
        if (receivedMsg != null) {
          finished = true;
        }
        else {
          block();
        }
      }

      public boolean done() {
        return finished;
      }
    } );

    addSubBehaviour(new OneShotBehaviour(a) {
      public void action() {
        // Process receivedMsg
      }
    } );
  }
}
```

In many cases, however, it is useful to create behaviours that can be reused in different contexts and are therefore not bound to a given agent or to a given parent composite behaviour. In these cases data shared among behaviours cannot be stored in agent or parent composite behaviour member variables. The `DataStore` class included in the `jade.core.behaviours` package provides a simple and generic solution to this problem. Each behaviour has its own data store (i.e. an instance of the `DataStore` class) accessible by means of the `getDataStore()` and `setDataStore()` methods of the `Behaviour` class. A data store is basically a map (actually `DataStore` extends `HashMap`) and provides the 'standard' mechanism by means of which behaviours designed to be reusable are supposed to share data. That is, by setting the same `DataStore` instance to two or more behaviours, these behaviours have a common space where they can store data that must be shared. For instance, let us assume we have a generic `MessageReceiver` behaviour that receives a message and we want to use it in step *n* of the above example. Here is how the `MessageReceiver` class could be coded:

```
public class MessageReceiver extends SimpleBehaviour {
  public static final String RECV_MSG = "received-message";

  private boolean finished = false;

  public void action() {
    ACLMessage msg = myAgent.receive();
    if (msg!= null) {
      getDataStore().put(RECV_MSG, msg);
      finished = true;
    }
    else {
      block();
    }
  }

  public boolean done() {
    return finished;
  }
}
```

And here is how we could modify the sequential behaviour to take advantage of the `MessageReceiver` class:

```
SequentialBehaviour sb = new SequentialBehaviour(anAgent);

Behaviour b = new MessageReceiver(anAgent);
b.setDataStore(sb.getDataStore());
sb.addSubBehaviour(b);

b = new OneShotBehaviour(anAgent) {
  public void action() {
    ACLMessage receivedMsg = getDataStore()
       .get(MessageReceiver.RECV_MSG);
    // Process receivedMsg
  }
};

b.setDataStore(sb.getDataStore());
sb.addSubBehaviour(b);
```

### *5.2.5 MORE ABOUT COMPOSITE BEHAVIOURS*

In this section more details about composite behaviours are described which will be useful to gain a deeper understanding of the behaviour composition support provided by JADE.

Besides the child scheduling policy a `CompositeBehaviour` subclass must also define a termination criterion and a blocking/restarting policy. The former specifies when the composite behaviour terminates. For instance, a `SequentialBehaviour` that (as described in Section 5.2.1) schedules its children one after the other, terminates when the last child is completed. The latter specifies how block and restart events (i.e. calls to the `block()` and `restart()` methods) in the composite behaviours are propagated to its children and vice versa.

A child scheduling policy is implemented (as already mentioned at the beginning of this section) by redefining the following abstract methods of the `CompositeBehaviour` class:

- `getCurrent()` – this method is intended to return the current child to run and is invoked each time the `CompositeBehaviour action()` method is executed.
- `scheduleFirst()` – this method is called just once as soon as the `CompositeBehaviour` starts and is intended to set the first child to be executed.
- `scheduleNext()` – this method has the same meaning of `scheduleFirst()`, but is called each successive time.

The termination criterion is implemented by redefining the `checkTermination()` abstract method of the `CompositeBehaviour` class. This method is invoked after the execution of

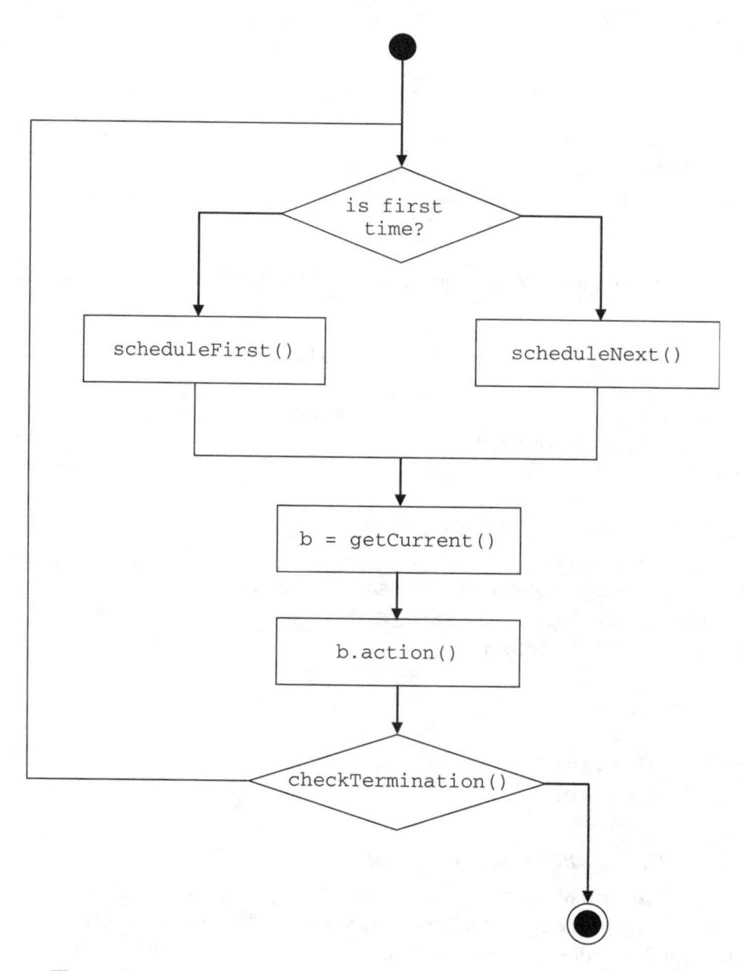


**Figure 5.7** `CompositeBehaviour action()` method operation flow

the action() method of the current child. When it returns true the CompositeBehaviour terminates too.

Figure 5.7 schematizes (in a simplified way) the flow of operations carried out by the CompositeBehaviour action() method.

The blocking/restarting policy is implemented by overriding the handle() method of the Behaviour class. Taking into account all block and restart propagation details is definitely not a trivial task and therefore two ready-to-use blocking/restarting policies, that cover the great majority of possible cases, are made available by means of the SerialBehaviour and ParallelBehaviour classes. In more detail, the SerialBehaviour extends CompositeBehaviour overriding the handle() method so that a block/restart event on the parent CompositeBehaviour is propagated downwards to the current child only and, similarly, only block/restart events occurring on the current child are propagated upwards to the parent. The ParallelBehaviour on the other hand propagates block/restart events downwards to all its children. As for upwards propagation, it becomes runnable when at least one of its children is runnable and becomes blocked when all its children are blocked. It should be noticed that while SerialBehaviour is an abstract class that only implements a blocking/restarting policy, the ParallelBehaviour is a ready-to-use class that also implements a children scheduling policy and a termination criterion as described in Section 5.2.3.

The last noteworthy bit of information about composite behaviours concerns the getParent() method of the Behaviour class. This method allows a sub-behaviour to get a pointer to its parent composite behaviour. If a behaviour is not part of any behaviour composition hierarchy (or if it is the root of a composition hierarchy) the getParent() method simply returns null.

## 5.3 THREADED BEHAVIOURS

As mentioned in Section 4.2.1, behaviour scheduling is performed in a non-pre-emptive way. That is, the action() method of a behaviour is never interrupted to allow other behaviours to step in. Only when the action() method of the currently running behaviour returns, is control given to the next behaviour. As already discussed, this approach has several advantages in terms of performance and scalability. However, when a behaviour needs to perform some blocking operation, it actually blocks the whole agent and not only itself. A possible solution to this is to use normal Java threads. JADE, however, provides a cleaner solution by means of threaded behaviours, i.e. behaviours that are executed in dedicated threads.

Any JADE behaviour (simple or composite) can be executed as a threaded behaviour by means of the jade.core.behaviours.ThreadedBehaviourFactory class. This class provides the wrap() method that wraps a normal JADE Behaviour into a threaded behaviour wrapper. This threaded behaviour wrapper is itself a behaviour. Adding it to the agent by means of the addBehaviour() method thus results in executing the original Behaviour object in a dedicated thread. It should be noted that developers only deal with the ThreadedBehaviourFactory class, while the actual class of threaded behaviour wrappers is private and not accessible.

The sample code below shows how to execute a JADE behaviour in a dedicated Java thread:

```
import jade.core.*;
import jade.core.behaviours.*;
public class ThreadedAgent extends Agent {
  private ThreadedBehaviourFactory tbf = new
     ThreadedBehaviourFactory();

  protected void setup() {
    // Create a normal JADE behaviour
    Behaviour b = new OneShotBehaviour(this) {
      public void action() {
```

```
            // Perform some blocking operation that can take a long time
        }
    };

    // Execute the behaviour in a dedicated Thread
    addBehaviour(tbf.wrap(b));
  }
}
```

Threaded behaviours can be mixed with normal behaviours inside composite behaviours. For example, a `SequentialBehaviour` may have two children executed as normal behaviours and a third child executed in a dedicated thread. The `ParallelBehaviour` class in particular can be used to assign a group of behaviours to a single dedicated thread.

There are some important points that must be taken into account when dealing with threaded behaviours:

- *The removeBehaviour() method of the Agent class has no effect on threaded behaviours.* A threaded behaviour is removed by retrieving its `Thread` object using the `getThread()` method of the `ThreadedBehaviourFactory` class and calling its `interrupt()` method.
- When an agent dies, moves or suspends, its active *threaded behaviours must be explicitly killed* using the technique described above.
- If a child of a parallel behaviour configured with the **`WHEN_ANY`** termination policy is a threaded behaviour, the termination of another child does not stop it. Once more the threaded child must be explicitly killed as described above.
- When a threaded behaviour accesses some agent resources that are also accessed by other threaded or non-threaded behaviours, proper attention must be paid to it synchronization issues.

## 5.4 INTERACTION PROTOCOLS

At this stage the reader should already be quite familiar with the FIPA-ACL language used by JADE agents to communicate. This language provides a standardized set of primitives (the performatives or communicative acts); each one with a well-defined semantics. One of the major advantages of this characteristic is the possibility to specify predefined sequences of messages that can be applied in several situations that share the same communication pattern regardless of the application domain. Such sequences of messages are known as interaction protocols.

Let us consider, for instance, task delegation. A coordinator must select an actor to delegate a task to, within a group of actors all able to perform the task. Each actor has different characteristics in terms of time, quality, cost and so on. This is a situation that programmers often have to face in many application domains. For example, a print manager connected to different printers must decide which printer to select depending on how many jobs each printer has in queue. Or, with respect to the book-trading case study, a buyer agent that must decide which seller agent to choose in case more than one has the requested book available for sale. Regardless of the application domain (trading books, printing documents, etc.), the task delegation problem can be solved by inducing the coordinator and the actors to follow a sequence of messages, such as that depicted in Figure 5.8, where the initiator corresponds to the coordinator and the participants correspond to the actor. This interaction protocol is known as the FIPA-Contract-Net (as described in Section 2.2.3.6).

Other examples of common situations that can be addressed by adopting standard interaction protocols are auctions, subscriptions to receive notifications, negotiations and so on. As described in Section 2.2, FIPA specifies several interaction protocols including ones that address these situations. For example, the FIPA-Request protocol can be used to request one or more agents to perform a given action and collect results; the FIPA-Subscribe protocol can be used to establish a notification

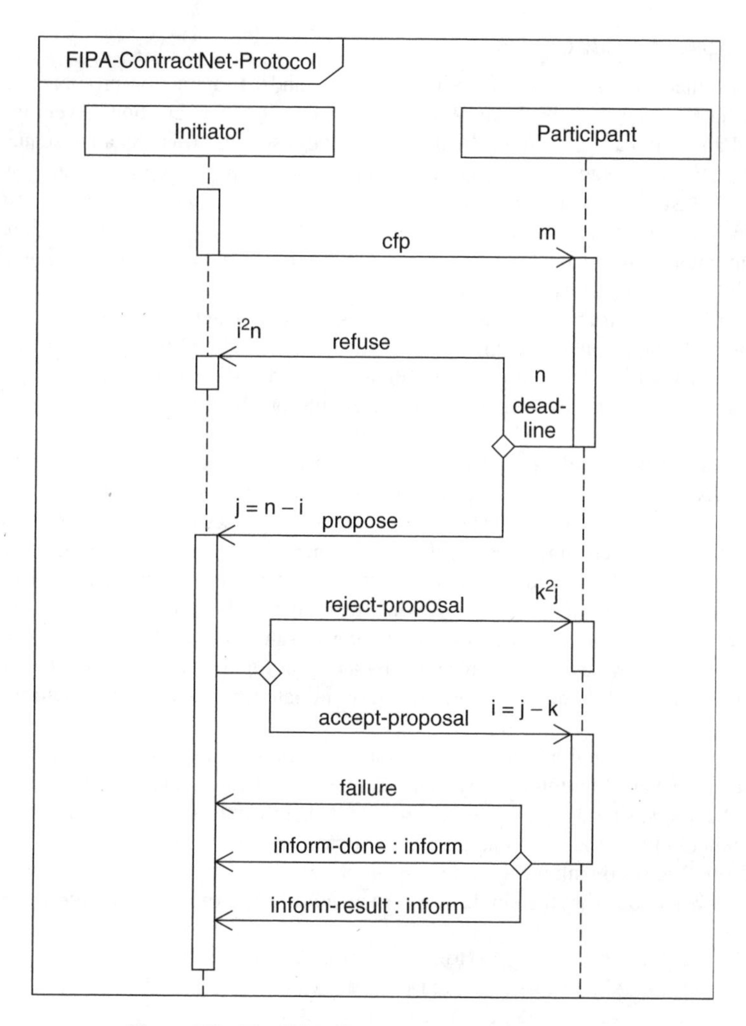


**Figure 5.8** The FIPA-Contract-Net interaction protocol

agreement with another agent to send an inform each time a given condition becomes true; the already mentioned FIPA-Contract-Net protocol and many more.

Of course, the added value in a library of standard interaction protocols is not limited to design (i.e. it is not just a suggestion about which messages agents should exchange in a given interaction). JADE in fact provides ready-made classes that free the programmer from the burden of implementing all the checks related to the flow of messages when two or more agents interact following a standard interaction protocol. These classes provide a number of callback methods that programmers are expected to redefine by inserting logic associated with the specific domain that cannot be generalized. For instance, in our book-trading scenario, the decision whether or not to accept an offer from a seller agent and which seller agent to select if more than one provided a suitable offer, strictly depends on the domain and must be implemented by the developer. All the checks related to the fact that each seller agent replies with either a `PROPOSE` or a `REFUSE` and that, if an offer is accepted, the selected seller agent confirms or disconfirms the trade with an `INFORM` or a `FAILURE`, are provided automatically if we use the `ContractNetInitiator` class.

### 5.4.1 *THE* JADE.PROTO *PACKAGE*

All the classes that provide support for implementing standard interaction protocols in JADE are included in the `jade.proto` package. When participating in a conversation driven by an interaction protocol an agent can play either the initiator or the responder role. As a consequence classes in the `jade.proto` package are divided into initiators and responders. For instance, we have the `ContractNetInitiator` and the `ContractNetResponder`, the `SubscriptionInitiator` and the `SubscriptionResponder` and so on. Playing a role in a conversation, no matter if it is the initiator or responder role, implies executing a task of some sort and thus all protocol classes (both initiators and responders) are JADE behaviours.

All initiator class constructors contain an `ACLMessage` parameter representing the message used to initiate the protocol. For instance, the `ContractNetInitiator` class gets the `CFP` message to be sent to responders to initiate the task delegation procedure. All protocol initiator classes support both one-to-one and one-to-many interactions depending on the number of receivers specified in the initiation message.

Responder classes are available in two versions. The cyclic version has a `MessageTemplate` parameter in its constructor, used to select protocol initiation messages from initiators. A cyclic protocol responder behaviour is typically added in the agent set-up and remains active for the whole agent lifetime. Each time a protocol initiation message matching the template is received, the protocol responder behaviour processes it, carries out the conversation and then goes back waiting for a new initiation message. Thus, when using the cyclic version of a protocol responder behaviour, an agent can be engaged in only one conversation driven by that protocol at a time. For instance, if two `CFP` messages are received by an agent running a `ContractNetResponder` behaviour, the latter will complete the conversation initiated by the first `CFP` before serving the second one.

The single session version has an incoming initiation message in its constructor, carries out the protocol-driven conversation initiated by that message and then terminates. Unlike cyclic protocol responder behaviours, single session protocol responder behaviours are not responsible for receiving the message that initiates the conversation they will carry out. There must be an external behaviour receiving them. The following piece of code exemplifies this in the case of the `SSContractNetResponder` behaviour (i.e. the single session version of the `ContractNetResponder`).

```
MessageTemplate template = MessageTemplate.and(
  MessageTemplate.MatchProtocol("fipa-contract-net"),
  MessageTemplate.MatchPerformative(ACLMessage.CFP) );

addBehaviour(new CyclicBehaviour(this) {
  public void action() {
    ACLMessage cfp = myAgent.receive(template);
    if (cfp != null) {
      myAgent.addBehaviour(new SSContractNetResponder(myAgent, cfp)
         {
        // Redefine callback methods to implement domain-dependent
        // logic
      } );
    }
    else {
      block();
    }
  }
} );
```

**Table 5.1** Supported interaction protocols

| Protocol | Initiator class | Responder class |
|---|---|---|
| FIPA-Request<br>FIPA-Query | AchieveREInitiator | AchieveREResponder |
| FIPA-Propose | ProposeInitiator | ProposeResponder |
| Iterated version of<br>FIPA-Request<br>FIPA-Query | IteratedAchieveREInitiator | SSIteratedAchieveREResponder |
| Contract-Net | ContractNetInitiator | ContractNetResponder<br>SSContractNetResponder |
| FIPA-Subscribe | SubscriptionInitiator | SubscriptionResponder |

When using the above pattern to play the responder role in a given protocol an agent can be engaged in several conversations driven by that protocol at the same time.

Table 5.1 summarizes the protocols currently supported together with the classes that can be used to implement them. It should be noted that we use a single class (`AchieveREInitiator/Responder` that implies 'achieve rational effect initiator/responder') for protocols like FIPA-Request and FIPA-Query that follow the same pattern after the initiation message. This also applies to the iterated version of these protocols which is useful when, for example, the requested action may produce a very large result that must be split into different `INFORM` messages. Receiving the first `INFORM`, the sender may request the responder to send the next trunk by sending another initiation message (and in this case the protocol goes on iteratively) or to close the session by sending a `CANCEL` message.

### *5.4.2 USING PROTOCOL CLASSES*

As mentioned, protocol classes provide a number of callback methods that programmers are expected to redefine by customizing them with logic that relates to the application domain. All these methods are declared protected and have a default (typically empty) implementation. In this way programmers can choose, depending on their specific requirements, which methods to implement and which ones to ignore. For both initiators and responders most callback methods are invoked following the reception of a protocol message and have the form

```
protected handle<message-performative>(ACLMessage receivedMessage)
```

For example, in the `ContractNetResponder` if an `ACCEPT_PROPOSAL` message is received, the `handleAcceptProposal(ACLMessage accept)` method is called. When the reception of a message terminates an interaction with the sender of that message (e.g. when a `REFUSE` message is received as a reply to a `CFP` in a `Contract-Net protocol` indicating that no further message need be sent back to the responder), the corresponding `handleXXX()` method returns `void`. On the other hand, if a reply must be sent back, we distinguish two cases. For responders that are always engaged in one-to-one interactions the `handleXXX()` method returns an `ACLMessage`. The returned value will be used as the reply. For instance, the `handleCfp()` method of the `ContractNetResponder` will typically be redefined as below.

```
protected ACLMessage handleCfp(ACLMessage cfp) {
  ACLMessage reply = cfp.createReply();
  // Evaluate the call
  if (call OK) {
    // Prepare a proposal
```

```
    reply.setPerformative(ACLMessage.PROPOSE);
  }
  else {
    reply.setPerformative(ACLMessage.REFUSE);
  }
  return reply;
}
```

For initiators that are designed to support one-to-many interactions, the `handleXXX()` method gets an additional argument of type `Vector` to which the reply must be added. For instance, the `handlePropose()` method of the `ContractNetInitiator` will typically be redefined as follows:

```
protected void handlePropose(ACLMessage propose, Vector acceptances)
   {
  ACLMessage reply = propose.createReply();
  // Evaluate the proposal
  if (proposal OK) {
    reply.setPerformative(ACLMessage.ACCEPT_PROPOSAL);
  }
  else {
    reply.setPerformative(ACLMessage.REJECT_PROPOSAL);
  }
  acceptances.add(reply);
}
```

Besides the methods triggered by the reception of protocol messages, the classes in the `jade.proto` package provide two other types of callback method when appropriate. Initiators that are designed to support one-to-many interactions provide methods that are invoked when the replies from all responders have been collected. These methods have the form `handleAllXXX(Vector v)` and allow the treatment of all replies at the same time. For instance, the `ContractNetInitiator` class provides the `handleAllResponses()` and the `handleAllResultNotifications()` methods. These are invoked when responses (i.e. `PROPOSE/REFUSE/NOT_UNDERSTOOD`) from all responders and result notifications (i.e. `INFORM/FAILURE`) from responders whose proposals were accepted are received respectively.

Finally, in some cases a protocol class must send one or more messages following a trigger not directly related to the reception of another message. For example, in a FIPA-Request protocol, if a responder replied with an `AGREE` to an incoming `REQUEST` message, it will have to successively send an `INFORM` or a `FAILURE` to notify the requester about the result of the agreed action. These cases are covered by callback methods that take the form `prepareXXX()` and return either an `ACLMessage` or a `Vector` of messages. For instance, the `AchieveREResponder` class (implementing the responder role in the FIPA-Request protocol) provides the `prepareResultNotification()` method to cover the case in the above example.

To further show the power of interaction protocol classes, here is how simple the `BookNegotiator` behaviour presented in Section 4.3.5 can exploit the `ContractNetInitiator` class:

```
public class BookNegotiator extends ContractNetInitiator {
  private String title;
  private int maxPrice;
  private PurchaseManager manager;
```

```
  public BookNegotiator(String t, int p, PurchaseManager m) {
    super(null, null);
    title = t;
    maxPrice = p;
    manager = m;
  }

  protected Vector prepareCFPs(ACLMessage cfp) {
    cfp = new ACLMessage(ACLMessage.CFP);
    cfp.setContent(title);
    for (int i = 0; i < sellerAgents.size(); ++i) {
      cfp.addReceiver((AID) sellerAgents.get(i));
    }
    Vector v = new Vector();
    v.add(cfp);
    return v;
  }

  protected void handleAllResponses(Vector responses
                                    Vector acceptances) {
    ACLMessage bestOffer = null;
    int bestPrice = -1;
    for (int i = 0; i < responses.size(); ++i) {
      ACLMessage rsp = (ACLMessage) responses.get(i);
      if (rsp.getPerformative() == ACLMessage.PROPOSE) {
        int price = Integer.parseInt(rsp.getContent());
        if (bestOffer == null || price < bestPrice) {
          bestOffer = rsp;
          bestPrice = price;
        }
      }
    }
    if (bestOffer != null) {
      ACLMessage accept = bestOffer.createReply();
      accept.setContent(title);
      acceptances.add(accept);
    }
  }

  protected void handleInform(ACLMessage inform) {
    // Book successfully purchased
    int price = Integer.parseInt(inform.getContent());
    myGui.notifyUser("Book "+title+" successfully purchased.
      Price = "+price);
    manager.stop();
  }
} // End of inner class BookNegotiator
```

The `prepareCFPs()` method is called as soon as the `ContractNetInitiator` behaviour starts. It is intended to adjust the CFP messages to be sent to responders. It is particularly useful either when the CFP message is not known at construction time or when we need to send customized messages to each responder. All protocol initiator classes have a similar method.

### *5.4.3 NESTED PROTOCOLS*

As we saw in previous sections, both initiator and responder classes invoke proper callback methods when protocol messages are received. If a reply has to be sent back the invoked callback method is responsible for creating that reply. There are, however, cases where, in order to create the reply it is necessary to execute a behaviour, e.g. to engage in a new protocol with another agent. Let us consider, for example, a broker agent that acts as a responder in a FIPA-Request protocol, but actually delegates the execution of the requested action to an executor selected from a pool of executor agents by means of a Contract-Net protocol. If we want to implement the FIPA-Request responder role in the broker agent using an `AchieveREResponder` we should be able to carry out the Contract-Net protocol with the executor agents within the `prepareResultNotification()` method. Clearly this prevents us from using the `ContractNetInitiator` class since it is not possible to execute a behaviour from within a method.

To overcome this limitation all JADE protocol classes are implemented as subclasses of `FSMBehaviour` (see Section 5.2.2) and each callback method is invoked in a dedicated state of the finite state machine. As an example, Figure 5.9 shows the finite state machine of the `AchieveREResponder` class. States dedicated to invoking callback methods are highlighted in grey. In general, for each callback method `mmm()` there is a `registerMmm()` method that allows overriding of the state that invokes the callback method `mmm()` with an application-specific behaviour. All `registerMmm()` methods take a `Behaviour` object parameter. For instance,

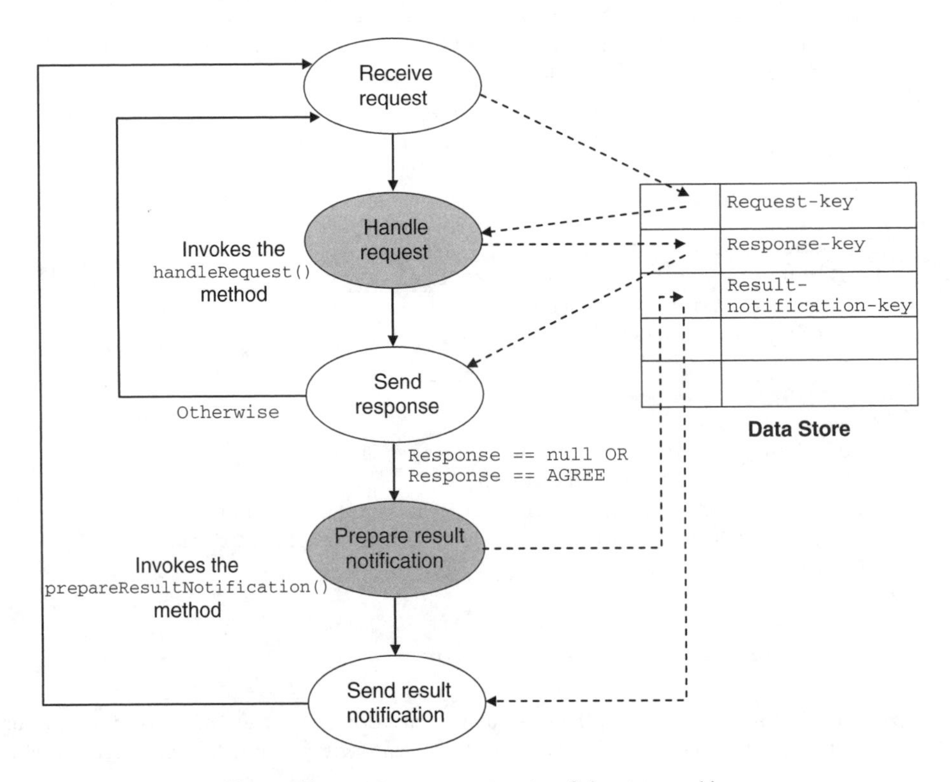


**Figure 5.9** `AchieveREResponder` finite state machine

with reference to Figure 5.9 and the broker example, the `registerPrepareResultNotification(Behaviour b)` method can be used to plug a `ContractNetInitiator` behaviour in the `Prepare-Result-Notification` state. It is important to note that, if we do this, the `prepareResultNotification()` callback method will no longer be invoked.

Besides states dedicated to invoking callback methods, there are typically other states that are responsible for receiving and sending messages and performing the checks related to the protocol flow. These states, however, are completely hidden and programmers need never know or care about them.

Dotted lines in Figure 5.9 also show how data is shared among the states of the protocol by means of the Data Store. For instance, the behaviour implementing the `Handle-Request` state of the `AchieveREResponder` class is as follows:

```
private class HandleRequest extends OneShotBehaviour {
  public void action() {
    ACLMessage request = getDataStore().get(REQUEST_KEY);
    ACLMessage response = handleRequest(request);
    getDataStore().put(RESPONSE_KEY, response);
  }
}
```

Thus, when registering an application-specific behaviour in the state of a protocol class, the registered behaviour is responsible for retrieving relevant previously received messages and storing produced replies in the Data Store using the correct keys. This is fully documented in the Javadoc of the corresponding protocol classes.

## 5.5 INTERACTING WITH THE AMS

The AMS (Agent Management System) was introduced in Section 2.2 when discussing the FIPA specifications. FIPA does not mandate the AMS to be an agent, but, since it must be able to send and receive ACL messages, that was the choice we made in JADE. However, the JADE AMS is a special agent. In fact, besides providing the white pages service as specified by FIPA, it also plays the role of authority in the platform. In this respect the AMS is the only agent able to perform platform management operations such as creating and killing agents, killing containers and shutting down the platform. Unlike the DF that is launched by default in the main container, but could live in any container, the AMS is tightly bound to the main container. As a result, when instantiated the AMS receives a pointer to the underlying main container in its constructor which it can then use to activate a number of features that are inaccessible to normal agents. Agents wishing to perform platform management actions must request the AMS to perform them.

### *5.5.1 REQUESTING PLATFORM MANAGEMENT OPERATIONS TO THE AMS*

All platform management operations supported by the JADE AMS are modelled by the actions of the `JADE-Agent-Management` ontology. This ontology is implemented by the `JADEManagementOntology` class included in the `jade.domain.JADEAgentManagement` package. The actions in the `JADE-Agent-Management` ontology are listed in Table 5.2 together with the classes implementing them (also included in the `jade.domain.JADEAgentManagement` package) and a brief description of the operations they represent. Besides these actions, the `JADE-Agent-Management` ontology also defines the following concepts:

- `location` representing a generic place where agents can live. This is implemented by the `Location` interface.

**Table 5.2** Actions of the `JADE-Agent-Management` ontology

| Action name | Class | Description |
|---|---|---|
| `create-agent` | `CreateAgent` | Create an agent of a given class with a given name in a given container |
| `kill-agent` | `KillAgent` | Kill a given agent |
| `kill-container` | `KillContainer` | Kill a given container |
| `shutdown-platform` | `ShutdownPlatform` | Shut down the whole JADE platform |
| `query-platform-locations` | `QueryPlatformLocationsAction` | Retrieve the list of all containers in the platform. This is returned as a `jade.util.leap.List` of `Location` (actually `ContainerID`) objects |
| `query-agents-on-location` | `QueryAgentsOnLocation` | Retrieve the list of agents currently living in a given container. This is returned as a `jade.util.leap.List` of `AID` objects |
| `where-is-agent` | `WhereIsAgentAction` | Retrieve the `ContainerID` of the container where a given agent currently lives |
| `install-mtp` | `InstallMTP` | Activates a new MTP of a given class in a given container. The URL of the installed MTP is produced as the result |
| `uninstall-mtp` | `UninstallMTP` | Deactivates an MTP with a given URL in a given container |

- `container-id` representing a JADE container. This is implemented by the `ContainerID` class.
- `platform-id` representing a generic FIPA platform. This is implemented by the `PlatformID` class.

Both `container-id` and `platform-id` are subtypes of `location` and, consistently, both the `ContainerID` and the `PlatformID` classes implement the `Location` interface. The `ContainerID`, `PlatformID` and `Location` classes are intensively used within the JADE kernel and are therefore included in the `jade.core` package.

All actions in the `JADE-Agent-Management` ontology must be requested of the AMS within the scope of a FIPA-Request interaction protocol and must be encoded using the FIPA-SL language. As an example, we show below the code to request that the AMS create an agent called 'john' of class '`myPackage.MyClass`' in a container called 'my-container'.

```
CreateAgent ca = new CreateAgent();
ca.setAgentName("john");
ca.setClassName("myPackage.MyClass");
ca.setContainer(new ContainerID("my-container", null));

Action actExpr = new Action(getAMS(), ca);

ACLMessage request = new ACLMessage(ACLMessage.REQUEST);
request.addReceiver(getAMS());
request.setOntology(JADEManagementOntology.getInstance().getName());
```

```
request.setLanguage(FIPANames.ContentLanguage.FIPA_SL);
request.setProtocol(FIPANames.InteractionProtocol.FIPA_REQUEST);

try {
  getContentManager().fillContent(request, actExpr);

  addBehaviour(new AchieveREInitiator(this, request) {
    protected void handleInform(ACLMessage inform) {
      System.out.println("Agent successfully created");
    }

    protected void handleFailure(ACLMessage failure) {
      System.out.println("Error creating agent.");
    }
  } );
}
catch (Exception e) {
  e.printStackTrace();
}
```

Note the use of the `jade.content.onto.basic.Action` class implementing the `action` operator required by the FIPA-SL language to express agent actions.

For all actions that do not imply any result to be sent back to the requester (e.g. the `create-agent` action), the content of the AMS reply makes use of the `done` operator of the FIPA-SL language implemented by the `Done` class of the `jade.content.onto.basic` package. For instance, if we sniff the reply sent back by the AMS to an agent using the code presented above we would see the following message content:

```
((done
    (action
      (agent-identifier :name ams@myPlatform)
      (create-agent :agent-name john
                    :class-name myPackage.myClass
                    :container (container-id :name my-container)
      )
    )
))
```

For all actions that imply a result is to be sent back to the requester, the content of the AMS reply makes use of the `result` operator of the FIPA-SL language implemented by the `Result` class of the `jade.content.onto.basic` package. As an example, if we request that the AMS perform the 'where-is-agent' action to identify where agent john currently lives, the code below shows how to retrieve the information from the AMS reply.

```
ACLMessage request = // Create a request to perform the
                     // where-is-agent action
addBehaviour(new AchieveREInitiator(this, request) {
  protected void handleInform(ACLMessage inform) {
    try {
      Result r = myAgent.getContentManager().extractContent(inform);
      ContainerID cid = (ContainerID) r.getValue();
    }
```

```
      catch (Exception e) {
        e.printStackTrace();
      }
    }
} );
```

If we sniff the reply received from the AMS in this case we would see the following:

```
((result
    (action
      (agent-identifier :name ams@myPlatform)
      (where-is-agent :agent (agent-identifier :name
         john@myPlatform))
    )
    (container-id :name my-container)
))
```

### *5.5.2 SUBSCRIBING TO PLATFORM EVENTS*

Besides serving requests to perform platform management operations, the JADE AMS also supports a subscription mechanism by means of which interested agents can be notified about platform events such as agent creations and terminations, container terminations and so on. Most of the JADE graphical tools such as the RMA and the Sniffer presented in Section 3.7 make use of this mechanism to keep their agent tree up to date.

All platform events are described by the concepts of the `JADE-Introspection` ontology implemented by the `IntrospectionOntology` class included in the `jade.domain.introspection` package. These concepts are listed in Table 5.3 together with the classes imple-

**Table 5.3** Platform events of the `JADE-Introspection` ontology

| Action name | Class | Description |
|---|---|---|
| added-container | AddedContainer | Indicates that a new container joined the platform |
| removed-container | RemovedContainer | Indicates that a container disappeared from the platform |
| born-agent | BornAgent | Indicates that an agent was born on a given container |
| dead-agent | DeadAgent | Indicates that an agent died on a given container |
| suspended-agent | SuspendedAgent | Indicates that an agent was suspended |
| resumed-agent | ResumedAgent | Indicates that an agent was resumed |
| moved-agent | MovedAgent | Indicates that an agent moved from one container to another |
| cloned-agent | ClonedAgent | Indicates that an agent was cloned on a given container |
| kill-container-requested | KillContainerRequested | Indicates that the AMS is about to serve a request to kill a given container |
| shutdown platform-requested | ShutdownPlatformRequested | Indicates that the AMS is about to serve a request to shut down the platform |

menting them (also included in the jade.domain.introspection package) and a brief description of the events they represent.

All the above concepts are subtypes of event and, consistently, all the related classes implement the jade.domain.introspection.Event interface. This interface includes a single method getName() that for each class returns the name of the concept representing the event implemented by that class. For instance, the BornAgent class implements the getName() method returning the String born-agent.

In order to subscribe to the AMS to receive notifications about platform events, agents should initiate a FIPA-Subscribe protocol with the AMS, specifying proper JADE-specific labels in the content slot and in the reply-with slot of the SUBSCRIBE message used to initiate the protocol. Programmers can, however, bypass these details by using the jade.domain.introspection.AMSSubscriber behaviour that manages them automatically. The AMSSubscriber class provides the installHandlers(Map handlers) method that should be redefined to associate a proper handler for each type of event an agent is interested in. These handlers must implement the AMSSubscriber.EventHandler interface which includes a single method handle(Event ev) that is invoked each time an event of the type associated with the handler is received from the AMS. The snippet of code below exemplifies usage of the AMSSubscriber class in the case of an agent that must detect agent creations and terminations.

```
AMSSubscriber myAMSSubscriber = new AMSSubscriber() {
  protected void installHandlers(Map handlers) {
    // Associate an handler to born-agent events
    EventHandler creationsHandler = new EventHandler() {
      public void handle(Event ev) {
        BornAgent ba = (BornAgent) ev;
        System.out.println("Born agent "+ba.getAgent().getName());
      }
    };
    handlers.put(IntrospectionVocabulary.BORNAGENT,
       creationsHandler);

    // Associate an handler to dead-agent events
    EventHandler terminationsHandler = new EventHandler() {
      public void handle(Event ev) {
        DeadAgent da = (DeadAgent) ev;
        System.out.println("Dead agent "+da.getAgent().getName());
      }
    };
    handlers.put(IntrospectionVocabulary.DEADAGENT,
       terminationsHandler);
  }
};
addBehaviour(myAMSSubscriber);
```

The names of all concepts included in the IntrospectionOntology are made available as constants in the IntrospectionVocabulary interface according to the vocabulary interface pattern described in Section 5.1.3.6.

## 5.6 STARTING JADE FROM AN EXTERNAL JAVA APPLICATION

Until now it has always been assumed that the JADE run-time is started from the command-line as shown in Section 3.4. In many cases, however, it is necessary to start one or more JADE agents from an external application, and therefore to create the JADE run-time to host them. To support this requirement, since JADE v2.3, an in-process interface has been implemented that allows JADE to be used as a kind of library to allow the run-time to be launched from within external programs.

The JADE run-time is implemented by the `jade.core.Runtime` class. According to the singleton pattern, a single instance of this class exists in a JVM and can be retrieved by means of the `instance()` static method. The singleton `Runtime` instance provides two methods: `createMainContainer()` to create a JADE main container and `createAgentContainer()` to create a JADE peripheral container (i.e. a container that joins an existing main container running somewhere in the network). Both methods require a `Profile` object as parameter that keeps the configuration options required to start the JADE run-time (e.g. the host name and port number of the main container). All options that can be specified when starting JADE from the command line are made available as constants in the `Profile` class and can be set in the profile using the `setParameter(String key, String value)` method.

Both the `createMainContainer()` and `createAgentContainer()` methods return a `jade.wrapper.ContainerController` object. This controller wraps the higher-level functionality of the agent container, such as installing and uninstalling MTPs, killing the container (the external application remains alive) and, of course, creating new agents. The `createNewAgent()` method of the `ContainerController` class returns an `AgentController` object that is also wrapped around some functionalities of the agent, while preserving its autonomy. In particular, the `AgentController` class provides methods with which the external application can control the life cycle of the wrapped agent but hides the reference to the `Agent` object so that it is not possible to perform direct method calls on it. Note that the `createNewAgent()` method creates the `Agent` instance but does not start it, as this is only possible via the `start()` method of the returned `AgentController` object.

To exemplify the usage of the JADE in-process interface let us go back to the book-trading case study and assume we want to integrate the buyer agent into a bigger system able to buy books over different channels. In this scenario the JADE book-trading environment becomes just one of the channels the bigger system can use to try to buy a book. Here is some sample code to activate the book-buyer agent that the external system will use to buy books in the JADE book-trading environment:

```
public AgentController startBuyerAgent(
    String host, // JADE Book Trading environment Main Container
       host
    String port, // JADE Book Trading environment Main Container
       port
    String name, // Book Buyer agent name
    ) {

  // Retrieve the singleton instance of the JADE Runtime
  Runtime rt = Runtime.instance();

  // Create a container to host the Book Buyer agent
  Profile p = new ProfileImpl();
  p.setParameter(Profile.MAIN_HOST, host);
  p.setParameter(Profile.MAIN_PORT, port);
  ContainerController cc = runtime.createAgentContainer(p);
  if (cc != null) {
    // Create the Book Buyer agent and start it
    try {
      AgentController ac = cc.createNewAgent(name,
                              "bookTrading.buyer.BookBuyerAgent",
```

```
                                  null);
        ac.start();
        return ac;
      }
      catch (Exception e) {
        e.printStackTrace();
      }
    }
    return null;
}
```

In this case a peripheral container is created that connects to an existing JADE platform (the JADE book-trading environment). We therefore we need to specify the host and port of the main container of that platform. Note also the use of the `ProfileImpl` class. This is required since `Profile` is an abstract class and therefore cannot be directly instantiated. Both `Profile` and `ProfileImpl` belong to the `jade.core` package.

### *5.6.1 OBJECT TO AGENT COMMUNICATION*

In many cases, besides starting one or more agents, an external application needs to interact with the agents to instruct them to perform some task. In the scenario considered above, for instance, the external system must notify the book buyer agent each time there is a new book to buy. Similarly the book buyer agent must notify the external system when it successfully purchases a book. As we saw in the previous section, however, the `AgentController` class does not disclose the reference of the wrapped `Agent` instance and therefore the external application cannot invoke any method of the agent directly. Interactions between an external application and an agent started by means of the in-process interface are made possible by the object-to-agent (O2A) communication mechanism. This is basically a synchronized FIFO queue where the external application can put Java objects that can be later extracted by the agent; each agent has its own O2A queue. The `AgentController` object wrapping an agent provides the `putO2AObject()` method that can be used by the external application to insert objects in the O2A queue of the agent. Similarly the `Agent` class provides the `getO2AObject()` method that can be used by the agent to read objects passed by the external application. The O2A communication mechanism is disabled by default and so an agent wishing to interact with external applications must enable it explicitly by means of the `setEnabledO2ACommunication()` method. Similarly to the standard message queue model, inserting an object into the O2A queue of an agent has the effect of restarting all behaviours of the agent to give them a chance to read and process the inserted object.

Returning to the example considered above, each time the external buying system is requested to buy a book it must notify its local book buyer agent. The following code used by the external system will achieve this:

```
BookInfo info = new BookInfo(title, maxPrice, deadline);
buyerAgentController.putO2AObject(info);
```

where `BookInfo` is a utility bean used to group the title of the book to buy, the maximum price the user is willing to pay and the deadline. `BuyerAgentController` is the `AgentController` object wrapping the book buyer agent as returned by the `startBuyerAgent()` discussed in the previous section. The following code is added to the `BookBuyerAgent setup()` method to process notifications from the external buying system (other modifications also need to be introduced to handle the successful purchase or deadline expiration).

```
// Enable O2A communication
setEnabledO2ACommunication(true, 0);

// Add the behaviour serving notifications from the external system
addBehaviour(new CyclicBehaviour(this) {
  public void action() {
    BookInfo info = (BookInfo) myAgent.getO2AObject();
    if (info != null) {
      purchase(info.getTitle(),
               info.getMaxPrice(),
               info.getDeadline());
    }
    else {
      block();
    }
  }
} );
```

Here the `purchase()` method is that shown in Section 4.2.5.1. Note also that a behaviour handling objects received via the O2A communication channel is similar to a behaviour handling incoming messages.

# 6

# Agent Mobility

Mobile agents (White, 1996) are a paradigm that derives from two different disciplines (Brown and Rossak, 2005). The first is artificial intelligence, which created the concept of an agent (Russell and Norvig, 1995), and the second is distributed systems, which defines the concept of code mobility (Picco, 2000).

## 6.1 AGENT MOBILITY

According to standard definitions, mobile agents are everything that a non-mobile agent is (i.e. autonomous, reactive, proactive and social), but in addition they are also moveable; they can migrate between platforms in order to accomplish assigned tasks.

From the distributed systems point of view, a mobile agent is a program with a unique identity that can move its code, data and state between networked machines. To achieve this, mobile agents are able to suspend their execution at any time and to continue once resident in another location.

We can position mobile agents in relation to other classical paradigms as follows (Picco, 2000):

- ***Client-server***: The most widely used paradigm where services are offered by a server and consumed by one or more, usually remote, clients.
- ***Remote execution***: One component sends code to another component for remote execution, either resulting from its own decision, a request from the remote component, or perhaps even as a part of a pre-existing contract. Once executed the executing component will typically return any result to the originating component.
- ***Mobile agents***: One component sends itself (or another, if allowed) to a remote host for execution. The component transitions with its code, data and perhaps state intact. Motivations may be similar to the previous case, but most typically result from the component (i.e. mobile agent) deciding for itself that it wishes to move to an alternate location.

A mobile agent, as depicted in Figure 6.1, consists of three parts: code, state and data. The *code* is that aspect of the agent that is executed when it migrates to a platform. In the simplest case there is a single code. The *state* is the data execution environment of the agent, including the program counter and the execution stack. This part is only found in agents that use strong migration (see Section 6.1.2). The *data* consists of the variables used by the agents, such as knowledge, file identifiers, etc. In weak migration (also see Section 6.1.2) this part is strictly necessary since agent code is constructed as a state machine with variables required to maintain state information.

*Developing Multi-Agent Systems with JADE* Fabio Bellifemine, Giovanni Caire, Dominic Greenwood

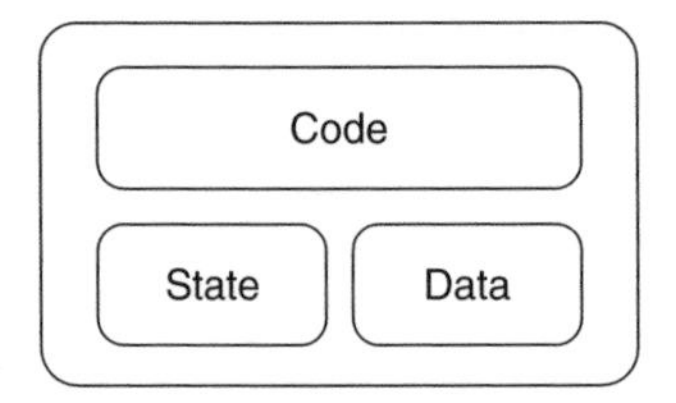


**Figure 6.1** Basic structure of a mobile agent

### *6.1.1 SOME ADVANTAGES AND DISADVANTAGES OF MOBILE AGENTS*

There have been many debates over the various advantages and disadvantages of mobile agents, usually in comparison with their non-mobile cousins. Some of the typical advantages are:

- ***Asynchronous and independent processing***: Once they have migrated to a new platform, agents do not have to contact their owner in order to perform their task. They may only need to send back the results. This is especially useful when considering mobile devices with limited resources; an agent can be migrated to another machine to perform complex tasks and periodically return results.
- ***Fault tolerance***: They can address and aid with fault conditions by moving to an alternative platform when problems are detected. Equally if a migration destination is down, an intermediary may be selected as a temporary host. This makes them quite suitable for hostile and disruptive environments.
- ***Sea of data applications***: Mobile agents are well suited to applications that need to process large amounts of remote data. Mobile agents can move to the data, rather than vice versa which in many cases is a much more efficient option.

But mobile agents also have some disadvantages. As described in Mir (2004), the most relevant of them are:

- ***Scalability and performance***: Even though mobile agents reduce network load, they also tend to increase processing load. This is because they are usually programmed with interpreted languages and also often need to observe rigorous interoperability standards that can incur data processing overheads.
- ***Portability and standardization***: Agents cannot interoperate if they do not follow common communication standards. Adoption of these standards, such as OMG MASIF (Mobile Agent System Interoperability Facility) or FIPA is typically necessary, especially for inter-platform mobility.
- ***Security***: The use of mobile agents can bring about security problems. Any mobile code offers a potential threat and should be carefully authenticated before invocation.

### *6.1.2 STRONG AND WEAK MIGRATION*

In mobile agents systems (Tanenbaum and Van Steen, 2001; Brown and Rossak, 2005) two primary types of migration can be distinguished: strong migration and weak migration.

Strong migration is more complex and is the case where an agent's execution is frozen, migration takes place, and then execution is restarted from the very next instruction. This technique requires the storage and protection of the agent state during the migration process. Implementation of this technique can be complex as it requires access to internal parameters of agent execution, generally only available to the operating system, and that can typically be very architecture dependent.

Weak migration, on the other hand, does not send the agent state and is therefore much simpler; agent execution always restarts from the beginning of the code. This kind of migration requires that the agent be implemented as a finite state machine if state is to be preserved.

### *6.1.3 MIGRATION ITINERARIES*

An itinerary defines the locations that a mobile agent must visit to complete a set of tasks. Two basic types of itineraries can be distinguished:

- ***Static itineraries*** are determined at agent creation time without any possibility of modification during agent execution.
- ***Dynamic itineraries*** are determined during agent execution according to needs and desires.

Additional hybrid methods can mix these two types (Mir, 2004).

## 6.2 INTRA-PLATFORM MOBILITY

JADE has a provided platform service called the *Agent Mobility Service* which implements intra-platform mobility. This provides software agents with the ability to move among different containers within the same platform. This mechanism, however, does not allow agents to move to containers belonging to other platforms.

### *6.2.1 MOBILITY ACCESSORS*

In JADE the agent mobility is simply controlled via the method `doMove()` in the `Agent` class:

```
void doMove(Location destination)
```

The `destination` parameter must be an object of a class implementing the `Location` interface. In the JADE platform there are two classes implementing this interface, both within the `jade.core` package. The first is `ContainerID`, which is used to specify that the destination of the agent will be a container of the platform where it is currently running. The second is `PlatformID`, which is used to indicate that the destination of the agent is the main container of a different platform. When a remote platform is indicated as destination of the mobile agent, inter-platform mobility mechanisms, described in Section 6.3, must be employed.

Once invoked this method initiates the process of moving the agent to the specified destination container. The majority of the code to achieve this is located in the `jade.core.mobility` package. The invocation of the `doMove()` method is forwarded to the Agent Mobility Service through the method `move()` of its helper. The first action that the helper carries out is to change the agent's state from ACTIVE to TRANSIT causing the agent to cease its current activities and suspend while the platform relocates it. Users can specify operations to be triggered when the mobility process is started, in order, for example, to save the agent state. These operations are specified by overriding the `Agent` class method:

```
void beforeMove()
```

A counterpart method called `afterMove()` can also be specified to trigger operations that must be executed just after the agent moves, *before* it recovers the ACTIVE state in the destination location.

### *6.2.2 AGENT SERIALIZATION*

After invocation of `beforeMove()`, the helper raises a vertical command (see Chapter 7) requesting that the agent be moved to the destination contained inside the command. If the command contains a destination object which is an instance of the `PlatformID` class, the Inter-Platform Mobility Service (see Section 6.3) will initiate the migration of the agent to the specified remote platform. When the destination object is an instance of the `ContainerID` class, the Agent Mobility Service will initiate the migration of the agent to the destination container within the current platform. The `ContainerID` of any container can alternatively be requested from the platform AMS agent, rather than constructing it from `ContainerID` class.

Migration of an agent must include the transmission of at least its code and data, and possibly also its state. In JADE, agents model their execution state as internal agent data, thus implying that it is only necessary to transfer the code and data. Agent data is contained in the Java object representing the agent, so transmitting this object together with its code will be sufficient to restore the agent at destination.

Java serialization is used to transmit an agent instance over a network connection by recursively recording the internal member values of the agent object into a byte stream. The user must specify which data members are to be transmitted together with the agent using the `transient` modifier. For example, if an agent has a `FileInputStream` member (which is not serializable) and which is not declared as transient, the mobility process will fail and throw a `NotSerializableException`.

### *6.2.3 MOBILE AGENT CLASSLOADER*

Once an agent instance has been serialized, it is transmitted to the destination container by means of a horizontal command (see Chapter 7). It is then deserialized to recover the original agent object.

However, JADE containers are often located in different Java virtual machines on different hosts. If an agent moves from one host to another, its original class files must also be made available as the deserialization process of an agent's object at a destination container requires the original class structure. To manage this, the Agent Mobility Service uses a built-in class loader capable of requesting classes from a remote container via horizontal service commands. Any container receiving such requests locates the required class file and sends it to the requesting container.

If the agent is successfully deserialized, a thread is created for the recovered object, and a 'power up' process is executed to restart the agent. All messages sent during migration are then removed from a temporary buffer at the originating container and redirected toward the new agent location. Finally, at the source container the agent moves into the GONE state, indicating that a migration was successful and that the local copy should now be terminated. The `afterMove()` method is invoked and the agent is removed from the container.

### *6.2.4 AGENT CLONING*

Thus far this chapter has described intra-platform mobility in terms of events occurring within JADE when an agent moves. However, JADE also provides a cloning mechanism that takes a snapshot copy of an existing agent using the method:

```
public void doClone(Location destination, String newName)
```

The `destination` parameter is used, as with mobility, to indicate the container at which the clone of the current agent will be created. The `newName` parameter is the name used to create the clone AID. The cloning process itself is the same as with mobility, except that the initiating agent is not terminated, thus resulting in two equal, executing clones similar in every way except their identities.

### *6.2.5 INDIRECT MOBILITY ASSERTION*

The process of moving an agent can actually be asserted either directly by the agent itself or indirectly by a platform AMS. The latter is initiated when an agent, or the user via the RMA GUI, sends an ACL message to the platform AMS requesting that a specific agent be migrated. If an

AMS receives such a request, and authority of the requestor is verified, the AMS automatically invokes the `doMove()` or `doClone()` method of the specified agent. For this reason, some care should be taken when creating the `afterMove()` and `beforeMove()` methods of agents that may be subject to indirectly asserted migration. Because the invocation of `doMove()` does not necessarily depend on the agent itself, it is prudent to ensure that actions, such as freeing resources and saving information, are taken to prevent problems occurring if an indirect migration call is initiated.

## 6.3 INTER-PLATFORM MOBILITY SERVICE

The Inter-Platform Mobility Service (IPMS) is a JADE add-on that has been created to provide platform-to-platform mobility for JADE agents, a feature which is not available with the built-in Agent Mobility Service.

The IPMS is specifically designed to be as transparent as possible to the agent programmer by ensuring that inter-platform migration is as straightforward as intra-platform migration. As much compliance as possible has been maintained with the deprecated (due to lack of initial implementation validation) FIPA Agent Management Support for Mobility Specification (this deprecated specification is available at http://www.fipa.org/specs/fipa00087/index.html). The current version is simple, but designed to be extendable with additional features such as security and support for fault-tolerant migration.

### *6.3.1 THE MIGRATION PROCESS*

The core mechanism of the IPMS is the movement of agents between platforms using FIPA-ACL messages as the transportation medium. These messages are sent between the AMSs of the endpoint platforms. As mentioned, a few alterations have been made to the original FIPA specification; the ontology now defines two actions, `move` and `power-up`. The first of these represents the movement of agent code and instance; the second represents the agent's activation once migration is complete. Additionally, several concepts are defined such as *mobile-agent-description* which contains all agent information, including its code, data and its *mobile-agent-profile*. This latter concept defines the basic agent characteristics to help ensure compatibility with receiving platforms, such as the name of its native agent system, the language in which it has been written, and the mobility protocol used. It is assumed that the messages containing this information, consistent with the mobility ontology, will be sufficient for a remote platform to decide whether it can execute an incoming agent.

Both of the actions, `move` and `power-up`, are effected using the standard FIPA Request interaction protocol. In each case a Request message is sent to the target platform with an Inform or a Failure message expected as a response. As illustrated in Figure 6.2, FIPA Request mobility protocol, the initiating agent (on the left) sends a Request to move the message to its platform which will in turn send a Request to move the message containing the agent code and data to the specified destination platform (on the right). If the destination platform successfully extracts the agent from the Request message, an Inform message is returned to the originating platform. When received, this Inform message triggers the originating platform to terminate the requesting agent and send a power up Request to the destination platform, which starts the agent. If something fails, the agent state in the originating platform is restored and any residual presence of the agent in the target platform is removed.

The advantage of using ACL messages to transport the agent from one platform to another is that no additional inter-platform communication channel is needed. The disadvantage is that performance is not particularly high due to the encoding and decoding processes intrinsic to ACL messaging. Standardization of MTPs designed to improve ACL message sending/receiving performance, perhaps using lightweight content encodings such as that expressed in (FIPA23), could significantly improve efficiency while retaining interoperability.

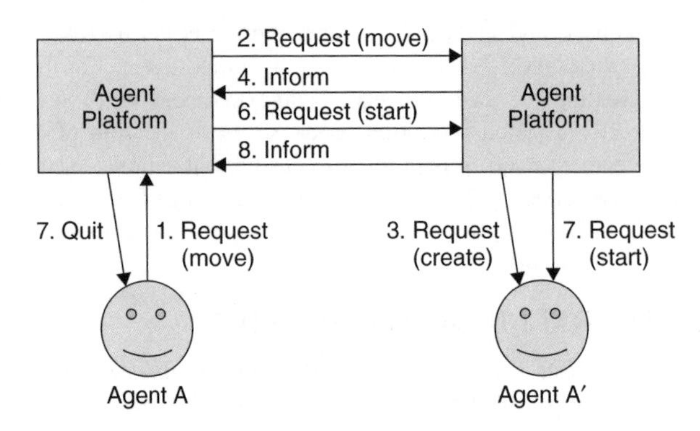


**Figure 6.2** FIPA Request mobility protocol

### 6.3.2 INTEGRATION OF MOBILITY SERVICES

The IPMS needs to take into account the Intra-Platform Mobility Service built into JADE. There are two reasons for this:

1. The method to migrate between platforms and the method to migrate between containers on the same platform are essentially the same from the user point of view. This method, `move()`, makes use of a new Location parameter in the inter-platform case called `PlatformID` (a variation of `ContainerID`) that symbolizes the target platform where the agent wants to migrate. As the IPMS is implemented using the JADE service architecture (see Chapter 7), it needs only wait to receive vertical commands from the Intra-Platform Mobility Service with a Location of `PlatformID` type. It must then simply cancel the command and begin its work. This scheme is described in Figure 6.3.
2. Because of intra-platform mobility, the executing agent code may not be resident in the same container from which the agent wants to migrate; thus a code retrieval mechanism is needed. As the intra-platform mobility service does not migrate agent code, the IPMS contains an additional mechanism layered over intra-platform mobility to ensure that code migration also takes place when an agent moves between containers.

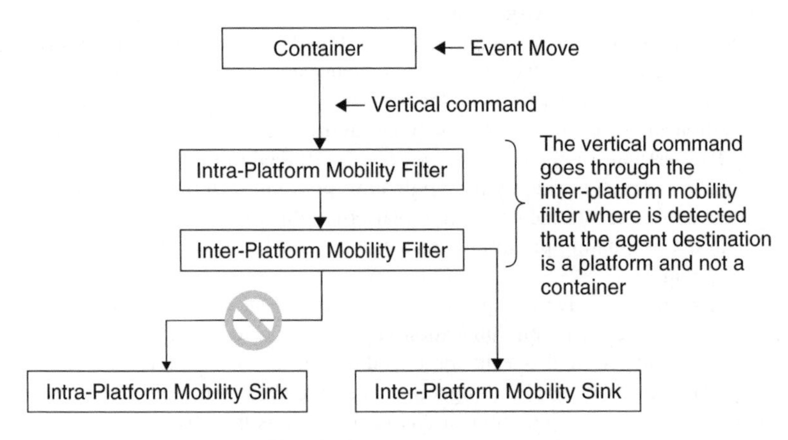


**Figure 6.3** The Move vertical command flow

### 6.3.3 CODE GROUPING

In order to move an agent from one location to another, all of its code must be grouped for placement into the ACL message that will transfer it. Locating this code is not necessarily a trivial task, as in JADE the Classpath tends to mix together references to platform code and libraries, and agent code. This makes it difficult to decide whether a library or a class in the Classpath may be used by the agent in the future.

Two different solutions are provided to this problem. The programmatically simpler method (although not so simple for the user) is for the programmer to manually pack the agent inside a Jar file with all the code and libraries it will require. This Jar file can then be transferred alongside the agent.

The Intra-Platform Mobility Service provides an initial mechanism in the migration system which automatically generates a Jar file given an agent. This Jar file contains all the agent's code and the programmer will never notice its existence. This mechanism is built by recursively examining the agent's main class looking for all dependent classes and building a Jar from the final set obtained. In the present version, it is possible that in some special cases not all agent classes are properly detected. This could happen in agents where, for example, reflection is used to load classes (e.g. `Class.forName("ClassName").newInstance()`). If problems occur with agent migration it is highly likely that some classes have not been included in the agent's Jar.

Currently, to avoid this problem, the Jar file must be manually created. The Jar should be named such that it corresponds to the agent's main class name. For example, the agent with a main class `org.example.MyAgent` must be packed in a Jar file named `org_example_MyAgent.jar`. The Jar must then be placed in a folder that is specified on the command line when launching JADE.

The second approach, programmatically more sophisticated but transparent to the user, consists of autonomously identifying all needed classes and constructing a Jar file containing them. Class selection is controlled by the service which may imply that non-optimal selections may be made.

### 6.3.4 THE JADE MOBILITY ONTOLOGY

The *jade-mobility-ontology* ontology contains all the concepts and actions needed to support agent mobility. JADE provides the class `jade.domain.mobility. MobilityOntology` which functions as a *singleton* giving access to a single, shared instance of the JADE mobility ontology through the `getInstance()` method. The ontology, which extends the `JADEManagementOntology`, contains five concepts and two actions, each associated with a class of the package `jade.domain.mobility`. The concept schemas are described in Tables 6.1–6.5. The actions are:

- `move-agent`. This is the action of moving an agent from one location to another. It is represented by the `MoveAction` class. The action has a single, unnamed slot of type `mobile-agent-description`. The argument is mandatory.
- `clone-agent`. This is the action of producing a copy of an agent, possibly running on another location. It is represented by the `CloneAction` class. The action has two unnamed slots: the first of type `mobile-agent-description` and the second of type `String`. Both arguments are mandatory.

Note that this ontology currently has no counterpart in the FIPA specifications.

## 6.4 USAGE OF THE JADE MOBILITY SERVICES

### 6.4.1 INTRA-PLATFORM MOBILITY SERVICE

The Intra-Platform Mobility Service is a JADE built-in service and thus requires no special installation steps to prepare it for use. The service is run by default when the JADE platform is started,

**Table 6.1** Mobile-agent-description: this describes a mobile agent. It is represented by the class MobileAgentDescription

| Slot name | Slot type | Mandatory/optional |
|---|---|---|
| Name | AID | **Mandatory** |
| Destination | Location | **Mandatory** |
| Agent-profile | mobile-agent-profile | Optional |
| Agent-version | String | Optional |
| Signature | String | Optional |

**Table 6.2** Mobile-agent-profile: this describes the computing environment needed by the mobile agent. It is represented by the class MobileAgentProfile

| Slot name | Slot type | Mandatory/optional |
|---|---|---|
| system | mobile-agent-system | Optional |
| language | mobile-agent-language | Optional |
| Os | Mobile-agent-os | **Mandatory** |

**Table 6.3** Mobile-agent-system: this describes the run-time system used by the mobile agent. It is represented by the class MobileAgentSystem

| Slot name | Slot type | Mandatory/optional |
|---|---|---|
| name | String | **Mandatory** |
| major-version | Integer | **Mandatory** |
| minor-version | Integer | Optional |
| dependencies | String | Optional |

**Table 6.4** Mobile-agent-language: this describes the programming language used by the mobile agent. It is represented by the class MobileAgentLanguage

| Slot name | Slot type | Mandatory/optional |
|---|---|---|
| name | String | **Mandatory** |
| major-version | Integer | **Mandatory** |
| minor-version | Integer | Optional |
| dependencies | String | Optional |

**Table 6.5** .Mobile-agent-os: this describes the operating system needed by the mobile agent. It is represented by the class MobileAgentOS

| Slot name | Slot type | Mandatory/optional |
|---|---|---|
| name | String | **Mandatory** |
| major-version | Integer | **Mandatory** |
| minor-version | Integer | Optional |
| dependencies | String | Optional |

but if some other services are running then it must be explicitly added to the list. It is run with the following command:

```
java jade.Boot -services jade.core.mobility.AgentMobilityService
```

### 6.4.2 INTER-PLATFORM MOBILITY SERVICE

The IPMS is not built into the platform and must be installed as an add-on. To install the add-on, the package distribution file must be unzipped inside the JADE folder from where the `ant lib` command is used to create a Jar file containing all compiled class files. To use the service, it must be explicitly specified on the command line, without forgetting to include the Intra-Platform Mobility Service first:

```
java jade.Boot -services
   jade.core.mobility.AgentMobilityService;jade.core.migration.
   InterPlatformMobilityService
```

### 6.4.3 SECURITY CONSIDERATIONS OF THE IPMS

It is very important to stress that the security implications of using inter-platform mobility are very important, typically more so than those of intra-platform mobility. Activating the IPMS naturally implies that the machine(s) hosting a JADE platform may execute code shipped from potentially any system reachable via a network connection. At the present time, the IPMS has no access control mechanism to decide which incoming agents will be executed and which will not. Currently, access control can only be performed by means of a secure MTP (for instance the HTTPS MTP provides some simple authentication features that can be used for this purpose), or by using network firewalls to prevent connections from specific platforms. Until a more complete access control system is implemented, it is recommended that platforms using IPMS take some basic security precautions:

1. Do not use the IPMS in an open environment or on machines containing sensitive information.
2. Try to restrict network access to platforms by using firewalls.
3. Try to use MTPs with some intrinsic form of authentication.

It is expected that forthcoming versions of the IPMS will provide comprehensive and configurable security mechanisms.

### 6.4.4 PROGRAMMING A MOBILE AGENT

A primary goal of the JADE mobility services is to be as simple to use as possible. In this vein, it is necessary to overload some methods and call the `doMove()` or `doClone()` methods of the `Agent` class to move or clone the agent (note that `doClone()` is only available with the Intra-Platform Mobility Service).

To indicate the destination of a mobile or cloned agent, JADE defines the `jade.core.Location` interface. Two implementations of this interface are provided: `jade.core.ContainerID` for intra-platform migration and `jade.core.PlatformID` for inter-platform migration. The `ContainerID` must be initialized with the name of the destination container and its transport address. The `PlatformID` must be initialized with the AID of the remote platform's AMS agent including its transport address.

To move an agent to another container or platform, an agent behaviour must include a call to the method `doMove()`. This method changes the agent to the TRANSIT state indicating that it is about to migrate. The following code illustrates an intra-platform migration:

```
// Create some variables
String containerName = "Container-1";
ContainerID destination = new ContainerID();

// Initialize the destination object
destination.setName(containerName);

// Change of the agent state to move
myAgent.doMove(destination);
```

In the case of inter-platform migration, a similar example is as follows:

```
// Build the AID of the corresponding remote platform's AMS
AID remoteAMS = new AID("ams@remotePlatform:1099/JADE", AID.ISGUID);

// Specify the MTP by setting the transport address of the remote
// AMS
remoteAMS.addAddresses("http://remotePlatformaddr:7778/acc");

// Create the Location object
PlatformID destination = new PlatformID(remoteAMS);

// Change of the agent state to move
myAgent.doMove(destination);
```

To clone an agent the `doClone()` method is used to change the agent state to COPY. To use it, two arguments must be passed: the container destination and the new agent name. This is an example:

```
// Create some variables
String containerName = "Container-1";
String newAgentName = "myClone";
ContainerID destination = new ContainerID();

// Initialize the destination object
destination.setName(containerName);

// Change of the agent state to clone
myAgent.doClone(destination, newAgentName);
```

As these mobility services implement weak migration, a code structure based on a finite state machine is required. This is because the program counter of the agent execution is not transmitted, making it is impossible to continue agent execution from the next line of the `doMove()` or `doClone()` instruction. Instead, agent execution can continue only from the beginning of the agent's behavioural code. Using a finite state machine representation, a structure can be created which segments the agent code into sections with assigned variables indicating the agent code execution state.

A switch statement is used, for example, an agent with both seller and buyer roles can have code with two states, one for selling and another for buying. The agent must set a state variable before leaving a container or a platform to indicate which role will be initially adopted in the next location. To create such a finite state machine representation. For example, in a two-container trip where the destination containers are in an array of locations, the behaviour code to be executed in each container is separated as in the following example:

```
addBehaviour(new CyclicBehaviour(this){

  public void action() {

    switch(_state){
      case 0:
        // Agent starts to migrate
        _state++;
        myAgent.doMove(_dests[0]);
        break;
      case 1:
        // Agent migrates to the second container
        _state++;
        myAgent.doMove(_dests[1]);
        break;
      case 2:
        // Agent dies
        myAgent.doDelete();
        break;
      default:
        myAgent.doDelete();
    }
  }

  private ContainerID[] _dests = ...;
  private int _state = 0;
});
```

This example shows how agent code must be structured in JADE to preserve its state by using a variable.

During the serialization and deserialization process of migrating agent code, some resources used by the agent will also be transferred, while others will be disconnected before migration of the agent and reconnected at the destination (this is the same distinction between transient and non-transient fields used in the Java Serialization API). JADE provides two matching methods in the `Agent` class for resource management that need only be overloaded by the programmer:

- **`beforeMove()`** – called at the source location when the move operation has successfully completed such that the moved agent instance is about to be activated on the destination container and the original agent instance is about to be terminated. This method is the correct place to release any local resource used by the original agent instance (e.g. closing open files and GUIs). If these resources were released before knowing if a migration was successful, they would just have to be reopened once again. However, a consequence of this is that any information that must be transported by the agent to the new location must be set before the `doMove()` method is called. Setting an agent attribute in the `beforeMove()` method will have no impact on the moved instance.
- **`afterMove()`** – called at the destination location as soon as the agent has arrived and its identity is in place, but before the behaviour scheduler has restarted.

For agent cloning, JADE provides a corresponding method pair, the `beforeClone()` and `afterClone()` methods. These are called in the same fashion as the `beforeMove()` and `afterMove()` methods. All four of these methods are protected members of the `Agent` class, defined as empty placeholders. User-defined mobile agents can override the four methods as needed.

### 6.4.5 ACCESSING THE AMS FOR AGENT MOBILITY

The JADE AMS has some extensions that support agent mobility. Each mobility-related action, as described in Section 6.3.4, can be requested of the AMS using a FIPA Request message, with the ontology attribute set to *jade-mobility-ontology* and the language set to *FIPA-SL0*.

The `move-agent` action takes a `mobile-agent-description` as its parameter. This action moves the agent identified by the name and address slots of the `mobile-agent-description` to the location specified in the destination slot. For example, if an agent wants to move the agent *Peter* to the location called *Front-End*, it must send the following ACL Request message to its platform AMS:

```
(REQUEST
 :sender (agent-identifier :name RMA@Zadig:1099/JADE)
 :receiver (set (agent-identifier :name ams@Zadig:1099/JADE))
 :content
  (
   (action (agent-identifier :name ams@Zadig:1099/JADE)
           (move-agent (mobile-agent-description
            :name (agent-identifier :name Johnny@Zadig:1099/JADE)
            :destination (location
              :name Main-Container
              :protocol JADE-IPMT
              :address Zadig:1099/JADE.Main-Container )
            )
           )
   )
  )
 :reply-with  Req976983289310
 :language  FIPA-SL0
 :ontology  jade-mobility-ontology
 :protocol  fipa-request
 :conversation-id  Req976983289310
)
```

Using JADE ontology support, an agent can easily add mobility to its capabilities, without needing to compose ACL messages manually. First, the agent must create a new `MoveAction` object and fill its argument with a suitable `MobileAgentDescription` object, which is in turn filled with the name and address of the agent to move (either itself or another mobile agent) and with the Location object for the destination. A call to the `Agent.getContentManager().fillContent(..,..)` method then converts the `MoveAction` Java object into a `String` and writes it into the content slot of a Request ACL message. The clone-agent action works in the same way, but has an additional `String` argument to hold the name of the new agent resulting from the cloning process. The AMS also supports the four mobility-related actions defined in the `JADEManagement Ontology`. The where-is-agent action has a single AID argument, holding the identifier of the agent to be located. This action has a result, namely the location for the agent that is put into the content slot of the Inform ACL message that successfully closes the protocol. For example, the Request message to ask for the location where the agent *Peter* resides would be:

```
(REQUEST
  :sender   (agent-identifier :name da1@Zadig:1099/JADE)
  :receiver (set (agent-identifier :name ams@Zadig:1099/JADE))
  :content  (( action
    (agent-identifier :name ams@Zadig:1099/JADE)
```

```
    (where-is-agent (agent-identifier :name Peter@Zadig:1099/JADE))
  ))
  :language  FIPA-SL0
  :ontology  JADE-Agent-Management  :protocol  fipa-request
)
```

The resulting Location would be contained within an Inform message such as:

```
(INFORM
  :sender (agent-identifier :name ams@Zadig:1099/JADE)
  :receiver  (set (agent-identifier :name da1@Zadig:1099/JADE))
  :content  ((result
    (action
      (agent-identifier :name ams@Zadig:1099/JADE)
      (where-is-agent (agent-identifier :name
         Peter@Zadig:1099/JADE))
    )
    (set (location
      :name Container-1
      :protocol JADE-IPMT
      :address Zadig:1099/JADE.Container-1
    ))
  ))
  :reply-with  da1@Zadig:1099/JADE976984777740
  :language  FIPA-SL0
  :ontology  JADE-Agent-Management
  :protocol  fipa-request
)
```

The query-platform-locations action takes no arguments but results in a set of all the Location objects available in the current JADE platform. The message for this action is very simple:

```
( REQUEST
 :sender (agent-identifier :name Johnny)
 :receiver (set (Agent-Identifier :name AMS))
 :content (( action (agent-identifier :name AMS)
                ( query-platform-locations ) ))
 :language FIPA-SL0
 :ontology JADE-Agent-Management
 :protocol fipa-request
)
```

If the current platform had three containers, the AMS would send back the following Inform message:

```
( INFORM
 :sender (Agent-Identifier :name AMS)
 :receiver (set (Agent-Identifier :name Johnny))
 :content (( Result ( action (agent-identifier :name AMS)
                             ( query-platform-locations ) )
              (set (Location
                  :name Container-1
```

```
                    :transport-protocol JADE-IPMT
                    :transport-address IOR:000....Container-1 )
                  (Location
                    :name Container-2
                    :protocol JADE-IPMT
                    :address IOR:000....Container-2 )
                  (Location
                    :name Container-3
                    :protocol JADE-IPMT
                    :address IOR:000....Container-3 )
            )))
  :language  FIPA-SL0
  :ontology  JADE-Agent-Management
  :protocol  fipa-request
)
```

The `Location` class implements the `jade.core.Location` interface and is passed into the `Agent.doMove()` and `Agent.doClone()` methods. A typical behaviour pattern for a JADE mobile agent will be to ask the AMS for locations (either the complete list or through multiple where-is-agent actions) and to then decide if, when and where to migrate.

### *6.4.6 EXAMPLES OF AGENT MOBILITY*

With respect to intra-platform mobility, the JADE examples package contains a demonstration application including a dedicated GUI to assist use. To run the demonstration a platform with two containers is needed. If all classes are in the Classpath, the application and the first container can be started as follows:

```
java jade.Boot test:examples.mobile.MobileAgent
```

The second container can then be started:

```
java jade.Boot -container
```

The GUI is shown in Figure 6.4. An intra-platform mobile agent example will appear. To run the example, it is only necessary to refresh the locations list, select a container and press the Move or Clone button. The entire application will then move or clone to the selected container.

With respect to inter-platform mobility, the add-on package of the IPMS contains two mobile agent examples. One is simple with easy to understand code, while the other has more complex code but provides a GUI similar to the intra-platform mobility example described above.

The first, simpler example is an application called MobileAgent. At least two platforms are required to run the example as it consists of an agent that travels through two or more platforms. To run the example, a properties file with the agent itinerary must first be created, with every platform to be visited added as a 'hop':

```
hop0=first_hostname_to_migrate
hop1=second_hostname_to_migrate
...
hopn=nth_hostname_to_migrate
```

Next the platforms which the agent will visit must be initiated, remembering to start an IPMS on each (see Section 6.4.2). Then, the source platform and the agent must be started with the

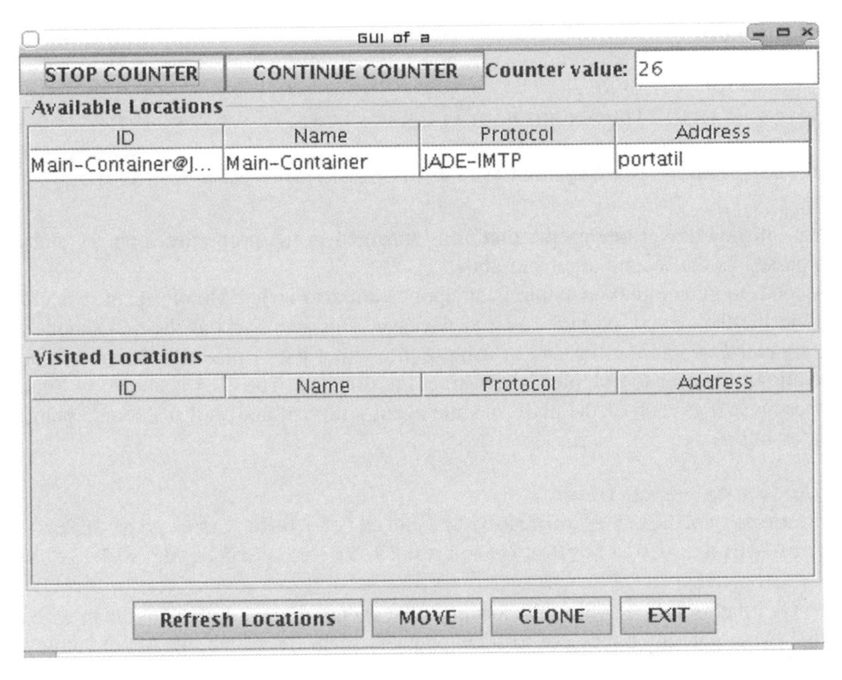


**Figure 6.4** Intra-platform mobile agent example

Itinerant Mobile Agent

Agent's Itinerary

| Machine Name / IP | AMS AID | Platform Transport Address |
|---|---|---|
| hostname | ams@hostname:1099/JADE | http://hostname:7778/acc |

New Hop

Display Message

Text to display:

Use GUI to display message

Move Exit

About Itinerant Agent

This Itinerant Agent is a kind of mobile agent that moves across a list of pre-established agent platforms showing some text in the console of each one.

In order to use this agent you must previously be sure that all the platforms of its itinerary have been correctly started and that inter-platform mobility service has also been loaded in each one.

In order to specify the agent itinerary you must create an entry for each agent platform you want the agent visit.

**Figure 6.5** Inter-platform mobile agent example

appropriate arguments:

```
java jade.Boot -services
   jade.core.mobility.AgentMobilityService;jade.core.migration.
   InterPlatformMobilityService
   test:samples.MobileAgent("file_properties","msg_to_display")
```

The agent will now travel among the platforms specified in the properties itinerary and print the message passed as the second argument above.

The second, more complex example is an application also called MobileAgent, but in this case is located inside the `samples.mobilegui` package. This agent acts as the first sample agent, as with the simpler example, but instead of using a file with a list of platforms to visit, it provides a GUI that allows the user to graphically construct the itinerary. The GUI is shown in Figure 6.5.

The user must start each of the platforms the agent will visit and start the source platform with the agent as follows:

```
java jade.Boot -services
   jade.core.mobility.AgentMobilityService;jade.core.migration.
   InterPlatformMobilityService test:samples.mobilegui.MobileAgent
```

Using the GUI the user then defines the locations the agent will visit by adding them with the New Hop button. After entering a message to show on every platform visited, the Move button will start the agent trip.

# 7

# JADE Internal Architecture

Up to this point only the features supported by the JADE run-time, and the API to access them, have been presented. In this section we now turn our attention towards the internal architecture of the JADE kernel, including how to modify and extend its behaviour.

## 7.1 DISTRIBUTED COORDINATED FILTERS

Prior to JADE v3.2, the run-time was implemented as a large monolithic kernel providing most of the features required by agents to live and communicate. This approach is rather inflexible as it requires modifications to the kernel every time new functionality is introduced, with the strong possibility of impacting other, apparently unrelated, features. Thus, in the v3.2 release, distributed in July 2003, the JADE run-time was completely restructured according to a new design known as the 'distributed coordinated filters architecture'.

### *7.1.1 IDEAS AND MOTIVATIONS*

The new JADE architecture was created by considering the following set of requirements:

- Implementation of platform features as separated modules.
- Extensibility to flexibly support the integration of new features and modification of existing ones.
- Easy deployment of features across a distributed platform.
- 'Deploy what you need' strategy to only start required features and use target-specific implementations (e.g. for mobile environments).

Taking into account these requirements, the architecture drew some inspiration from *aspect oriented programming* (AOP), of which the main tenet is the promotion of clean *separation of concerns*. This is achieved by writing software code as a collection of independent *aspects*, each expressing a different concern. Aspects are then carefully combined according to some rules in a process called *aspect weaving*. Several approaches to realize aspect weaving are available, with probably the most popular one (in the Java world) being classload-time aspect weaving. In this approach an ad hoc ClassLoader is used to merge the different aspects together when loading Java classes. However, in an attempt to keep things as simple as possible, and considering the requirement of running JADE on mobile devices (see Chapter 8) where sophisticated classloading techniques are generally not available, the approach proposed by Aksit *et al.*, 1993) was selected. In this approach, known as *composition filters*, each object is provided with two *filter chains*: an *incoming chain* whose filters are invoked whenever the object receives a method call and an *outgoing chain* whose filters

*Developing Multi-Agent Systems with JADE* Fabio Bellifemine, Giovanni Caire, Dominic Greenwood

are invoked whenever the object is about to call some other object's method. The distributed coordinated filters architecture results from a blend of the composition filters approach and the intrinsic distribution of JADE platforms across multiple containers.

### *7.1.2 MAIN ELEMENTS*

The main elements in the distributed coordinated filters architecture are depicted in Figure 7.1. Each JADE container resides on top of a *node*. While a container is designed to host agents, a node hosts *services*. A service implements a set of platform-level features that can be grouped together according to their conceptual cohesion. It is the primary abstraction in the distributed coordinated filters architecture. Examples of services are the Messaging Service dealing with the delivery of ACL messages exchanged by agents, the Agent Management Service which controls all operations related to agents' activation and termination and the Mobility Service which implements the mechanisms that allow agents to migrate between containers.

The element that manages the activation of services inside a node is called the *Service Manager*. Actually a single Service Manager exists within the node hosting the main container. This Service Manager keeps the table of all nodes in the platform and the list of all services that are active in each node. Service Managers on peripheral containers are just proxies to the real Service Manager in the main container.

### *7.1.3 SERVICE COMPONENTS*

As mentioned, the main abstraction in the JADE distributed coordinated filters architecture is the service. All agent-level operations, such as sending messages, moving and even starting and terminating, are implemented by a JADE service which will contain several components that are described in this section.

#### 7.1.3.1 Vertical Commands, Filters and Sinks

All agent-level operations are forwarded by the underlying container to the service responsible for implementing that operation. For instance, an attempt by an agent to migrate to a remote container

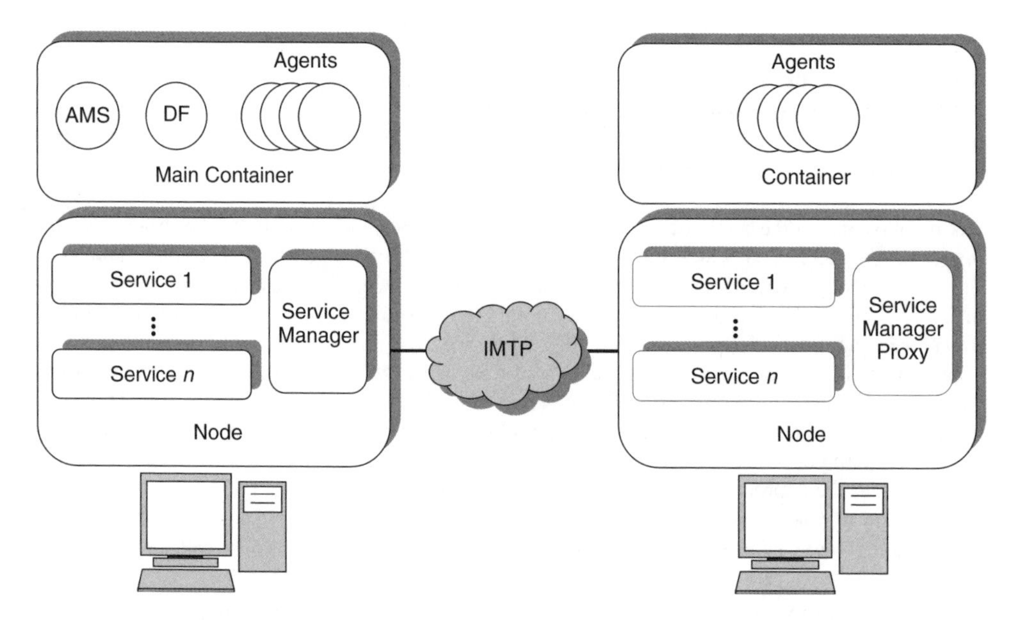


**Figure 7.1** Distributed coordinated filters architecture main elements

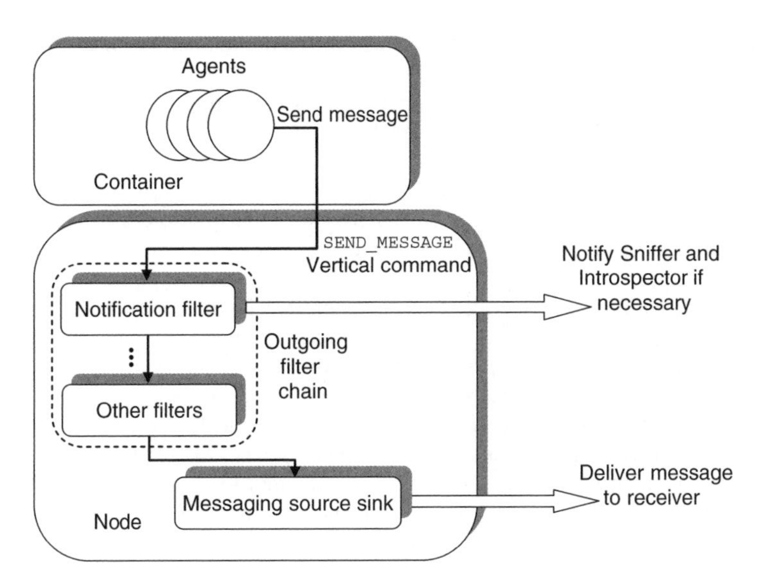


**Figure 7.2** The outgoing filter chain

(i.e. by calling the `doMove()` method of the `Agent` class) is redirected to the Mobility Service by the underlying container. The service may either directly implement the operation or issue a *vertical command*. A vertical command is an instance of the `jade.core.GenericCommand` class and embeds the operation to be performed along with any parameters. When issuing a vertical command the actual implementation of the operation to be performed is delegated to a component of the service itself called an *outgoing (or source) sink*. Before reaching the service source sink, however, a vertical command traverses a chain of *outgoing filters* (according to the composition filters approach). All services installed in a node may provide an outgoing filter with which they can react in a service-specific way to all vertical commands issued by all services in that node. For instance, as depicted in Figure 7.2, when an agent sends a message, the container passes the message to the Messaging Service. The latter issues a `SEND_MESSAGE` vertical command embedding the message to be delivered and the intended receiver as parameters. The Notification Service (the service used to notify the Sniffer Agent and the Introspector Agent described in Section 3.7) outgoing filter processes the command by checking if the sender or the receiver are currently sniffed or introspected, and issuing a notification if so. Other filters belonging to other services not interested in the `SEND_MESSAGE` vertical command simply ignore it. Finally the messaging sink actually delivers the message to the intended receiver.

When a service implements an agent-level operation issuing a vertical command, it creates an extension/modification point in the platform so that other services (possibly created by application developers described in Section 7.2) can intercept the command and add further processing. A filter may modify a command and even block it. For instance, by intercepting the `SEND_MESSAGE` vertical commands and blocking non-compliant messages, a filter can easily prevent all messages that do not respect given application-specific constraints from being delivered.

#### 7.1.3.2 Horizontal Commands and Slices

A service can also be distributed across several platform nodes with each part often needing to interact. For instance, the Messaging Service is installed in all nodes that compose a platform. When the messaging service sink on a node needs to process a `SEND_MESSAGE` vertical command carrying a message directed to an agent living on a remote container, it must detect where the receiver lives and then pass the message to the destination container. The first operation may involve contacting the Messaging Service running on the main container to obtain the receiver location. The second operation certainly requires contacting the Messaging Service running on the destination node to actually transfer the message and post it in the receiver's message queue.

These node-to-node interactions are carried out by means of *horizontal commands*. As with vertical commands, horizontal commands embed the operation to be performed along with any parameters it may have. The same class, `jade.core.GenericCommand`, is used to implement them.

The component, within every service, that is responsible for accepting horizontal commands from remote nodes is called a *slice*. A slice may perform the operation embedded in a received horizontal command directly, or it can issue a new vertical command. In the latter case, the actual operation is delegated to another service component called the *incoming (or target) sink*. Before reaching the service target sink, however, a vertical command issued by a slice traverses, according to the composition filters approach, a chain of *incoming filters*. All services installed in a node may provide an incoming filter with which they can react in a service-specific way to all vertical commands issued by all service slices in that node. It should be noted that both the source and target sinks and the outgoing and incoming filters must be different entities (even if they are instances of the same class).

As an example, Figure 7.3 shows the complete path followed by an ACL message exchanged between two agents. It can be noticed that this path has the form of a schematized 'U' where the vertical edges correspond to the vertical commands traversing the outgoing and incoming filter chains, and the horizontal edge corresponds to the horizontal command sent by the messaging source sink to the messaging slice on the destination node.

When a service slice serves a horizontal command by just issuing an incoming vertical command and delegating the actual operation to the service target sink, it creates an extension/modification point in the platform as other services can intercept the command and add further processing.

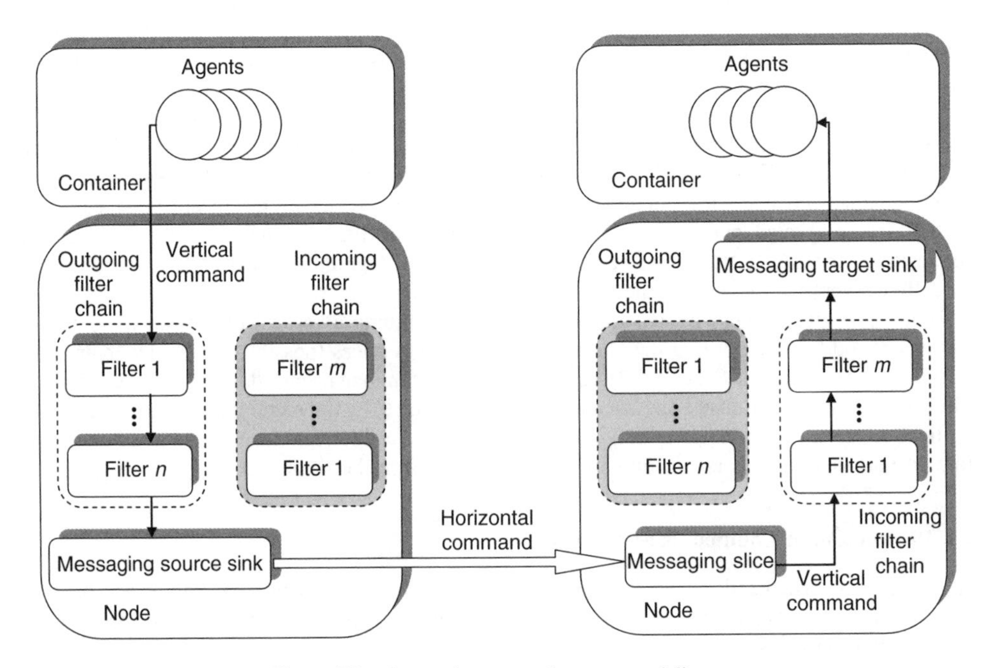


**Figure 7.3** A complete example: message delivery

#### 7.1.3.3 Service Helpers

For JADE built-in services, agent-level operations are typically handled by the underlying container that is responsible for passing them to the proper service for processing, as discussed in Section 7.1.3.1. For extended and user-defined services this approach would require modifying the container code each time a new service supporting new agent-level operations is added. Therefore, the distributed coordinated filters architecture includes the Service Helper abstraction that allows an agent to directly access the features provided by a service. The Service Helper is the mechanism by which the JADE kernel API can be extended in a non-intrusive and backward compatible way. An agent can retrieve the helper of a given service by means of the `getHelper()` method of the `Agent` class. This method gets the service name as parameter and returns an object implementing the `jade.core.ServiceHelper` interface. That object must be properly downcast to the correct service helper before its business methods can be invoked.

For instance, assuming the JADE kernel must be extended to give agents the ability to send emails, an `EMailService` should be developed (as will be detailed in Section 7.2). Clearly the sending feature of the service would not be made available through a new method of the `Agent` class (as, for example, in the case of sending messages or migrating), since this would imply modifying the JADE kernel classes. On the other hand a proper `EMailHelper` would have to be defined, providing, for instance, the `sendMail()` method. Code similar to the following could be used.

```
1: ...
2: EMailHelper emHelper = (EMailHelper) getHelper
   (EMailService.NAME) ;
3: emHelper.sendMail(...) ;
4: ...
```

The class diagram depicted in Figure 7.4 summarizes all the components that make up a kernel-level JADE service. It should be noted that none of these components are mandatory; a minimal service need not have any filters, sinks or slices.

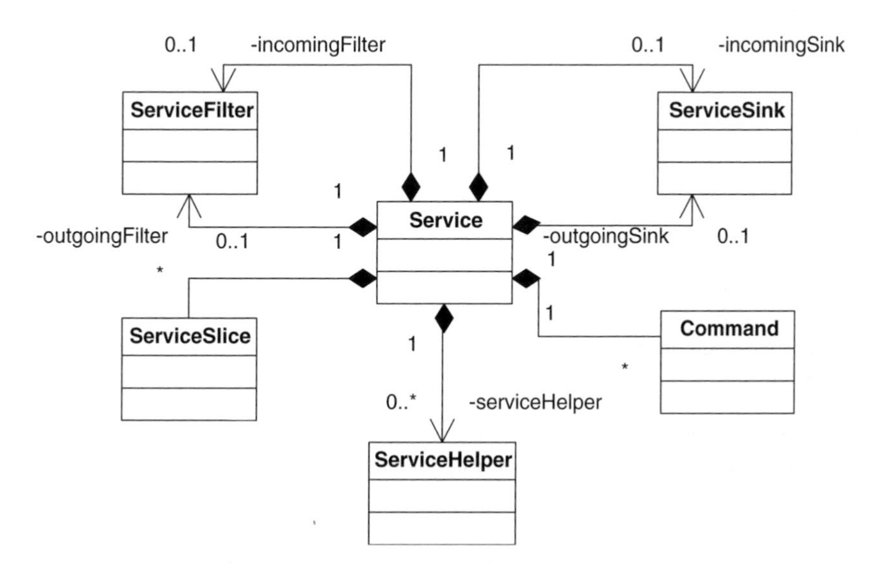


**Figure 7.4** Components of a JADE kernel-level service

### 7.1.4 SELECTING SERVICES TO ACTIVATE

When starting a JADE container, the services to be activated in the underlying node are specified by means of the -services option. The value for this option is a list of semicolon (';') separated, fully qualified class names, each implementing a service as described later in Section 7.2. It is important to note that the Messaging Service and the Agent Management Service, which are mandatory in every JADE platform, are always activated regardless of the value of the -services option. Besides the two mandatory services, JADE also activates the Mobility Service (see Chapter 6) and the Notification Service by default. Thus, launching JADE without the -services option, as shown below, is equivalent to not specifying it at all.

```
java jade.Boot ... -services jade.core.mobility
   .AgentMobilityService;jade.core.event.NotificationService
```

It should also be noted that when an additional service is to be started, launching JADE with a command line such as

```
java jade.Boot ... -services myPackage.myService
```

will start a container that does not support agent mobility and agent sniffing/introspection because, since the default is overridden, the Mobility Service and the Notification Service are no longer activated.

## 7.2 CREATING A JADE KERNEL SERVICE

Having described the components that make up a JADE kernel service and illustrated their roles and interactions, this section focuses on the classes that implement them and how to create a user-defined service.

A simple logging service for the book-trading case study used throughout this book will be used as an example application which will simply trace all messages exchanged by agents. The service will be built incrementally. In the first phase messages will be printed on the local standard output which illustrates that a service can simply provide filters to intercept certain outgoing and incoming vertical commands (as the majority of user-defined services do). In the second phase the service will be modified to print all messages on the standard output of the main container. This phase illustrates that a service can require some degree of coordination between different nodes. Finally the service will be extended by adding a Service Helper that allows agents to directly interact with it.

### 7.2.1 IMPLEMENTING THE SERVICE CLASS

A service can be created either by implementing the jade.core.Service interface or, more conveniently, by subclassing the jade.core.BaseService class. The latter provides default implementations for most methods of the Service interface. The only method that must be implemented is the getName() method which returns a string used as an identifier for the service being implemented. Although this is not strictly mandatory, the recommended convention for service names is as follows:

- Use the 'Service' suffix in all classes extending BaseService.
- Use the fully qualified service class name without the 'Service' suffix as the name of the service returned by the getName() method.

During service start-up two methods of the Service interface are called by the underlying node. The init() method is called first to make passive initializations before the service is actually installed in the node; this can be compared to a constructor. The `boot()` method is then called after the service has been activated. The service is then able to use features of the `ServiceManager` and the local container. The `boot()` method can be used to implement, for example, a distributed initialization protocol. The code below shows the skeleton of the `bookTrading.logging.LoggingService` class that implements the sample Logging Service described above.

```
 1: package bookTrading.logging;
 2:
 3: import jade.core.*;
 4:
 5: public class LoggingService extends BaseService {
 6:    // Service name
 7:    public static final String NAME =
   "bookTrading.logging.Logging";
 8:
 9:    // Service parameter names
10:    public static final String VERBOSE =
   "bookTrading_logging_LoggingService_verbose";
11:
12:    private boolean verbose = false;
13:
14:    public String getName() {
15:      return NAME;
16:    }
17:
18:    public void boot(Profile p) throws ServiceException {
19:      super.boot(p);
20:      verbose = p.getBooleanProperty(VERBOSE, false);
21:      System.out.println("VERBOSE = "+verbose);
22:    }
23: }
```

As can be seen at line 18, the `boot()` method takes a `jade.core.Profile` object as a parameter. This `Profile` object holds all JADE start-up configurations as described in Section 5.6. Service-specific configurations can therefore be specified as command line parameters following the usual `-key value` format. In this way they can be retrieved by the service itself through the `Profile` object passed as a parameter to the `boot()` method. With respect to the example, we assume the Logging Service accepts a *verbose* parameter that tells it to print all fields of a message (if verbose is `true`) or just the performative (if verbose is `false`). A good naming convention used in JADE to avoid confusion between parameters related to different services, is to build parameter names by concatenating the service class name (substituting dots with underscores) and the parameter identifier. The `VERBOSE` static constant defined at line 10 provides an example. At line 20 the `getBooleanProperty()` method of the `Profile` class is used to get the value (already as a `boolean`) of the `bookTrading_logging_LoggingService_verbose` parameter.

### *7.2.2 STARTING THE SERVICE*

A Main Container hosting the Logging Service created above can be started by typing the following command line:

```
java -cp <JADE classes + bookTrading classes> jade.Boot -gui
```

```
-services jade.core.event.NotificationService;bookTrading.logging.
LoggingService -bookTrading_logging_LoggingService_verbose true
```

The service will work in verbose mode. Note also that, besides the Logging Service, only the Notification Service is started, while the Mobility Service is not. As a consequence the started platform will of course not support mobility.

This is the output produced by JADE from the above command line:

```
prompt> java -cp <jade classes + bookTrading classes> jade.Boot -gui
-services bookTrading.logging.LoggingService;jade.core.event.NotificationService
-bookTrading_logging_LoggingService_verbose true
29-dic-2005 16.41.11 jade.core.Runtime beginContainer
INFO: ----------------------------------
    This is JADE snapshot - revision 5752 of 2005/07/15 14:22:11
    downloaded in Open Source, under LGPL restrictions,
    at http://jade.tilab.com/
----------------------------------------
29-dic-2005 16.41.14 jade.core.BaseService init
INFO: Service jade.core.management.AgentManagement initialized
29-dic-2005 16.41.15 jade.core.BaseService init
INFO: Service jade.core.messaging.Messaging initialized
29-dic-2005 16.41.15 jade.core.BaseService init
INFO: Service bookTrading.logging.Logging initialized
29-dic-2005 16.41.15 jade.core.BaseService init
INFO: Service jade.core.event.Notification initialized
29-dic-2005 16.41.16 jade.core.messaging.MessagingService boot
INFO: MTP addresses:
http://anduril:7778/acc
VERBOSE = true
29-dic-2005 16.41.19 jade.core.AgentContainerImpl joinPlatform
INFO: ----------------------------------
Agent container Main-Container@JADE-IMTP://NBNT2004130496 is ready.
```



By comparing this output with that presented in Figure 3.4, it can be noted that the Logging Service is initialized and that the *verbose* parameter has been set to `true` as expected.

### *7.2.3 USING FILTERS TO INTERCEPT VERTICAL COMMANDS*

Thus far the Logging Service does not provide any functionality, so the next step is to make it actually log messages exchanged by agents in the book-trading scenario. To do this some service filters must be implemented, specifically an outgoing filter is needed to log messages once they are sent. Service filters are implemented by subclassing the `jade.core.Filter` class and redefining the `accept()` and the `postProcess()` methods. Both methods receive a `jade.core.VerticalCommand` parameter. When a vertical command traverses a filter chain, the following sequence of invocations occurs:

```
Filter-1.accept()
...
Filter-n.accept()
Sink.consume()
Filter-n.postProcess()
...
Filter-1.postProcess()
```

Therefore a filter can operate on a given vertical command both before (method `accept()`) and after (method `postProcess()`) the operation corresponding to the vertical command is actually carried out by the service sink. The `accept()` method returns a `boolean` value. By returning `false` a filter can block a vertical command.

A service provides its outgoing and incoming filters (if any) by means of the `getCommandFilter()` method of the `Service` interface. This method receives a parameter indicating the direction of the filter to be returned. Possible values for this parameter are `Filter.INCOMING` and `Filter.OUTGOING`.

Even if this is not mandatory, a typical programming style when developing JADE kernel services is to implement service components such as filters, sinks and slices as inner classes of the main service class. The code below shows how the Logging Service can print all sent messages on the standard output.

```
 1: ...
 2: private outFilter = new OutgoingLoggingFilter();
 3: ...
 4: public Filter getCommandFilter(boolean direction) {
 5:   if (direction == OUTGOING) {
 6:     return outFilter;
 7:   }
 8:   else {
 9:     return null;
10:   }
11: }
12: ...
13: private class OutgoingLoggingFilter extends Filter {
14:   public boolean accept (VerticalCommand cmd) {
15:     if (cmd.getName().equals (MessagingSlice.SEND_MESSAGE) ) {
16:       object[] params = cmd.getParams();
17:       AID sender = (AID) params [0];
18:       GenericMessage gMsg = (GenericMessage) params [1];
19:       ACLMessage msg = gMsg.getACLMessage();
20:       AID receiver = (AID) params [2];
21:       System.out.println("Message from "+sender+" to
  "+receiver+:");
22:       if (verbose) {
23:         System.out.println (msg) ;
24:       }
25:       else {
26:         System.out.println(ACLMessage.getPerformative
  (msg.getPerformative()));
27:       }
28:     }
29:     // Never block a command
30:     return true;
31:   }
32: }
```

As can be seen at lines 17 to 20, the `SEND_MESSAGE` vertical command has three parameters representing the sender of the message, the message itself and the intended receiver. Actually the `ACLMessage` object is incapsulated in a `jade.core.messaging.GenericMessage` object. The `GenericMessage` class embeds, in addition to the real ACL message, other information that is used by the Messaging Service sink to handle special delivery cases.

### *7.2.4 IMPLEMENTING A DISTRIBUTED JADE SERVICE*

In this section the sample Logging Service is extended to show all logging printouts on the main container standard output regardless of the container in which they are generated. To achieve this, a Logging Service slice must be implemented. Furthermore, the OutgoingLoggingFilter presented in Section 7.2.3 must be modified to send a horizontal command to the Logging Service slice on the main container each time it intercepts a message.

Implementing a slice for a service is slightly more complex than implementing a filter, as it implies developing three classes.

#### 7.2.4.1 The *Horizontal Interface*

This interface declares all the methods that can be invoked on a remote slice and must extend the `Service.Slice` interface. All methods in the horizontal interface must throw `jade.core.IMTPException` if some network problem occurred at the IMTP level, e.g. the remote slice is not reachable. In the ongoing example the horizontal interface can be as follows:

```
1: package bookTrading.logging;
2:
3: import jade.core.*;
4:
5: public interface LoggingSlice extends Service.Slice {
6:   public static final String H_LOGMESSAGE = "log-message";
7:
8:   public void logMessage (String s) throws IMPException;
9: }
```

The horizontal interface is also a good place to define the constants representing the names of the horizontal commands that can be served by the service slice. In particular there must be one such horizontal command for each method in the horizontal interface.

#### 7.2.4.2 The *Slice Proxy*

This is a class whose instances are proxies to a remote slice. It extends the `jade.core.SliceProxy` class and must implement the horizontal interface. When a service needs to interact with a slice on a remote node it first retrieves a proxy to that slice and then invokes the required methods. The proxy has the main purpose of converting method calls into proper horizontal commands that will be sent to the remote slice. Proxy classes are loaded dynamically when needed and must comply with a well-defined naming convention: they must be named `<service name>Proxy`. The `bookTrading.logging.LoggingProxy` class is as follows:

```
1: package booTrading.logging;
2:
3: import jade.core.*;
4:
5: public class LoggingProxy extends SliceProxy implements
   LoggingSlice {
6:   public void logMessage (String s) throws IMTPException {
7:      GenericCommand cmd = new GenericCommand(H_LOGMESSAGE,
   LoggingService.NAME, null);
8:      cmd.addParam(s) ;
9:      getNode().accept(cmd);
```

```
10:     }
11: }
```

As can be seen at line 7, the `jade.core.GenericCommand` class is used to implement horizontal commands. These commands are then filled with the necessary parameters (in this case, just the string to be logged) and forwarded to the remote node. Note that a slice proxy embeds a proxy to the node the proxied slice resides in.

#### 7.2.4.3 The *Slice Implementation*

This is a class (typically implemented as an inner class of the main service class) implementing the `Service.Slice` interface. It is responsible for actually serving incoming horizontal commands. The slice implementation does not need to implement the horizontal interface as it receives horizontal commands and not method calls. The slice implementation class for the Logging Service is as follows:

```
 1: ...
 2: private class LoggingSlicelmpl implements Service.Slice {
 3:
 4:   public Service getService() {
 5:     return LoggingService.this;
 6:   }
 7:
 8:   public Node getNode() throws ServiceException {
 9:     try {
10:       return LoggingService.this.getLocalNode() ;
11:     }
12:     catch (IMTPException imtpe) {
13:       // Should never happen as this is a local call
14:       throw new ServiceException("Unexpected error retrieving
    local node") ;
15:     }
16:   }
17:
18:   public VerticalCommand serve (HorizontalCommand cmd) {
19:     String cmdName = cmd.getName() ;
20:     if (cmd.getName().equals (LoggingSlice.H_LOGMESSAGE)) {
21:        Object[] params = cmd.getParams () ;
22:        System.out.println (params [0]) ;
23:     }
24:   }
25: }
26: ...
```

The `getService()` and `getNode()` methods are quite straightforward and their bodies are almost identical in all slice implementation classes. The `serve()` method on the other hand is the central point of a slice implementation class and is invoked each time a horizontal command is received. A slice may serve a horizontal command directly as in the example above, or alternatively, as mentioned in Section 7.1.3.2, if developers want other services to react to the reception of a horizontal command, they can make the `serve()` method return a `VerticalCommand` that will traverses the incoming filter chain up to the service sink.

A service declares its ability to perform inter-node coordination by implementing the `getHorizontalInterface()` and the `getLocalSlice()` methods so that they return the horizontal interface class and the slice implementation instance respectively. To do this, the `LoggingService` class should be modified as follows:

```
 1: ...
 2: private ServiceSlice localSlice = new LoggingSliceImpl() ;
 3: ...
 4: public Class getHorizontalInterface() {
 5:   return LoggingSlice.class;
 6: }
 7:
 8: public Service.Slice getLocalSlice() {
 9:   return localSlice;
10: }
11: ...
```

At this point the Logging Service is ready to forward logging printouts across platform nodes. The last thing to do is modify the OutgoingLoggingFilter, presented in Section 7.2.3, to forward all logging printouts to the logging slice on the main container. The modified code is as follows:

```
 1: ...
 2: private class OutgoingLoggingFilter extends Filter {
 3:   public boolean accept (VerticalCommand cmd) {
 4:     if (cmd.getName().equals (MessagingSlice.SEND_MESSAGE)) {
 5:       Object[] params = cmd.getParams() ;
 6:       AID sender = (AID) params [0] ;
 7:       GenericMessage gMsg = (GenericMessage) params[1] ;
 8:       ACLMessage msg = gMsg.getACLMessage() ;
 9:       AID receiver = (AID) params[2] ;
10:       // Prepare the log record
11:       String logRecord = "Message from "+sender+" to
   "+receiver+": \n" ;
12:       if (verbose) {
13:         logRecord = logRecord+msg;
14:       }
15:       else {
16:         logRecord = logRecord+ACLMessage.getPerformative
   (msg.getPerformative());
17:       }
18:
19:       // Send the log record to the logging slice on the Main
   Container
20:       try {
21:         LoggingSlice mainSlice = (LoggingSlice)
   getSlice(MAIN_SLICE) ;
22:         mainSlice.logMessage(logRecord) ;
23:       }
24:       catch (ServiceException se) {
25:         System.out.println("Error retrieving Main Logging
   Slice") ;
26:         se.printStackTrace() ;
```

```
27:       }
28:        catch (IMTPException impte) {
29:          System.out.println ("Error contacting Main Logging
   Slice") ;
30:          se.printStacktrace() ;
31:       }
32:     }
33:     // Never block a command
34:     return true;
35:   }
36: }
```

The `getSlice()` method of the `Service` interface (already implemented in the `BaseService` class) is used to retrieve a proxy to a remote slice. This method gets the name of the node (that coincides with the name of the container) that the remote slice resides in. The `MAIN_SLICE` constant defined in the `BaseService` class always indicates the slice residing on the main container regardless of the actual name of the main container node.

### *7.2.5 AGENT AND SERVICE INTERACTIONS*

As described in Section 7.1.3.3, agents can interact directly with services that provide a Service Helper. To exemplify how a service can provide a helper, the Logging Service is further extended to allow agents to change its verbose attribute.

A recommended programming style here consists of defining a `LoggingHelper` interface extending `ServiceHelper` and a `LoggingHelperImpl` inner class of the main service class. The `LoggingHelper` interface may then appear as follows:

```
1: package bookTrading.logging;
2:
3: import jade.core.*;
4:
5: public interface LoggingHelper extends ServiceHelper {
6:   public void setVerbose(boolean verbose);
7: }
```

A service provides a helper by implementing the `getHelper()` method of the Service interface. This method is invoked whenever an agent retrieves the helper for that service for the first time and obtains the agent as parameter. This allows services to use a single helper object for all agents, one helper object for each agent, or even to give access to its helper only to agents of given types depending on the specific scenario. The code of the `LoggingService` class related to the provision of the Service Helper is shown below. A single helper object for all agents is used in this case.

```
1: ...
2: private ServiceHelper helper = new LoggingHelperImpl();
3: ...
4: public ServiceHelper getHelper (Agent a) {
5:   return helper;
6: }
7: ...
8: public class LoggingHelperImpl implements LoggingHelper {
9:   public void init(Agent a) {
```

```
10:   }
11:
12:   public void setVerbose (boolean v) {
13:     verbose = v;
14:   }
15: }
16: ...
```

The `init()` method defined at line 9 is the only method included in the `ServiceHelper` interface and is invoked just before returning a helper object to an agent requesting it for the first time. When using a helper object for each agent, it allows linking a helper object to the associated agent. Since in this case a single helper object is used for all agents, its body is left empty.

# 8

# Running JADE Agents on Mobile Devices

The introduction of always connected wireless networks (GPRS, UMTS, WLAN), combined with the continuous growth in power and resources of hand-held devices such as PDAs and cellphones, has directly resulted in the progressive integration of wireless and wire-line environments. In turn, the need to deploy applications distributed partly in the fixed network and partly on hand-held devices has become increasingly important. It is possible to run JADE agents on MIDP-enabled mobile devices thanks to the LEAP add-on component developed in 2002 by Motorola, British Telecommunications, Broadcom Eireann, Siemens AG, Telecom Italia and the University of Parma, as part of the partially European Commission financed LEAP (IST 1999–10211) project. This chapter focuses on this feature of JADE.

By means of the LEAP add-on it is also possible to execute JADE agents on a Microsoft.NET framework. This functionality is not addressed in this book, however. Interested readers should refer directly to the JADE website or to the LEAP User Guide included in the LEAP add-on. ·

## 8.1 MAIN LIMITATIONS OF THE MOBILE ENVIRONMENT

When deploying agents on mobile devices, such as PDAs and phones, a number of constraints must be properly taken into account. These constraints are related to the limitations of the devices themselves, the limited features supported by the Java Virtual Machines on these devices, and the nature of the wireless network(s) used such as GPRS or UMTS. This section describes these issues, which are the basis for all the architectural choices that guided the design of the LEAP add-on.

### *8.1.1 HARDWARE LIMITATIONS*

Hand-held devices typically have reduced processing power with 16-bit processors and, in general, clock rates lower than 200 MHz. They are also characterized by their memory limitations which varies from less than 64 kb in SIM cards to approximately 16 MB in phones and 64 MB in PDAs. Other important constraints include persistent storage with generally no file system available and limited battery life. Though next generation phones are progressively overcoming these limitations and the gap between laptop and desktop computer capability is narrowing, in order meet future user expectations, developers will be required to treat memory and processing power as valuable resources.

*Developing Multi-Agent Systems with JADE* Fabio Bellifemine, Giovanni Caire, Dominic Greenwood

### *8.1.2 JAVA LIMITATIONS*

Java Virtual Machines available on hand-held devices do not provide the full set of features available in laptop and desktop computers. In particular, the MIDP specification (the only one commonly available on the majority of mobile phones) does not provide support for accepting incoming connections. Even in PDAs, where Java Virtual Machines conforming to either the Personal Profile of the CDC configuration or the old Personal Java specification are typically available, the opportunity of opening a connection towards a device is not guaranteed. This is due to the mobile carriers' common practice of establishing firewalls. Other noteworthy limitations of the MIDP specification concern the lack of Java Reflection, Java Serialization and the Collection Framework. It also has only limited support for graphics and user interfaces.

### *8.1.3 NETWORK LIMITATIONS*

JADE assumes full and continuous connectivity between containers. That is, each container must be able to open and accept network connections to and from all other containers in the platform at any time. Moreover, as described in Section 3.6.2.1, all container-to-container interactions in JADE occur using the RMI protocol, which is not particularly efficient in terms of the number of bytes transmitted over the network. Unfortunately wireless networks such as GPRS and UMTS do not meet these assumptions as they are characterized by:

1. Intermittent connectivity due to areas with poor coverage, shielded environments and the need to turn off devices to conserve battery power;
2. Volatile IP addresses due to a mobile device's IP address not necessarily remaining the same after a temporary disconnection;
3. High network latency and low bandwidth ranging from 9.6 kbps (GSM) to 128 kbps (GPRS) and 1 Mbps (HSDPA) and beyond.

## 8.2 THE LEAP ADD-ON

Towards the end of 1999 a group of major players in the mobile telecommunications sector decided to combine their efforts to develop a common platform supporting the deployment of FIPA-compliant agents on mobile devices. The group was led by Motorola and included two other major manufacturers, Siemens and Ericsson (represented by Broadcom, Ericsson's Irish centre of research), and two major carriers, British Telecommunications and Telecom Italia. As a result of this initiative, a project in the IST Programme of the European Commission was launched at the beginning of 2000. The name of the project was LEAP, Lightweight Extensible Agent Platform, as its primary goal was to create a middleware sufficiently lightweight to run on resource-constrained devices such as mobile phones. The system was also designed to be fully extensible to provide scaled-up functionality when executed on larger devices.

The LEAP project team quickly realized that the optimum route was not to develop a complete platform from scratch, but to choose an existing one and customize it to fit the project requirements. After some months of evaluation the project selected JADE as the basis for LEAP. This choice was motivated by a number of reasons:

- The Java 2 Micro Edition was, at the time, quickly becoming a de facto standard to develop mobile client-based applications.
- JADE is written completely in Java.
- JADE is distributed open source thus giving the project free access to all its sources. Moreover, by this time JADE had already gathered a considerable community of users that could form a good basis for the adoption of the new lightweight extensible agent platform.

- Some features of JADE, such as the possibility of executing multiple concurrent tasks (behaviours) in a single Java thread, matched well the constraints imposed by devices with limited resources.
- Finally, the development of JADE was led by Telecom Italia, a member of the LEAP project.

The implementation produced by the LEAP project was configured as a JADE add-on and made publicly available in 2003. The add-on actually replaces certain parts of the JADE kernel, forming a modified run-time environment that is identified as JADE-LEAP ('JADE powered by LEAP') and which can be deployed on a wide range of devices varying from servers to Java-enabled cellphones. To manage the scaling, JADE-LEAP can be shaped in three different ways corresponding to the three main types of Java environments (editions, configurations and profiles) found on the devices considered:

- *J2SE*: to execute JADE-LEAP on PCs and servers in the fixed network running JDK1.4 or later.
- *pJava*: to execute JADE-LEAP on hand-held devices supporting J2ME CDC or Personal Java. [1]
- *MIDP*: to execute JADE-LEAP on hand-held devices supporting MIDP1.0 (or later) such as the great majority of Java-enabled cellphones.

Though different internally, the three versions of JADE-LEAP provide the same API subset to developers, thus offering a homogeneous layer over a diversity of devices and network types as depicted in Figure 8.1.

Only a few features that are available in JADE-LEAP for J2SE and pJava are not supported in JADE-LEAP for MIDP, as they are related to Java classes that are not supported in MIDP. The pJava version completely removes all graphical administration tools while retaining all the APIs of the J2SE version.

JADE-LEAP for J2SE, pJava and MIDP can be directly downloaded in binary form from the 'download' area of the JADE website. However, when developing JADE-LEAP MIDP-based applications it is typically useful to download the JADE sources plus the LEAP add-on from the 'add-ons' area of the JADE website. The add-on includes some tools that can be helpful to correctly package JADE-LEAP based applications for the MIDP environment. This is addressed in more detail in Section 8.4.

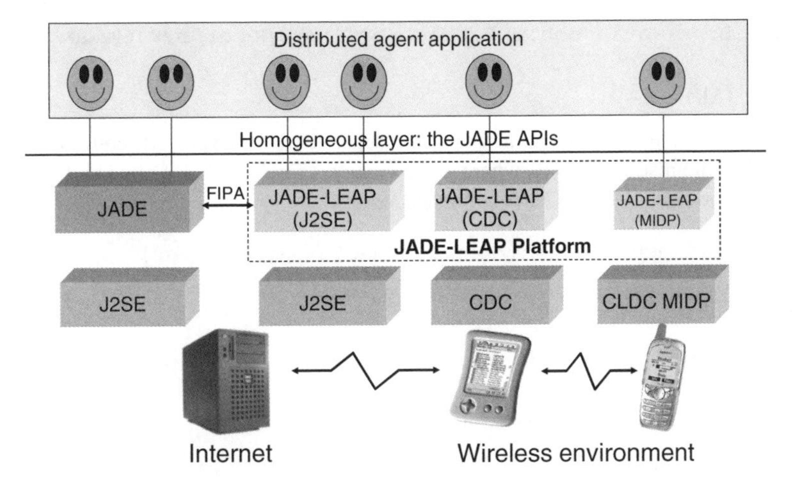


**Figure 8.1** The JADE-LEAP run-time environment

[1] In 2003 the Personal Java specification was declared obsolete by Sun and replaced with the Personal Profile of the CDC configuration of the J2ME edition. From the point of view of running JADE-LEAP there is no difference, hence throughout this book the terms 'CDC' and 'Personal Java' are considered synonymous.

### *8.2.1 JADE AND JADE-LEAP*

From the perspective of application developers and users, JADE-LEAP for J2SE is identical to JADE in terms of APIs, and is almost identical as far as the run-time administration is concerned (see Section 8.2.1.1 for a summary of the differences). Developers can generally deploy their JADE agents on JADE-LEAP and vice versa without changing a single line of code. There is no need for a JADE-LEAP guide or API documentation, as those provided with JADE also apply to JADE-LEAP.

However, it should be kept in mind that *JADE containers and JADE-LEAP containers cannot be mixed within a single platform*. This is the main impetus for having a J2SE version of JADE-LEAP. Without this it would be impossible to deploy a single JADE-LEAP platform spawning both the wireless and the wire-line environment.

As depicted in Figure 8.1, a JADE-LEAP platform and a JADE platform can, of course, communicate as specified by FIPA, e.g. by using the HTTP MessageTransportProtocol.

#### 8.2.1.1 Differences between JADE and JADE-LEAP for J2SE

As previously mentioned, JADE-LEAP for J2SE provides exactly the same API as JADE. There are, however, a few differences in terms of run-time administration.

***Jar files***: The JadeLeap Jar file that is produced when building JADE-LEAP for J2SE also includes the classes related to both the administration tools (that in JADE are included in the jadeTools.jar file) and to the two ready-made MTPs for HTTP and IIOP (that in JADE are respectively included in the http.jar and iiop.jar files).

***Specification of bootstrap agents on the command line***: When starting JADE from the command line, agents that must be activated at bootstrap time must be separated by blanks (‘ ’). On the other hand, the semicolon character (‘;’) is used as separator when launching JADE-LEAP. For instance, the JADE command line

```
java jade.Boot -gui john:myPackage.MyClass peter:myPackage.MyClass
```

is substituted with

```
java jade.Boot -gui john:myPackage.MyClass;peter:myPackage.MyClass
```

when starting JADE-LEAP.

***Agent arguments***: Agent arguments in JADE are separated by blanks (‘ ’), both in the command line and in the administration GUI; whereas the comma character (‘,’) is used as the separator in JADE-LEAP. For example, the JADE command line

```
java jade.Boot -gui john:myPackage.MyClass(arg1 arg2)
```

is substituted with

```
java jade.Boot -gui john:myPackage.MyClass(arg1,arg2)
```

when starting JADE-LEAP. In JADE-LEAP, agent arguments cannot contain blanks.

***Command line options***:

- If JADE-LEAP must be started without options and only the specification of bootstrap agents, the `-agents` option must be used, since a command line such as `java jade.Boot xxxx` is interpreted as if xxxx represented the name of a configuration file. Therefore, for example, `java jade.Boot john:MyClass` will not work; `java jade.Boot -agents john:MyClass` must be used instead.

- The `-nomobility` and `-dump` options are not available in JADE-LEAP.
- The LEAP IMTP (see Section 8.2.2) can be configured by specifying a number of options that are ignored when launching pure JADE.

### *8.2.2 THE LEAP IMTP*

Though JADE and JADE-LEAP are almost identical from an external perspective, they are quite different internally. One critical change is that the IMTP, i.e. that part of the platform dealing with container-to-container interactions as described in Section 3.6.2, is completely different. The normal JADE IMTP is based on Java RMI and this is not suited for mobile devices. JADE-LEAP therefore uses an alternative IMTP based on a proprietary protocol called JICP (Jade Inter Container Protocol). The difference between the two protocols (RMI and JICP) is the main reason why a JADE-LEAP container cannot register with a JADE main container and vice versa.

The LEAP IMTP is composed of a so-called 'Command Dispatcher' and one or more Internal Communication Peers (ICPs). The former is a singleton and is shared among all containers in the same JVM. It is responsible for:

- serializing and deserializing JADE horizontal commands (see Section 7.1.3.2) exchanged among containers;
- routing serialized commands from the local containers to the proper ICP (according to the addresses of the destination container) for transmission over the network; and
- passing commands received by the underlying ICPs to the local containers.

The ICPs are responsible for sending and receiving serialized commands (i.e. raw sequences of bytes) over the network using a given protocol. Figure 8.2 depicts the main components of the LEAP IMTP.

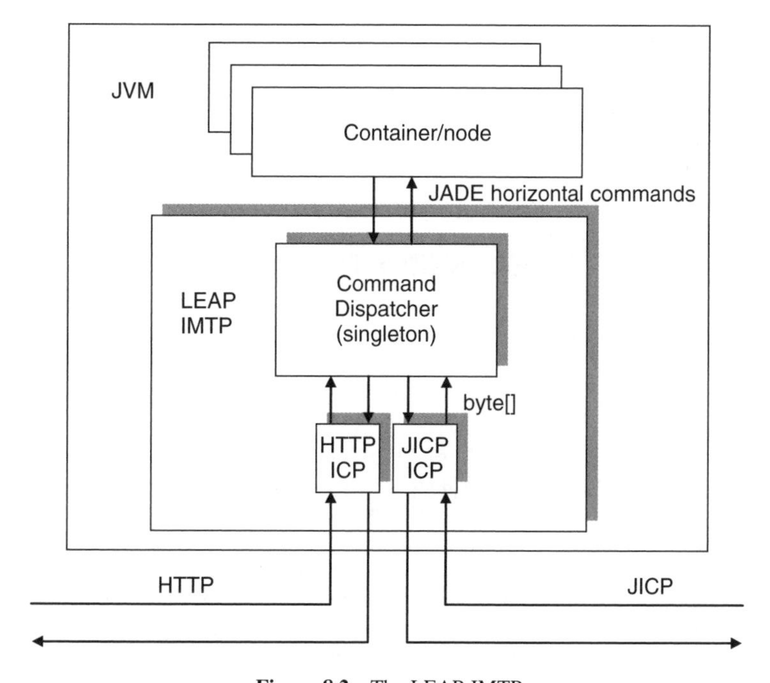


**Figure 8.2** The LEAP IMTP

The LEAP add-on includes three ICP implementations: the default JICP-based ICP implemented by the `jade.imtp.leap.JICP.JICPPeer` class, the SSL version of the JICP ICP implemented by the `jade.imtp.leap.JICP.JICPSPeer` class, and an HTTP-based ICP implemented by the `jade.imtp.leap.http.HTTPPeer`. This means that a JADE-LEAP platform can use JICP, JICP over SSL, or HTTP as the container-to-container communication protocol.

It should be noted that, since more than one ICP can be activated in the Command Dispatcher, all containers that live in the local JVM, and therefore share the Command Dispatcher, can be reachable at multiple addresses. For instance, in Figure 8.2 all containers are reachable from other containers at an address of type `jicp://...` and at another address of type `http://...`

#### 8.2.2.1 Configuring the LEAP IMTP

The main configuration attribute accepted by the LEAP IMTP is `icps` which specifies which ICP to activate on a given container. The value of the attribute is a semicolon (';') separated list of ICP specifiers (in the same format as agent specifiers or MTP specifiers), such as:

```
java jade.Boot -gui -icps
   jade.imtp.leap.JICP.JICPPeer;jade.imtp.leap. http.HTTPPeer(2000)
```

The above command line starts a JADE-LEAP main container with two ICPs; one using the JICP protocol and listening for incoming horizontal commands on the default port 1099 (for consistency with JADE) and the other using the HTTP protocol and listening for horizontal commands on port 2000. The default value for the `icps` option is a single JICP ICP using the default port 1099. Therefore, the command line

```
java jade.Boot -gui
```

is equivalent to

```
java jade.Boot -gui -icps jade.imtp.leap.JICP.JICPPeer
```

When using the HTTP protocol (i.e. when activating the `HTTPPeer` ICP) the `-proto http` option must be specified on peripheral containers in addition to the usual `-host` and `-port` attributes.

## 8.3 THE SPLIT CONTAINER EXECUTION MODE

It should now be clear that JADE agents require an underlying run-time that provides them with the necessary functionalities to live and communicate. Normally JADE run-times are always implemented with containers, but the LEAP add-on provides an alternative means of implementing the run-time known as the split execution mode and is specifically designed to meet the requirements of mobile devices.

When launching the JADE run-time using the split execution mode, the user is not creating a normal container, but a very thin layer called the *front-end*. The front-end provides agents with exactly the same features of a container, but only implements a small subset of them directly, while delegating the others to a remote process called the *back-end*. If the front-end looks like a normal container from the point of view of the agents residing on top of it, the back-end, in turn, looks like a normal container from the perspective of the other containers in the platform, including the main container. The result is that the union of the front-end and the back-end forms a container that is split into two parts, resulting in the term 'split execution mode'. The front-end and the back-end communicate via a dedicated connection as depicted in Figure 8.3.

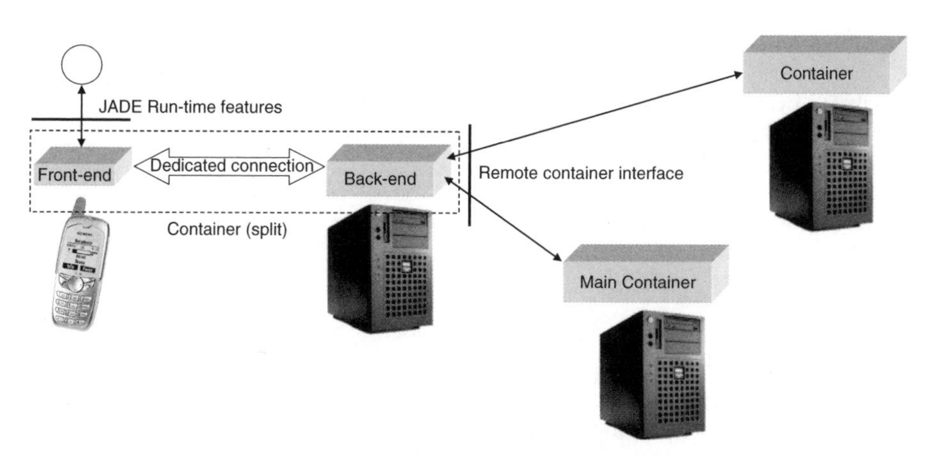


**Figure 8.3** The split execution mode

The split execution mode is particularly suited for resource-constrained and wireless devices since:

- The front-end that runs in the device is considerably more lightweight than a complete container.
- The bootstrap phase is much faster since all communications with the main container required to join the platform are performed by the back-end and therefore not carried out over a wireless connection.
- The usage of the wireless link is optimized.
- Both the front-end and the back-end embed a store and forward mechanism to make the potential loss of connection transparent to applications. That is, if the connection between the front-end and the back-end goes down (e.g. because a cellphone host enters a dead spot), messages to and from agents living in the front-end are buffered in both the back-end and in the front-end. As soon as the front-end re-establishes the connection, buffered messages are delivered to their intended receivers.
- The IP address of the mobile device is never seen by other containers in the platform, since they always interact with the back-end. It may even change without any impact to the application.

Table 8.1 summarizes how the two execution modes (i.e. the normal or 'stand-alone' mode and the split mode) are supported in the different environments targeted by JADE-LEAP. Note that the split execution mode is *mandatory* in MIDP and strongly suggested in pJava. It is also important to note that agent developers do not have to be concerned with whether their agents will run on a stand-alone container or on the front-end of a split container, as the APIs they provide are exactly the same.

There are, however, three issues of which developers must be aware:

1. It is not possible to execute a main container using the split execution mode.
2. The `jade.core.Runtime` class and the classes in the `jade.wrapper` package described in Section 5.6 are designed to control a stand-alone container. External applications can control

**Table 8.1** JADE-LEAP Execution modes

| | J2SE | pJava | MIDP |
|---|---|---|---|
| Stand-alone | Suggested | Supported | Not supported |
| Split | Supported | Suggested | Mandatory |

a split container by means of a minimal in-process interface provided by the `jade.core.MicroRuntime` class (see Section 8.5.3). This class is clearly only available when working with JADE-LEAP, since the split execution mode does not exist in pure JADE.
3. Agent mobility and cloning are not supported on a split container.

When running JADE-LEAP on a PersonalJava/CDC device and mobility features are needed, it is recommended to try the split execution mode plus the dynamic behaviour loading mechanism supported by the `jade.core.behaviours.LoaderBehaviour` class first. Only if this does not fit application needs should the stand-alone execution mode be used.

### *8.3.1 THE MEDIATOR*

When considering a real application with several users each of whom is running agents on his own mobile phone, there must be a proper mechanism to automatically manage the back-ends of all front-ends active on user devices. This mechanism is implemented by the JADE-LEAP architectural element called a *mediator*, depicted in Figure 8.4.

The mediator must be up and running on a host with a known, fixed and visible address before front-ends can be initiated on any mobile device. When the user activates JADE-LEAP using the split execution mode on his phone, the front-end starts up and connects to the mediator requesting the creation of a back-end. This is accomplished by sending a `CREATE_MEDIATOR` request over the newly opened connection using the JICP protocol mentioned in Section 8.2.2. The mediator instantiates a back-end and passes to it the connection opened by the front-end. The front-end and the back-end will then use this connection for all successive interactions. The back-end, in turn, registers with the main container, allowing the newly born split container to join the platform. Only at this time will the mediator send back a response to the front-end that includes information such as the name assigned to the container, the platform-name and the platform addresses, if any. The sequence diagram in Figure 8.5 shows the split container start-up procedure.

After the start-up procedure, all successive interactions between the front-end and the back-end do not involve the mediator. However, the mediator maintains a table that maps container names with the corresponding back-ends. If for some reason (e.g. a mobile phone goes out of coverage)

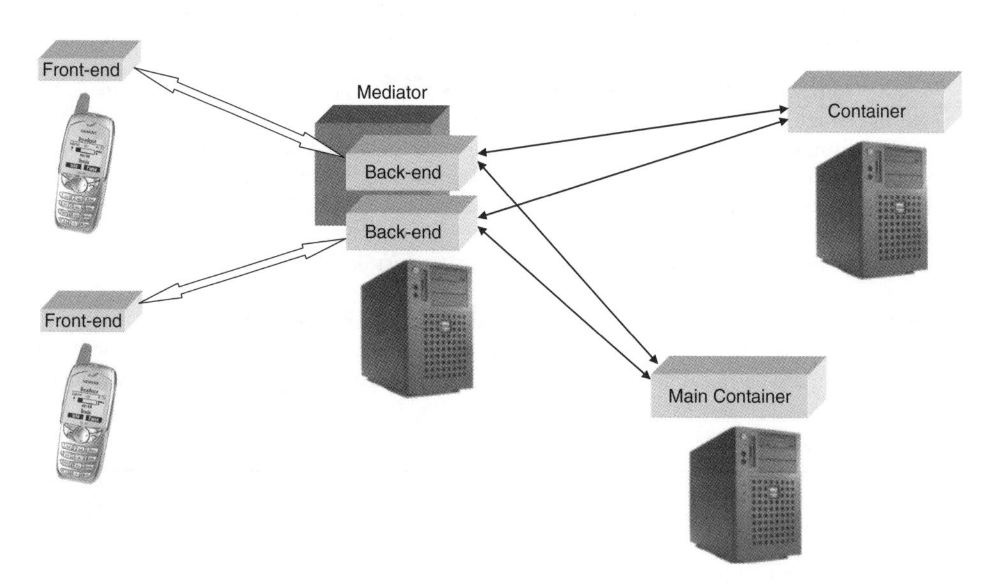


**Figure 8.4** The mediator

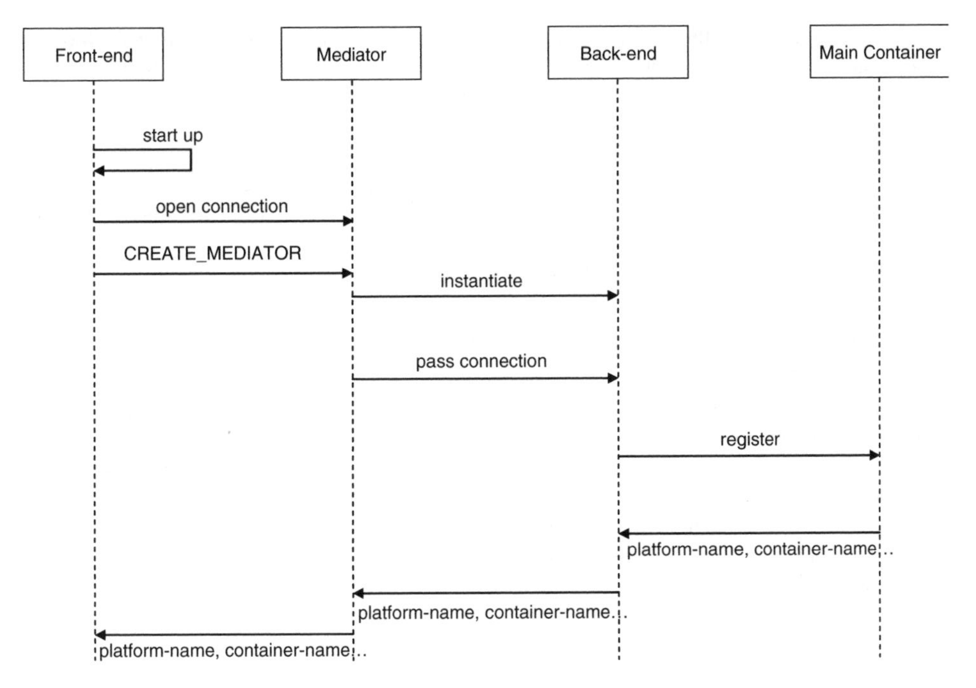


**Figure 8.5** Split container start-up procedure

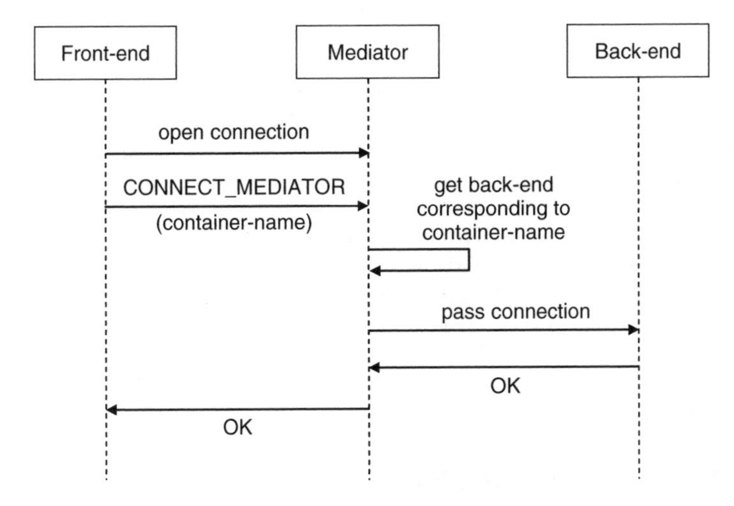


**Figure 8.6** Reconnection procedure

the connection between a front-end and its back-end goes down, the front-end periodically attempts to re-establish the connection via the mediator, as shown by the sequence diagram in Figure 8.6. The `CONNECT_MEDIATOR` JICP request is used in this scenario. The request embeds the container-name, thus allowing the mediator to pass the connection to the correct back-end.

It should be noted that in both the start-up and the reconnection procedures the only element accepting connections (i.e. embedding a server socket) is the mediator. Therefore, if the mediator, the main container, and other normal containers, if any, are running behind a firewall, the only requirement is to open the port used by the mediator to listen for incoming connections.

JADE-LEAP provides two entities that are capable of playing the mediator role:

1. The ICPs of the LEAP IMTP. It follows that in order to have a mediator up and running, it is sufficient to start a normal JADE-LEAP container (including of course the main container). If a port different than the default 1099 is needed to accept connections from front-ends, the usual `-local-port` option can be specified.
2. The `BEManagementService` included in the LEAP add-on. This is described in Section 8.5.5.

### *8.3.2 SPLIT CONTAINER CONFIGURATION OPTIONS*

A number of options can be specified when starting the front-end of a split container, the most important of which are `-host` and `-port`. However, these options do not indicate the host and port of the main container, as is the case when starting a normal container, but instead indicate the host and port of the mediator. Only in the case where the Main Container (or more precisely its ICP) acts as mediator do the `-host` and `-port` options for split and normal containers coincide. This configuration is frequently used when experimenting with the split execution mode, but, in general, it is not suited for real deployment scenarios where it is typically preferable to protect the main container behind a firewall.

Other noteworthy options that can be specified to control the start-up of a split container include:

- `-agents <list of agent specifiers>` As with a normal container, this option specifies the agents that must be activated at front-end bootstrap.
- `-exitwhenempty [true | false]` When set to `true`, this option tells the front-end to terminate as soon as there are no more agents living on it. The default is `false`.
- `-msisdn <telephone number>` This option tells JADE that the phone hosting the starting front-end has a given phone number (msisdn is the telecom term to indicate the telephone number of a mobile phone). As a consequence, JADE will use the specified phone number as name for the newly born split container. Section 8.5.5.3 explains how to connect the JADE-LEAP mediator to an external system that is capable of automatically detecting the mobile phone msisdn. In this configuration the `-msisdn` option is ignored.
- `-connection-manager <connection-manager class name>` This option allows the specification of a class that implements the `jade.core.FEConnectionManager` interface. This is used to manage (at the front-end) the connection between the front-end and the back-end. This option is further described in Section 8.5.5.1.
- `-mediator-class <mediator class name>` This option allows the specification of a class implementing the `jade.imtp.leap.JICP.JICPMediator` interface that is used to manage (at the back-end) the connection between the front-end and the back-end. This option is further described in Section 8.5.5.1.

## 8.4 DEVELOPING MIDP AGENTS

This section assumes that the reader is familiar with the MIDP profile of the Java Micro Edition. Nevertheless, a short overview is provided the MIDP specification elements most relevant to developing MIDP applications based on the JADE-LEAP platform.

Java applications intended to be deployed on a MIDP-enabled phone must be packaged as MIDlet suites. A MIDlet suite is basically a Jar file containing:

- Some code including at least one class extending the `javax.microedition.midlet.MIDlet` class. Each of these classes corresponds to a MIDlet (i.e. a launchable MIDP application) in the MIDlet suite.

- A manifest file specifying, among others, one `MIDlet-<n>` (where `n` indicates a progressive number starting from 1) entry for each MIDlet in the MIDlet suite.
- Other optional resources such as images, data files, etc.

The following issues should be taken into account when developing a MIDlet suite:

1. The code for all MIDlets included in a MIDlet suite must be packaged into the MIDlet suite Jar file. It is not possible to use external libraries.
2. Code must be preverified for it to be executed on an MIDP JVM, Java byte. This process mainly consists of verifying that the byte code does not make use of Java API methods that are not available in MIDP. Moreover, preverification also manipulates the byte code to assist in its execution by the JVM on the device. This allows the footprint of MIDP JVMs to be kept very small.
3. MIDlets are not started from the command line as is typically the case for desktop Java application. Instead they are initiated using some device-specific mechanism, e.g. by clicking on an icon on the phone display.
4. A MIDlet suite is described by an external descriptor called the Java Application Descriptor (JAD). The JAD has the same format as a manifest file and is almost identical to the manifest file included in the MIDlet suite itself. It is used to download a MIDlet suite over the air. The Java Wireless Toolkit, i.e. the suite of tools provided by Sun to support the development and testing of MIDP applications, uses the JAD to start a MIDlet in the embedded emulated environment. With respect to the MIDlet suite manifest, the JAD file includes two additional mandatory entries: the `MIDlet-Jar-URL`, whose value must be the URL where the MIDlet suite Jar file can be downloaded, and the `MIDlet-Jar-Size`, whose value must be the size in bytes of the MIDlet suite Jar file.

### *8.4.1 BUILDING A JADE-LEAP-BASED MIDLET SUITE*

When working on cellphones running MIDP1.0 (or higher), the MIDP version of JADE-LEAP must be used. In order to be deployed on MIDP devices, JADE-LEAP for MIDP is configured as a MIDlet suite and includes the following MIDlets:

- `jade.MicroBoot`. This MIDlet starts the front-end of a split container (remember that the split execution mode is the only one supported in MIDP).
- `jade.util.leap.Config`. This MIDlet allows manual editing of the JADE-LEAP configuration properties as described in Section 8.4.2.
- `jade.util.leap.OutputViewer`. This MIDlet allows reviewing of the output printed during the previous JADE-LEAP execution session as described in Section 8.5.1.

The requirement to include all classes that make up an application (both application-specific classes and library classes if any) within a single Jar file does not make using the JADE-LEAP MIDlet suite 'as is' particularly beneficial, as it does not include any application-specific code. To resolve this issue, the typical build process to create a JADE-LEAP-based MIDlet suite goes through the following steps:

1. Compile application-specific code putting JADE-LEAP for MIDP in the classpath;
2. Unjar all JADE-LEAP classes;
3. Preverify;
4. Create a single Jar file including both application-specific classes and JADE-LEAP classes with a proper manifest file;
5. Modify the JAD file to include the correct Jar URL and size.

### 8.4.2 SETTING CONFIGURATION OPTIONS

JADE-LEAP-based MIDlets typically require some configuration options such as `-host` and `-port`, which are described in Section 8.3.2. However, since a MIDP phone does not have a command line, an alternative way of specifying these properties is required. JADE-LEAP provides two possibilities. Configuration options can be specified as entries in the MIDlet suite manifest with the following syntax:

```
LEAP-<option>: <value>
```

For example, the following manifest tells JADE-LEAP to connect to the mediator on host `avalon.tilab.com`, port `1099` and to start an agent called john of class `myPackage.MyAgent`.

```
MicroEdition-Configuration: CLDC-1.0
MicroEdition-Profile: MIDP-1.0
MIDlet-Name: JADE-LEAP
MIDLet-Vendor: TILAB
MIDlet-Version: 3.4
MIDlet-1: JADE, , jade.MicroBoot
MIDlet-2: View-output, , jade.util.leap.OutputViewer
LEAP-host: avalon.tilab.com
LEAP-port: 1099
LEAP-agents: john:myPackage.MyAgent
```

This method is quite straightforward, but must be done during packaging of the MIDlet suite, i.e. before uploading a JADE-LEAP-based MIDlet onto the cellphone.

If configuration options are not known at MIDlet suite packaging time, the `jade.util.leap.Config` MIDlet included in the JADE-LEAP MIDlet suite provides an alternative mechanism that allows them to be set after the MIDlet is installed on the device. Basically, the `Config` MIDlet provides the user with a simple form in which a key can be specified (without the 'LEAP' prefix) and a related value as shown in the screenshot in Figure 8.7. This must be repeated for each option that the user wants to set.

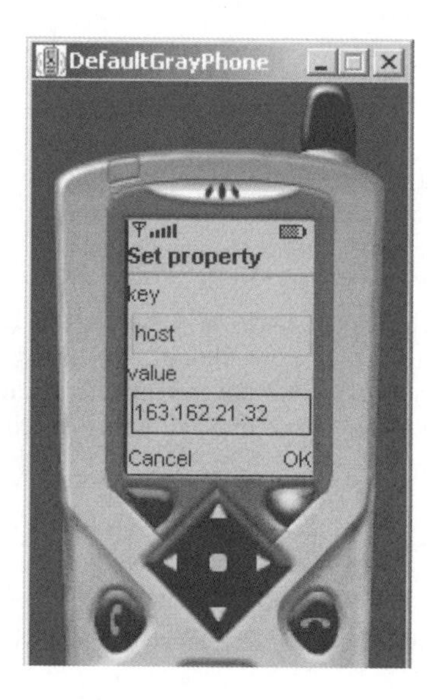


**Figure 8.7** Setting configuration options by means of the `Config` MIDlet

To force JADE-LEAP to read the configuration options set through the `Config` MIDlet, the `LEAP-conf: conf` entry must be specified in the manifest as follows:

```
MicroEdition-Configuration: CLDC-1.0
MicroEdition-Profile: MIDP-1.0
MIDlet-Name: JADE-LEAP
MIDlet-Vendor: TILAB
MIDlet-Version: 3.4
MIDlet-1: JADE, , jade.MicroBoot
MIDlet-2: Config, , jade.util.leap.Config
MIDlet-3: View-output, , jade.util.leap.OutputViewer
LEAP-conf: conf
```

The `Config` MIDlet must be included in the MIDlet suite in this case.

### *8.4.3 BUILDING THE BOOK TRADER EXAMPLE FOR THE MIDP ENVIRONMENT*

In order to review all the steps required to create a JADE-LEAP-based MIDP application, we return once again to the book-trading case study and modify it to run on a MIDP phone. To simplify things we compile the whole project, but package a MIDlet suite for the book buyer agent only. The process to package a MIDlet suite for the book seller agent is very similar. Figure 8.8 shows how the project directory structure should be modified with respect to the example presented in Section 4.1.5.

The following differences can be observed:

- The JADE-LEAP jar files (in the MIDP version for the agent that will run on the device and in the J2SE version for the main container and the mediator in the fixed network) are used instead of the normal JADE Jar files.
- In addition to the classes directory used to place compiled byte code, there is also the verified directory that is used to store preverified byte code.
- Two new files have been added into the bookTrading project root directory. The buyer.dlc file used to drive the minimization process is described in Section 8.4.4. The buyer.manifest file is a manifest file ready to be inserted into the MIDlet suite Jar file. The manifest appears as follows:

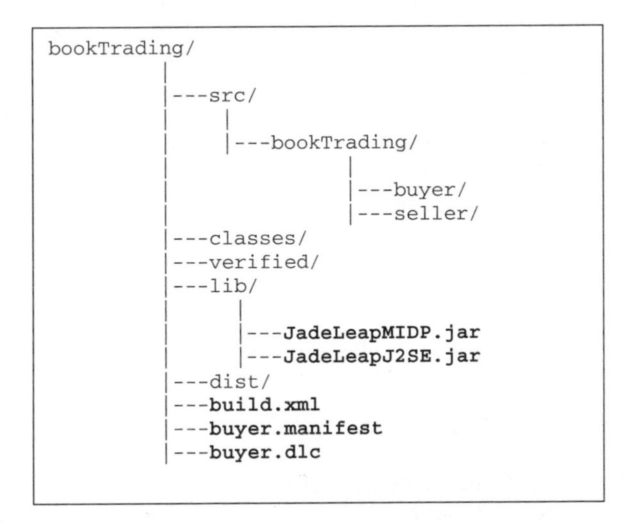
```
bookTrading/
          |
          |---src/
          |     |
          |     |---bookTrading/
          |                |
          |                |---buyer/
          |                |---seller/
          |---classes/
          |---verified/
          |---lib/
          |     |
          |     |---JadeLeapMIDP.jar
          |     |---JadeLeapJ2SE.jar
          |---dist/
          |---build.xml
          |---buyer.manifest
          |---buyer.dlc
```

**Figure 8.8** Book-trading project directory structure for the MIDP environment

```
MicroEdition-Configuration: CLDC-1.0
MicroEdition-Profile: MIDP-1.0
MIDlet-Name: Book-Buyer
MIDlet-Vendor: TILAB
MIDlet-Version: 1.0
MIDlet-1: Book-Buyer, , jade.MicroBoot
MIDlet-2: View-output, , jade.util.leap.OutputViewer
LEAP-host: <mediator-host>
LEAP-port: <mediator-port>
LEAP-agents: %C-book-buyer:bookTrading.buyer.BookBuyerAgent
LEAP-exitwhenempty: true
```

Three interesting details should be noted in the above manifest file:

1. The MIDlet that is started to activate the book buyer application on the device is `jade.MicroBoot`, i.e. the JADE-LEAP front-end. The `agents` option tells JADE-LEAP to launch a book buyer agent at bootstrap time. The `exitwhenempty` option tells JADE to terminate as soon as the user closes its book buyer agent.
2. The `%C` wildcard is used to assign a name to the book buyer agent that is going to be launched. At bootstrap time JADE will replace this wildcard with the name of the (split) container so that the book buyer agent will be named `<container-name>-book-buyer`. This approach ensures consistent agent names and is particularly useful when connecting the mediator to a system able to detect the msisdn (telephone number) of the device where the front-end is starting, as described in Section 8.5.5.3. In this case the book buyer agent will be named `<user-phone-number>-book-buyer`.
3. The MIDlet suite also includes the `jade.util.leap.OutputViewer` MIDlet that is a JADE-LEAP-specific utility to review debugging printouts, as described in Section 8.5.1.

The same sources for both the buyer and the seller agents can be reused precisely as if they are components from the GUI-related code. Both the AWT and the Swing API are not supported in MIDP. The choice of hiding the actual GUI implementation classes behind the `BookBuyerGui` and `BookSellerGui` interfaces allows their easy replacement with others based on the `javax.microedition.lcdui` package that provides support for creating user interfaces in MIDP. Although the details of these classes are outside the scope of this book, there is one issue that is worth mentioning related to the retrieval of the `Display` object necessary to show graphic elements on the screen in a MIDP environment. This is achieved by means of the static method `getDisplay()` of the `javax.microedition.lcdui.Display` class that requires the current `MIDlet` object to be passed as parameter. The current MIDlet in this case is `jade.MicroBoot` and its reference is made available to application-specific code by means of the public static field `midlet` of the `Agent` class. Therefore the `Display` object can be retrieved by simply calling

```
Display d = Display.getDisplay(Agent.midlet);
```

The ANT build file used to compile the sources and package the book buyer MIDlet suite Jar file is shown below and reflects the steps listed in Section 8.4.1.

```
1: <project name="Book-Trading ">
2:
3:   <! -- Where to find MIDP classes -->
4:   <property name="j2mewtk.home" value=" Wireless Toolkit home
   dir."/>
```

```
 5:  <property name="midplib value="${j2mewtk.home} /lib"/>
 6:  <property name="boot-midp-classes"
 7:            value="${midplib}/midpapi10.jar;${midplib}
   /cldcapi10.jar"/>
 8:
 9:  <! -- Additional ANT tasks Require the StampysoftAntTasks.jar
   in the ANT lib directory -->
10:  <! -- Preverifier -->
11:  <taskdef name="preverify"
   classname="cpm.stampysoft.ant.j2me.PreverifyTask"/>
12:  <! -- Jad Updater -->
13:  <taskdef name="updatejad"
   classname="com.stampysoft.ant.j2me.UpdateJARSizeTask"/>
14:
15:  <! -- Targets -->
16:
17:  <target name="init">
18:    <tstamp/>
19:  </target>
20:
21:  <target name="compile" depends="init">
22:    <javac srcdir="src"
23:           destdir="classes"
24:           classpath="lib/JadeLeapMIDP.jar"
25:           bootclasspath="${boot-midp-classes}"
26:           source="1.1"
27:           target="1.1"/>
28:  </target>
29:
30:  <target name="jar" depends="compile">
31:    <! -- Mix JADE classes and book-trading-specific classes -->
32:    <unjar src="lib/JadeLeapMIDP.jar" dest="classes"/>
33:
34:    <! -- Preverify the whole -->
35:    <preverify unverified="classes" verified="verified"/>
36:
37:    <! -- Create the JAR -->
38:    <delete file="dist/buyer.jar" quiet="true"/>
39:    <jar jarfile="dist/buyer.jar "
40:         basedir="verified"
41:         manifest="buyer.manifest"/>
42:
43:    <! -- Create the JAD -->
44:    <copy file="buyer.manifest" toFile="dist/buyer.jad"
   overWrite="yes"/>
45:    <manifest file="dist/buyer.jad" mode="update">
46:      <attribute name="MIDlet-Jar-URL" value="buyer.jar"/>
47:    </manifest>
48:    <updatejad jad="dist/buyer.jad"/>
49:   </target>
50: </project>
```

Some issues of note in the above build file are as follows:

- Lines 3–7 tell ANT where to find the Java Wireless Toolkit. This must be properly installed prior to the development of any MIDP project. In this section a MIDP1.0 compliant MIDlet suite is built; therefore, the midpapi10.jar library file is used.
- Lines 9–13 declare two additional ANT tasks: preverify and updatejad. The former allows the preverify process to be launched as an ANT task (line 35). The latter is used to automatically set the value of the `MIDlet-Jar-Size` entry in the JAD file (line 48). Neither of these tasks are directly included in the ANT distribution but are instead included in a library of additional ANT tasks (StampysoftAntTasks.jar) that can be downloaded from http://www.jeckels.com/ant/. This library must be placed in the ANT_HOME/lib directory.
- Lines 43–48 create a proper JAD file by simply copying the MIDlet suite manifest and setting the two additional entries `MIDlet-Jar-URL` (lines 45–47) and `MIDlet-Jar-Size` (line 48).

Moving to the `bookTrading` directory and typing `ant jar` will create the book buyer MIDlet suite Jar file (buyer.jar) and the related JAD file (buyer.jad) in the `dist` directory of the bookTrading project.

### 8.4.4 MINIMIZATION

JADE-LEAP also includes several library classes in the MIDP version that are available when the programmer needs to use them. For this reason, a JADE-LEAP-based MIDlet Jar file, produced by the ANT build script presented in the previous section, is typically quite large even for very simple applications. In MIDP, however, it is desirable to remove all unused classes and reduce the size of the MIDlet suite Jar file as much as possible. This operation can be performed by means of the minimization utility included in the LEAP add-on. This utility is configured as a target of the ANT build script included in the LEAP add-on and prompts the user for:

- The MIDlet suite Jar file to be minimized.
- The `.dlc` file that specifies the dynamically loaded classes; these are used as starting point for the minimization process.
- The manifest file to be included in the minimized Jar file.

The result of the minimization process is a new Jar file placed in the same directory as the original Jar file and named as original jar but appended with the '-min' suffix. This new Jar file will include only those classes that are referenced, starting from the classes listed in the `.dlc` file.

The `.dlc` file is a text file that includes just one fully qualified class name per line *with no blank lines and no comments*. The buyer.dlc file mentioned in the previous section is presented below as an example:

```
jade.MicroBoot
jade.util.leap.OutputViewer
jade.imtp.leap.JICP.BIFEDispatcher
bookTrading.buyer.BookBuyerAgent
```

In general, the following classes are listed in a dlc file:

- All MIDlet classes to be included in the MIDlet suite (`jade.MicroBoot` and `jade.util.leap.OutputViewer` in the example above);
- All agent classes to be included in the MIDlet suite (`bookTrading.buyer.BookBuyerAgent` in the example above);

- The front-end connection management class controlling (front-end side) the connection between the front-end and the back-end, mentioned in Section 8.3.2 and described in Section 8.5.5.1 (`jade.imtp.leap.JICP.BIFEDispatcher` in the example above);
- All application-specific classes (if any) loaded by means of the `Class.forName()` instruction (none in the example above).

In order to minimize the book buyer MIDlet created in Section 8.4.3 the following steps should be performed:

1. Move to the LEAP add-on directory and type `ant minimize;`
2. The minimization utility prompts for the Jar file to minimize: type the full pathname of the buyer.jar file;
3. The minimization utility prompts for the dlc file from which to start the minimization process: type the full pathname of the buyer.dlc file;
4. The minimization utility prompts for the manifest file to include in the minimized jar file: type an asterisk ('*') to tell the minimization utility to reuse the manifest of the original Jar file.

As a result of the minimization process two new files will be created in the bookTrading/dist directory: buyer-min.jar (the minimized book buyer MIDlet suite Jar file) and buyer-min.jad (the corresponding JAD file). A comparison of the size of the buyer.jar and buyer-min.jar files reveals that the minimization process reduced the MIDlet suite size from ~600 kbyte to ~150 kbyte.

## 8.5 LEAP ADD-ON ADVANCED

### *8.5.1 THE LOGGING API AND THE OUTPUTVIEWER MIDLET*

Normally MIDP phones do not have any standard output. As a consequence, debugging and logging printouts produced by means of the typical `System.out.println()` instruction are lost when a MIDlet runs on a device. Given that on a real device there are, in general, no opportunities to perform any kind of step-by-step execution, the need to produce debugging printouts is extremely important, especially when investigating problems.

To overcome this limitation JADE-LEAP provides an ad hoc logging API that is homogeneous across both the standard JADE and the three versions (J2SE, CDC/pJava and MIDP) of JADE-LEAP, but generates logs differently according to the underlying environment. The central element of this API is the `jade.util.Logger` class whose instances can be obtained by means of the static method `getMyLogger(String loggerName)`. The `jade.util.Logger` class has three different implementations.

In JADE and JADE-LEAP for J2SE it simply extends `java.util.logging.Logger`, thus providing exactly the same features both in terms of API and configuration mechanisms. In order to have a homogeneous set of APIs across the J2SE, CDC/pJava and MIDP environments, the logging levels defined as static constants in the `java.util.logging.Level` class are also provided as static constants (with the same names) in the `jade.util.Logger` class. The following code is a typical example of how JADE logging API is used:

```
1: import jade.util.Logger
2:
3: public class Foo {
4:    private Logger myLogger = Logger.getMyLogger (getClass
  ().getName());
5:
6:    public Foo() {
7:      myLogger.log(Logger.INFO, "Creating a Foo instance");
```

```
 8:
 9:       try {
10:          myLogger.log (Logger.FINE, "Entering try block...");
11:          ...
12:       }
13:       catch (Exception e) {
14:         myLogger.log(Logger.WARNING, "Unexpected error", e);
15:       }
16:     }
17:   }
```

In JADE-LEAP for CDC/pJava, where the `java.util.logging` package is not available, a call to one of the overloaded versions of the `log()` method results in a call to `System.out.println()`. Alternatively, it is possible to redirect logging printouts to a text file by setting the `jade_util_Logger_logfile` option. Note that, in order to cope with resource limitations it is not possible to redirect logging printouts produced by different `Logger` objects to different files.

In the MIDP implementation of the `jade.util.Logger` class, logging printouts are redirected to a MIDP RecordStore (the only support for persistent storage available in the MIDP specification). These printouts can be viewed at a later time by means of the `jade.util.leap.OutputViewer` MIDlet included in the LEAP add-on. The latter, of course, must be included in the same MIDlet suite as the MIDlet that uses the `jade.util.Logger` class to produce the printouts.

The default logging level is set to `Logger.INFO`. All logs produced using this level or higher levels (`Logger.WARNING` and `Logger.SEVERE`) will be produced by default. Also, in the vice versa case, all logs produced with lower levels will not be produced by default.

In both MIDP and pJava, the logging level for a `Logger` object registered with name `x.y.z` can be configured by setting the configuration option `x_y_z_loglevel` to one of the following: `severe`, `warning`, `info`, `config`, `fine`, `finer`, `finest`, `all`. If, for example, a MIDlet uses the `Foo` class presented above, assuming there are no exceptions, only the log produced at line 7 is actually generated. In contrast, setting the entry `LEAP-Foo_loglevel: fine` in the MIDlet suite manifest will result in generating both the log produced at line 7 and that produced at line 10.

### *8.5.2 MANAGING COMPLEX CONTENT EXPRESSIONS IN THE MIDP ENVIRONMENT*

Though compatible with MIDP1.0 in terms of API, JADE support for content languages and ontologies presented in Section 5.1 is not particularly lightweight and may cause memory problems on devices with limited resources. The ability to automatically convert between Java beans and strings or sequences of bytes relies on Java reflection, and as a consequence, developers wishing to use it in MIDP are required to write a significant amount of code to perform these conversions by hand. These two issues make the normal JADE support for content languages and ontologies unsuitable for MIDP.

For this reason the LEAP add-on includes an alternative, lightweight support comprising five classes that are only included in `jade.content.frame`. This support allows the creation of complex content expressions as `Frame` objects and the conversion of them to/from strings or sequences of bytes encoded according to the SL or the LEAP language. More specifically, each frame has a type-name and a number of composing elements that can be either other frames or primitive elements (`Integer`, `Boolean`, `String`, `Date`, `byte[]` and `AID`). There are two types of frames:

- Ordered frames whose composing elements can be retrieved by an index.
- Qualified frames whose composing elements can be retrieved by a unique name.

These are respectively represented by the `OrderedFrame` class (that extends `java.util.Vector`) and `QualifiedFrame` class (that extends `java.util.Hashtable`). It is important to note that, in order to keep it as simple and lightweight as possible, the content expression management framework included in the `jade.content.frame` package does not perform any validity checks. It is thus the responsibility of the user to ensure that elements in a `Frame` object are instances of one of the supported types mentioned above.

Having built a content expression as a tree of `Frame` objects, the `encode()` method of the `SLFrameCodec` or the `LEAPFrameCodec` classes can be used to convert it into a string or a sequence of bytes that can be respectively set in the `:content` slot of an `ACLMessage` by means of the `setContent()` or `setByteSequenceContent()` methods. Similarly, when processing an incoming message, the `decode()` method can be used to reconstruct a tree of `Frame` objects.

The lightweight content expression management framework included in the `jade.content.frame` package is fully interoperable with the 'standard' JADE support for content languages and ontologies, provided, of course, that the same vocabulary of symbols is used.

To exemplify the usage of the `jade.content.frame` package the following code describes how the `CallForOfferServer` inner class (presented in Section 5.1.3.5) should be written to support lightweight content expression management.

```
 1: private class CallForOfferServer extends CyclicBehaviour {
 2:   public void action() {
 3:     ACLMessage msg = myAgent.receive();
 4:     if (msg != null) {
 5:       // Message received. Process it
 6:       ACLMessage reply = msg.createReply();
 7:       try {
 8:         OrderedFrame actFrame = (OrderedFrame)
   slFrameCodec.decode(msg.getContent());
 9:         OrderedFrame sellFrame = (OrderedFrame)
   actFrame.elementAt(1);
10:         QualifiedFrame bookFrame = (QualifiedFrame)
   sellFrame.elementAt (0);
11:         String title = (String)
   bookFrame.get(BookTradingVocabulary.Book_TITLE);
12:         PriceManager pm = (PriceManager) catalogue.get(title);
13:         if (pm != null) {
14:           // The requested book is available for sale
15:           reply.setPerformative (ACLMessage.Propose);
16:           OrderedFrame listFrame = new OrderedFrame (null);
17:           listFrame.addElement (actFrame);
18:           OrderedFrame costsFrame = new OrderedFrame
   (BookTradingVocabulary.COSTS);
19:           costsFrame.addElement(bookFrame);
20:           costsFrame.addElement(new Integer
   (pm.getCurrentPrice()));
21:           listFrame.addElement(costsFrame);
22:           reply.setContent (slFrameCodec.encode(listFrame));
23:         }
24:         else {
25:           // The requested book is NOT available for sale.
26:           reply.setPerformative(ACLMessage.REFUSE);
27:         }
28:       }
```

```
29:         catch (FrameExpection fe) {
30:           fe.printStackTrace();
31:           reply.setPerformative(ACLMessage.NOT_UNDERSTOOD);
32:         }
33:         myAgent.send(reply);
34:       }
35:     }
36:   } // End of inner class CallForOfferServer
```

In the above code, the `slFrameCodec` variable used at lines 8 and 22 to decode the content of the incoming CFP message and to encode the content of the reply, is an instance of the `SLFrameCodec` class. It should be defined at the level of the enclosing `BookSellerAgent` class. Note also that the lightweight content expression management framework included in the `jade.content.frame` package does not require any initialization (such as registration of codecs and ontologies in the agent's ContentManager).

### *8.5.3 STARTING JADE FROM USER-DEFINED MIDLETS*

Section 8.4 presented the simplest approach to creating JADE-LEAP-based MIDlets. This approach does not involve the development of any user-defined class extending `javax.microedition.midlet.MIDlet`, since `jade.MicroBoot` is used. Developers are only requested to create their agent classes and to tell JADE-LEAP which agent to start by means of the `agents` configuration attribute.

There are, however, many instances in which it is necessary to perform application-specific operations before starting JADE-LEAP and the agents running on it. Suppose that the MIDP book-trading application built in Section 8.4.3 must be modified so that it first requests the user to select whether he wants to sell or buy books and then starts the seller or the buyer agent according to the user selection. In order to implement this it is necessary to create an ad hoc MIDlet class (e.g. `bookTrading.BookTradingMIDlet`) that allows the user to make his choice before starting JADE-LEAP and the correct agent.

The `jade.core.MicroRuntime` class can be used in these cases. It provides a minimal in-process interface to control JADE-LEAP in the split execution mode. More specifically, the `startJADE()` static method allows starting the front-end of a JADE-LEAP split container. It takes two parameters:

1. A `jade.util.leap.Properties` object including the configuration options used to tune the JADE-LEAP start-up process. The names of all options described in Section 8.3.2 are made available as static constants of the `MicroRuntime` class.
2. A `Runnable` object (possibly `null`) that will be executed just after the termination of the split container that is going to be launched. This can be used, for instance, to terminate the MIDlet when JADE-LEAP terminates.

The `jade.util.leap.Properties` class offers methods very similar to the `java.util.Properties` class, but, unlike the latter, it is also available in MIDP and therefore provides a uniform way to pass configuration options to JADE-LEAP across all environments. The `jade.util.Logger` class presented in Section 8.5.1, though presenting the same API, has three different implementations depending on the environment in which it is used. For example, when in J2SE it simply extends `java.util.Properties`, and when in MIDP it extends `java.util.Hashtable` and reimplements the `setProperty()` and `getProperty()` methods. The `jade.util.leap.Properties` object including JADE-LEAP configuration options can be initialized by hand, as shown in the code below:

```
1: Properties pp = new Properties();
2: pp.setProperty (MicroRuntime.HOST_KEY, "avalon.tilab.con");
3: pp.setProperty (MicroRuntime.PORT_KEY, "1099");
4: pp.setProperty (MicroRuntime.AGENTS_KEY,
   "%C-book-buyer:bookTrading.buyer.BookBuyerAgent");
```

Alternatively, the configuration options set in the MIDlet suite manifest or set by means of the `Config` MIDlet as described in Section 8.4.2 can be automatically read by means of the `load()` method of the `jade.util.leap.Properties` class as exemplified below:

```
1: Properties pp = new Properties();
2: pp.load("jad");
```

Specifying the `'jad'` string constant as a parameter to the `load()` method indicates that properties set in the MIDlet suite manifest must be loaded. Specifying the `'conf'` string constant, on the other hand, indicates that properties set by means of the `Config` MIDlet must be loaded.

When starting JADE-LEAP from a user-defined MIDlet it is *strongly recommended* to set the `Agent.midlet` static field mentioned in Section 8.4.3 to point to the current MIDlet object before doing any JADE-LEAP specific operation. Not doing this may lead to unexpected behaviour.

The following illustrates the `BookTradingMIDlet`:

```
 1: public class BookTradingMIDlet extends MIDlet implements
   CommandListener {
 2:   ChoiceGroup choices;
 3:   Command okCommand = new Command ("OK", Command.OK, 1);
 4:
 5:   // MIDlet startup method
 6:   public void startApp() throws MIDletStateChangeException {
 7:     // Set the Agent.midlet static field
 8:     Agent.midlet = this;
 9:
10:     // Show a form to let the user select the operation he wants
   to do
11:     Form f = new Form ("Book Trading");
12:     choices = ChoiceGroup ("Select an operation",
   Choice.EXCLUSIVE);
13:     choices.append("BUY", null);
14:     choices.append("SELL", null);
15:     f.append(choices);
16:     f.addCommand(okCommand);
17:     f.setCommandListener(this);
18:     Display.getDisplay(this).setCurrent(f);
19:   }
20:
21:   // This is executed when the user clicks OK
22:   public void commandAction (Command c, Displayable d) {
23:     if (c == okCommand) {
24:       int i = choices.getSelectedIndex();
25:       String agentsOption = null;
26:       if (i == 0) {
27:         // The user selected Buy --> start a BookBuyerAgent
```

```
28:           agentsOption = "%C-book-buyer:bookTrading.buyer
   .BookBuyerAgent";
29:         }
30:         else {
31:           // The user selected SELL --> start a BookSellerAgent
32:           agentsOption = "%C-book-seller:bookTrading
   .seller.BookSellerAgent";
33:         }
34:
35:         // Launch JADE with the proper agent
36:         Properties pp = new Properties();
37:         pp.load("jad");
38:         pp.setProperty(MicroRuntime.AGENTS_KEY, agentsOption);
39:         MicroRuntime.startJADE(pp, null);
40:         if (!MicroRuntime.isRunning()) {
41:           // JADE startup failed for some reason. Handle the
   failure. . .
42:         }
43:       }
44:     }
45:   }
```

This code assumes that the `host` and `port` options telling JADE-LEAP where to find the mediator are properly set in the MIDlet suite manifest file. Furthermore, the minimization process for this MIDlet suite requires the following. dlc file:

```
bookTrading.BookTradingMIDlet
jade.util.leap.OutputViewer
jade.imtp.leap.JICP.BIFEDispatcher
bookTrading.buyer.BookBuyerAgent
bookTrading.seller.BookSellerAgent
```

Note that `bookTrading.BookTradingMIDlet` replaced `jade.MicroBoot` and both the `BookBuyerAgent` class and the `BookSellerAgent` class are now listed, as it is not known a priori which of them will have to be loaded.

Besides `startJADE()`, the `MicroRuntime` class provides four other notable methods:

- `isRunning()` (used at line 40 in the `BookTradingMIDlet` class presented above) returns a Boolean value indicating whether the JADE-LEAP split container is up and running.
- `stopJADE()` makes the JADE-LEAP split container kill all its agents and terminate.
- `startAgent(String name, String className, String[] arguments)` allows an agent of a given class to be started and passed some arguments.
- `killAgent(String name)` allows an agent living in the front-end of the local split container to be killed.

### *8.5.4 CONTROLLING THE STATUS OF THE CONNECTION BETWEEN THE FRONT-END AND THE BACK-END*

As mentioned in Section 8.3, the front-end and the back-end of a split container keep a permanent connection used to transfer all data that must be exchanged, such as ACL messages sent and received by agents resident in the front-end. This connection is fully managed by JADE-LEAP in a manner that is completely transparent to developers. Even in the case of temporary disconnections

**Table 8.2** ConnectionListener events

| Event code | Event description |
|---|---|
| BEFORE_CONNECTION | Raised each time the front-end attempts to open a network connection to the mediator. A common use is reacting to the set-up of an appropriate PDP context if not already in place |
| DISCONNECTED | Raised whenever a temporary disconnection is detected |
| RECONNECTED | Raised whenever the connection between the front-end and the back-end is successfully re-established after a temporary disconnection |
| RECONNECTION_FAILURE | Raised when the front-end detects that it is no longer possible to reconnect (e.g. because the maximum disconnection timeout expired) |
| BE_NOT_FOUND | Raised when the mediator replies with a 'BE Not Found' error response to a CONNECT_MEDIATOR request |
| NOT_AUTHORIZED | Raised when the mediator replies with a 'Not Authorized' error response to a CREATE_MEDIATOR or a CONNECT_MEDIATOR request |

and changes of the hand-held device IP address, agent messages are automatically buffered and delivered as soon as the connection is restored.

There are, however, cases in which the developer is interested in being notified when events related to the management of the connection occur, such as a disconnection or a reconnection. This is accomplished by setting a proper `ConnectionListener` object among the start-up properties passed in the `startJADE()` method described in Section 8.5.3. The `jade.imtp.leap.ConnectionListener` interface provides a single method `handleConnectionEvent()` with two parameters:

1. An integer code specifying the type of event that occurred. Possible events with the related code (available as constants defined in the `ConnectionListener` interface itself) are summarized in Table 8.2.
2. An `Object` parameter carrying additional event-specific information. This parameter is typically `null` and is there mainly for future extensions.

The following piece of code shows how to launch a JADE-LEAP split container, setting a ConnectionListener that logs all disconnections and reconnections.

```
 1: private Logger myLogger =
   Logger.getMyLogger(getClass().getName());
 2: ...
 3: Properties pp = new Properties();
 4: pp.load("jad");
 5: pp.put(MicroRuntime.CONNECTION_LISTENER_KEY, new
   ConnectionListener() {
 6:   public void handleConnectionEvent(int code, Object info) {
 7:     switch (code) {
 8:     case ConnectionListener.DISCONNECTED:
 9:       myLogger.log(Logger.WARNING, "Connection down");
10:       break;
11:     case ConnectionListener.RECONNECTED:
12:        myLogger.log(Logger.INFO, "Connection reestablished");
13:       break;
14:     }
15:   }
```

```
16:   } );
17:   MicroRuntime.startJADE(pp, null);
18:   if (!MicroRuntime.isRunning()( {
19:     // JADE startup failed for some reason. Handle the failure...
20:   }
```

Note that, since the `ConnectionListener` object is not a String, at line 5 the `put()` method (inherited from `java.util.Hashtable`) is used to set it in the configuration properties.

### *8.5.5 THE BACK-END MANAGEMENT SERVICE*

As mentioned in Section 8.3.1, two elements in the LEAP add-on can play the role of the mediator, accepting incoming connections from front-ends and serving back-end creation and (re)connection requests. Typically, when deploying JADE-LEAP-based applications in a laboratory setting, a normal JADE-LEAP container (its local ICP) is used as mediator. Though very simple, this approach has scalability limitations and does not allow more than about 1000 devices to be connected to a single mediator. In a real scenario it is suggested to use the `jade.imtp.leap.nio.BEManagementService` as mediator, since it is able to simultaneously manage up to 10 000 devices. The `BEManagementService` is a JADE kernel service (as described in Chapter 7) which makes use of the `java.nio` package to manage the connections with all front-ends. This approach allows the reception of data from a large number of connections by means of a pool of threads instead of using one thread per connection. When using the `BEManagementService`, the deployment architecture becomes similar to that depicted in Figure 8.9.

As in the case where an ICP acts as mediator, several JADE nodes live in the same JVM: Container1 (i.e. the container hosting the BEManagementService) and all back-ends. All share the same singleton CommandDispatcher and its installed ICPs. ICPs are only used to transfer data between such nodes and remote non-split containers. Incoming connections from front-ends are accepted by the BEManagementService and installed in Container1.

Another difference is that the BEManagementService is not simply responsible for connection acceptance and the back-end creation phases. In this scenario, all data from front-ends are always

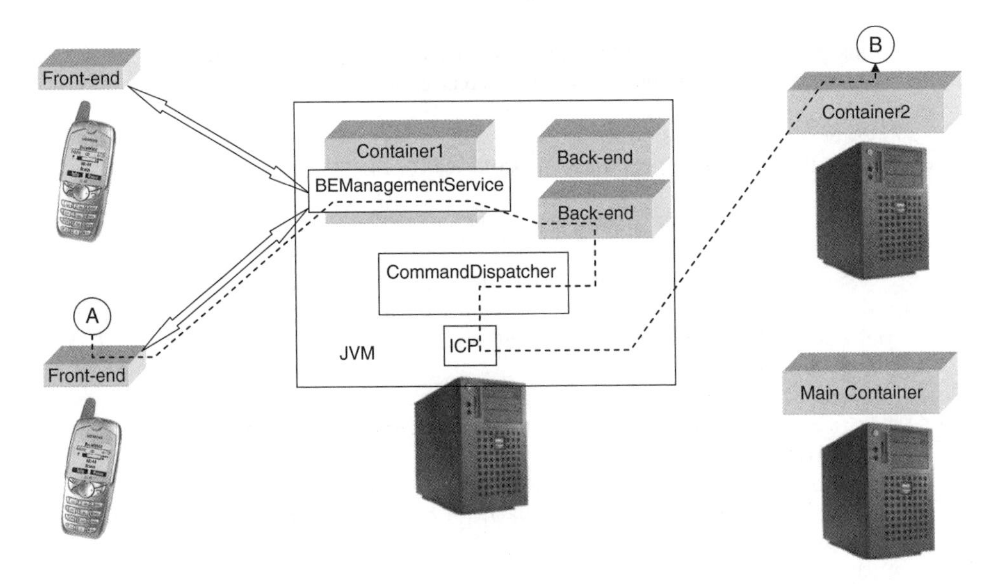


**Figure 8.9** Deployment architecture using the BEManagementService as mediator

received by the BEManagementService that then dispatches them to the relevant back-ends for processing. The dotted line in Figure 8.9 shows the path followed by a message sent by agent A (living in a front-end) to agent B (living in Container2).

#### 8.5.5.1 Connection Management Classes

Taking into account the different behaviour of the two types of mediators described above, it is clear that the back-end component responsible for exchanging data with the front-end must be different depending on whether an ICP or the BEManagementService is used as mediator. In the first case, it must directly read/write data from/to the connection with the front-end. In the second case, it needs only get/pass data from/to the BEManagementService, since the connection is fully managed by the latter. The actual class implementing that component is specified in a proper field of the `CREATE_MEDIATOR` JICP command sent by the front-end to the mediator at bootstrap time. Each front-end side connection management class indicates by default (when available) the back-end side data exchange class, assuming an ICP is used as mediator. When using the BEManagementService, therefore, it is necessary to specify the `mediator-class` option among the front-end start-up arguments to override this default. Table 8.3 summarizes the available front-end side connection management classes with the related back-end side data exchange classes to be used with the two mediator types.

As stated in Section 8.3.2, the `connection-manager` option must be specified to select the front-end side connection management class to use. The main characteristics of the connection management classes listed in Table 8.3 are described hereunder.

The `jade.imtp.leap.JICP.BIFEDispatcher` is the default front-end connection management class. It can be used with both a JICP ICP and the BEManagementService as mediator, provided that the right back-end side data exchange class is specified (see Table 8.3). The `BIFEDispatcher` class actually keeps two network sockets open with the back-end: one for commands sent by the front-end to the back-end (and related responses) and the other for commands sent by the back-end to the front-end. This behaviour was implemented to overcome a limitation that is present in many Symbian devices: the same socket can not be used for writing and reading data at the same time.

The `jade.imtp.leap.JICP.FrontEndDispatcher` class, a modified version of the `BIFEDispatcher` class described above, uses a single full duplex network socket. As indicated in Table 8.3 at the time of writing, it can only be used when the BEManagementService plays the mediator role. The consequence is that it does not require specification of the `mediator-class` configuration option. When working with devices supporting full duplex network connections, usage of the `FrontEndDispatcher` class is preferable to the default `BIFEDispatcher` class. In fact,

**Table 8.3** Connection management classes

| Front-end side connection management class | Back-end side data exchange class when an ICP is used as mediator | Back-end side data exchange class when the BEManagementService is used as mediator |
|---|---|---|
| `jade.imtp.leap.JICP.BIFEDispatcher` | `jade.imtp.leap.JICP.BIBEDispatcher` | `jade.imtp.leap.nio.NIOBEDispatcher` |
| `jade.imtp.leap.JICP.FrontEndDispatcher` | NOT AVAILABLE | `jade.imtp.leap.nio.BackEndDispatcher` |
| `jade.imtp.leap.http.HTTPFEDispatcher` | `jade.imtp.leap.http.HTTPBEDispatcher` | NOT AVAILABLE |

the latter implies keeping $2 \times N$ sockets (where $N$ is the number of devices) simultaneously open in the JVM, where the mediator is running.

Unlike the `BIFEDispatcher` and the `FrontEndDispatcher` classes that use the JICP protocol to communicate with their back-end side data exchange class, the `jade.imtp.leap.http.HTTPFEDispatcher` class transfers data over the HTTP 1.0 protocol. This is significantly less efficient, and its usage is not encouraged. There are, however, two cases in which it is the only option:

1. When starting the front-end on a device with a MIDP implementation that does not provide support for plain sockets, but HTTP only. Note that HTTP is the only protocol that a JVM is required to support in order to be MIDP-compliant.
2. When connecting to a mediator in the Internet from within a LAN through an HTTP proxy as depicted in Figure 8.10.

As indicated in Table 8.3, the `HTTPFEDispatcher` class can only be used when an ICP plays the role of the mediator. However, it should be noted that, since it uses the HTTP protocol to transfer data, the ICP must be a `jade.imtp.leap.http.HTTPPeer` that also uses the HTTP protocol.

#### 8.5.5.2 BEManagementService Activation and Configuration

Being a JADE kernel service, the BEManagementService is activated by specifying the `-services` configuration option as shown:

```
java jade.Boot ... -services jade.imtp.leap.nio.BEManagementService
```

See Chapter 7 for a detailed description of the `-services` option.

Three main configuration options can be specified to control the BEManagementService:

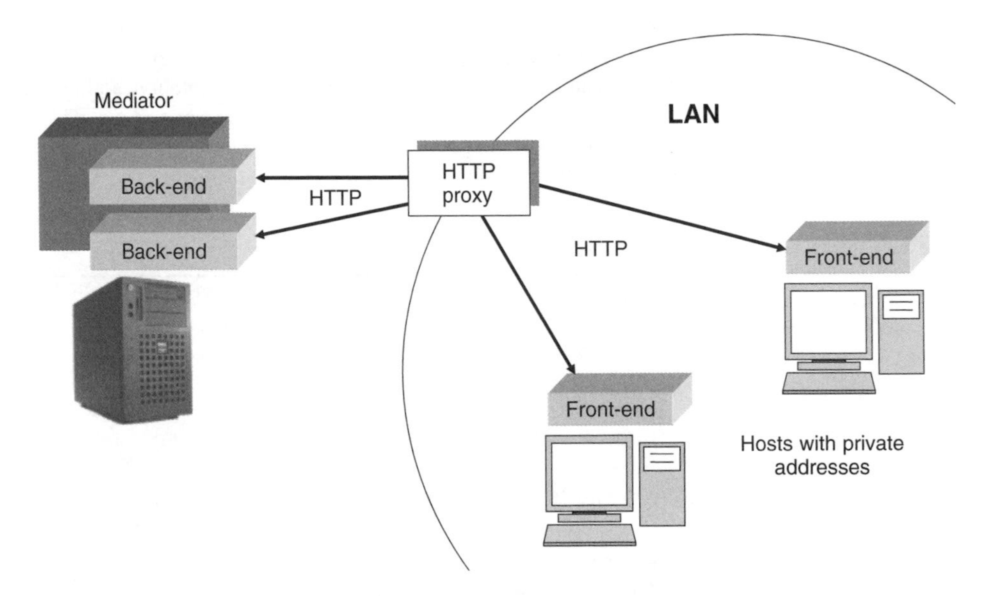


**Figure 8.10** Connecting to a mediator through an HTTP proxy

- `jade_imtp_leap_nio_BEManagementService_local-port`. This option allows specification of the port used by the BEManagementService to listen to incoming connections from front-ends. The default value for this option is 2099.
- `jade_imtp_leap_nio_BEManagementService_poolsize`. This option allows the thread pool size to be set which is used by the BEManagementService to read data from all the connections. The default for this option is 5.
- `jade_imtp_leap_nio_BEManagementService_leap-property-file`. This option specifies the name of a property file, including configuration options to be applied to all back-ends that will be created by the BEManagementService. By default, a file called 'leap.properies' is searched in the current directory. The leap property file also allows specification of a `PDPContextManager` implementation to be connected to the BEManagementService, as described in the subsequent section.

#### 8.5.5.3 Using Telephone Numbers as Container and Agent Names

In a JADE platform both agent and container names must be unique. Mobile phones intrinsically embed a very secure and intuitive identification system: telephone numbers. It is therefore desirable, when working with mobile phones, to use phone numbers as the basis of agent and container names. For instance, in the book-trading case study a good naming convention could be that, for a phone whose msisdn (the telecommunication term to indicate the telephone number) is 1234567890, the split container is called 1234567890, the buyer agent is called book-buyer-1234567890 and the seller agent is called book-seller-1234567890.

Assigning agent names on the basis of the container name can be easily achieved by means of the %C wildcard, as discussed in Section 8.4.3. Unfortunately reading the msisdn on a phone to use it as container name is generally quite difficult. In fact, the msisdn is associated with the SIM card, not the phone, and is generally not stored in the SIM card itself.

By means of the `jade.imtp.leap.JICP.PDPContextManager` interface, the BEManagementService provides a hook to be connected to an external system able to detect the msisdn of a mobile phone on the basis of the IP address assigned to that phone. The most important method of the `PDPContextManager` interface is `getPDPContextInfo()`. When the BEManagementService is configured to use a `PDPContextManager`, each time an incoming connection is received from a front-end, it invokes the `PDPContextManager` `getPDPContextInfo()` method and passes the IP address of the connecting front-end as a parameter. A typical `PDPContextManager` implementation is expected to connect to the proper element of the wireless access infrastructure (typically the Radius server) and retrieve the msisdn corresponding to that IP address. The `getPDPContextInfo()` method returns a `Properties` object that is expected to include the msisdn of the connecting phone as the value corresponding to the `PDPContextManager.MSISDN` key. The choice of having a `Properties` object as a return value is mainly to allow `PDPContextManager` implementations to provide more information (e.g. the GPRS username and password) in those cases where they are able to retrieve them without having to change the `getPDPContextInfo()` method signature. The `PDPContextManager` implementation class (if any) to use is declared in the LEAP properties file passed to the BEManagementService specifying the `pdp-context-manager` property. When a `PDPContextManager` is attached to the BEManagementService, all split containers managed by the BEManagementService are automatically named using the msisdn values retrieved by that PDPContextManager at front-end start-up.

# 9

# Deploying a Fault-Tolerant JADE Platform

When deploying real-world applications fault tolerance is often one of the main requirements that must be met. JADE is a distributed platform but nevertheless relies on a main container to house key platform services such as the AMS and DF. This is clearly a potential single point of failure that must be effectively managed to ensure the platform remain fully operational even in the event of a main container failure. This is achieved by combining the following two features:

- *Main container replication*. This feature, presented in Section 9.1, allows the main container and the AMS agent within it to be replicated, to keep all replicas fully synchronized and to ensure that if fails, another can take over.
- *DF persistence*. This feature, presented in Section 9.2, allows the catalogue of the DF agent to be recorded in a relational DataBase (DB). In the case of main container failure a new DF agent is automatically started on the new master main container and can recover its catalogue from the DB.

## 9.1 THE MAIN REPLICATION SERVICE

The JADE main container can be replicated to deploy a fault-tolerant platform with the Main Container Replication Service (MCRS) implemented by the `jade.core.replication.MainReplicationService` class. As the name suggests, this is a JADE kernel-level service according to the distributed coordinated filters architecture discussed in Chapter 7.

The MCRS can launch any number of *logical* main containers in a platform. Only one can be appointed as the master while the others must serve as replica backups. All active main containers constitute a logical ring within which they can monitor one another. If one fails, the others detect the failure event and take the appropriate recovery actions, as described later. Non-main containers can join the platform through any of these active main container nodes as all the replicas are kept coherent and synchronized with one another via cross-notification.

As shown in Figure 9.1, the MCRS changes the topology of a JADE platform from a star into a ring of stars. Using the MCRS, the administrator can control the level of fault tolerance of the platform, the level of platform scalability, and the level of platform distribution. A control layer composed of several distributed instances of the main container can be configured to implement a distributed bootstrapping system (i.e. several bootstrap points can be passed, explicitly or implicitly,

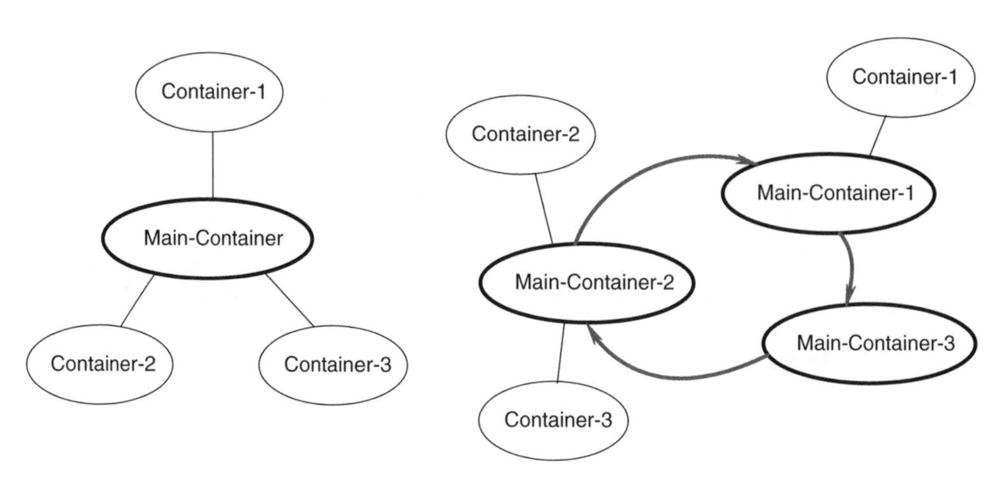


**Figure 9.1** Topology of a JADE platform without (left) and with (right) the Main Container Replication Service

to each container), and a distributed control system. In the extreme case, each container can be instructed to join the Main Replication Service and act as part of the control layer.

In the example shown on the right of Figure 9.1, three main containers are arranged in a unidirectional ring with each node monitoring its next neighbour. If *main-container-1* fails, its monitoring node *main-container-2* informs all the other main containers (in this case just *main-container-3*) of the failure, allowing the survivors to reorganize into a new, smaller ring. When a main container fails, all its directly connected *normal* containers become orphans (in the current example, *container-1*), because their main-container is no longer available. When a container detects that it has been orphaned it must register with an alternative main container node, implying that every non-main container must retain a list of all the main containers present in its platform. This is achieved through two alternative, though non-exclusive, means:

1. For each container, the list of main containers can be directly specified on the command line. This can of course only be used when the list of main containers is fixed and known a priori.
2. An additional kernel-level service known as the *Address-Notification Service* (ANS) can be initiated on both main container and non-main container nodes. This service automatically detects additions and removals to the main container ring and keeps the address list of all platform nodes up to date. This mechanism is particularly suitable for dynamic systems where the topology of the platform and its containers can vary over time.

To launch a platform configured as that in the right part of Figure 9.1, we will first assume that each of the six main and non-main container nodes are to be initiated on six different network hosts. The following command launches a main container on host1 that will act as the master and starter of the sample initialization sequence. The command also requests activation of an MCRS and an ANS:

```
prompt@host1> java jade.Boot -services
   jade.core.replication.MainReplicationService;jade.core.repli-
   cation.AddressNotificationService
```

This next command launches a backup main container on host1 (`-backupmain` command-line argument) with, again, an MCRS and an ANS. The `-host host1` argument specifies the host running the master main container. Finally, the argument `-local-port 1102` specifies that this backup main container will listen on local port 1102, although this is actually superfluous when each container is executed on a different host. If six different hosts are not available, select different port numbers for each container running on the same host machine.

```
prompt@host2> java jade.Boot -backupmain -host host1 -local-port
   1102 -services jade.core.replication.MainReplicationService;
   jade.core.replication.AddressNotificationService
```

This next command launches a backup main container on host3 that connects to main-container-2, as specified by `-host host2 -port 1102`:

```
prompt@host3> java jade.Boot -backupmain -host host2 -port 1102
   -local-port 1103 -services jade.core.replication.
   MainReplicationService; jade.core.replication.
   AddressNotificationService
```

Now we can start container-1 that registers directly with the main container on host1. A GUI can also be initiated on this container in order to observe platform recovery from main container faults.

```
prompt@host4> java jade.Boot -gui -container -host host1 -services
   jade.core. replication.AddressNotificationService
```

Finally, it is time to start container-2 and container-3 that will attempt to register with the backup main-container-2 on host2. This registration request will be *automatically redirected to the current actual main container* running on host1. The command to launch container-3 uses the argument `-smaddrs` (service manager addresses) to specify the list of main container nodes, as an alternative to launching an ANS.

```
prompt@host5> java jade.Boot -container -host host2 -port 1102
-services jade.core.replication.AddressNotificationService
```

```
prompt@host6> java jade.Boot -container -host host2 -port 1102
-smaddrs host1:1099;host2:1102;host3:1103
```

We now have a ring of three main containers. If the master main container on host1 is now abruptly killed (e.g. by pressing Ctrl-C in its command shell window), we can observe, using the GUI, that the AMS and DF agents are automatically recreated at main-container-2 on host2. The logs on the remaining containers will indicate that main-container-2 was notified of the failure of main-container-1 and automatically assumed the role of master main container. Container-1 also reregistered with main-container-2.

The main container on host1 can now be restarted. Note that the command-line arguments differ slightly from the original inception of this main container, as it is now simply joining an existing platform, rather than starting a new one.

```
prompt@host1> java jade.Boot -backupmain -host host2 -port 1102
   -local-port 1099 -services jade.core.replication.
   MainReplicationService;
jade.core.replication.AddressNotificationService
```

Notice that the MCRS ensures that each backup main container is updated with all relevant information, except the DF repository. For maximum reliability, administrators should use a persistent DF with an underlying database to ensure that all stored data are recoverable – Section 9.2 describes how to configure such a persistent DF repository.

## 9.2 ATTACHING THE DF TO A RELATIONAL DB

By default the DF stores its catalogue in memory. This guarantees short access times but at the same time leads to linearly increasing memory consumption. For applications based on a great number of agents this can lead to scalability problems.

It is therefore also possible to configure the DF to store its catalogue in a relational database. This has the advantage of a nearly constant level of memory consumption. Moreover it ensures DF fault-tolerance and, used in combination with the Main Container Replication Service, allows the deployment of a fully fault-tolerant JADE platform. Of course in terms of performance, storing the DF catalogue into a database leads to an increase in access times, especially when transferring large data sets from the database to the DF.

Which variant, in-memory or database persistence, fits best depends mainly on application requirements; optimal behaviour often lies between the two extremes. The JADE administrator must determine the best compromise, taking into account performance scalability and fault tolerance. The administrator can choose between two persistence modes, default and custom, which are described in the following two sections.

### *9.2.1 DEFAULT PERSISTENCE*

JADE provides a default implementation for DF persistence, based on the Java DataBase Connectivity (JDBC) interface. To guarantee compatibility with most database engines the implementation is kept as general as possible and thus does not use any non-standardized SQL features. With the following start-up parameter the DF is instructed to store its catalogue into a database, reachable under a specific JDBC-URL:

```
-jade_domain_df_db-url  JDBC-URL
```

The database access can be further configured with the parameters described in Table 9.1, DF DB configuration options.

Since version 3.3, JADE also provides a specialized implementation for the HSQL database engine. HSQL is a lightweight, open source SQL database engine which is 100% implemented in Java. One of its main advantages is the flexible way in which it caches data in memory, including detailed configuration of how much data should be held in memory and how much should be stored externally.

**Table 9.1** DF DB configuration options

| | |
|---|---|
| `jade_domain_df_db-driver` | Indicates the JDBC driver to be used to access the DF database (defaults to the ODBC-JDBC bridge) |
| `jade_domain_df_db-username` | Indicate the user name to be used to access the DF database (defaults to null) |
| `jade_domain_df_db-password` | Indicate the password to be used to access the DF database (defaults to null) |
| `jade_domain_df_db-cleantables` | If set to `true`, indicates that the DF will clean the content of all pre-existing database tables, used by the DF. This parameter is ignored if the catalogue is not stored in a database (defaults to `false`) |

Since DF persistence is an optional feature and the majority of the JADE users will probably not use it, the HSQL database is not included with the JADE distribution. Therefore in order to use the default persistence for HSQL the binaries must be downloaded from the project homepage (http://hsqldb.sourceforge.net) and added to the class path.

After adding the binaries to the class path, HSQL support can be activated with the following start-up parameter:

```
-jade_domain_df_db-default true
```

With this parameter the JADE platform starts together with an instance of the HSQL database engine within the same Java Virtual Machine. The engine is configured to allow reasonable access times with relatively constant memory consumption.

### *9.2.2 CUSTOM PERSISTENCE*

As described previously, default persistence provides two different means of persisting DF descriptions. While the HSQL-specific implementation provides a good compromise between high performance and low memory consumption, the more general implementation guarantees the usability of most relational database engines.

But in some situations neither of the two implementations can fulfil all the requirements of the administrator. For this reason JADE also allows the integration of custom implementations. With this it is possible to optimize the persistence mechanisms for specific database engines or to introduce a custom database schema.

For implementation the developer has two options illustrated in Figure 9.2: either adapt one of the base implementations by inheriting from the class `jade.domain.DFDBKB` or from the class `jade.domain.DFHSQLKB`, or start an implementation from scratch by inheriting from the abstract base class, `jade.domain.DBKB`.

By default the DF uses the factory `jade.domain.DFKBFactory` to instantiate the correct persistence implementation according to the start-up parameters. When integrating a custom implementation a new factory must be implemented by subclassing the default factory. To load the custom implementation the method `getDFDBKB()` must be overwritten so that it returns the newly implemented class. Finally the DF must be instructed to use the new factory. This is achieved with the following, additional, start-up parameter:

```
jade_domain_df_kb-factory  Factory Class Name
```

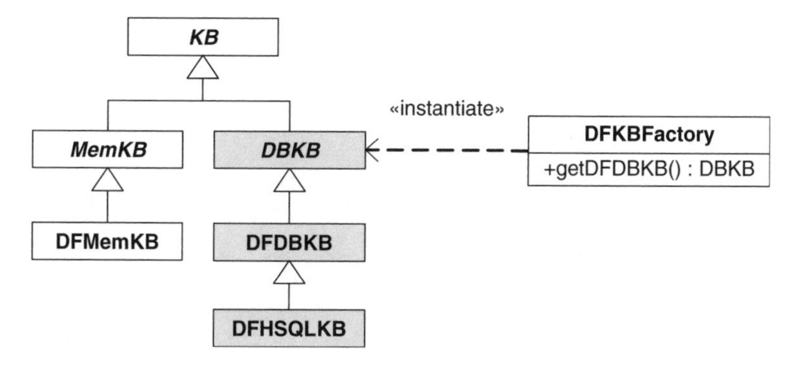


**Figure 9.2** The classes of the default implementation and the factory class used by the DF. The marked classes are potential base classes for custom implementations

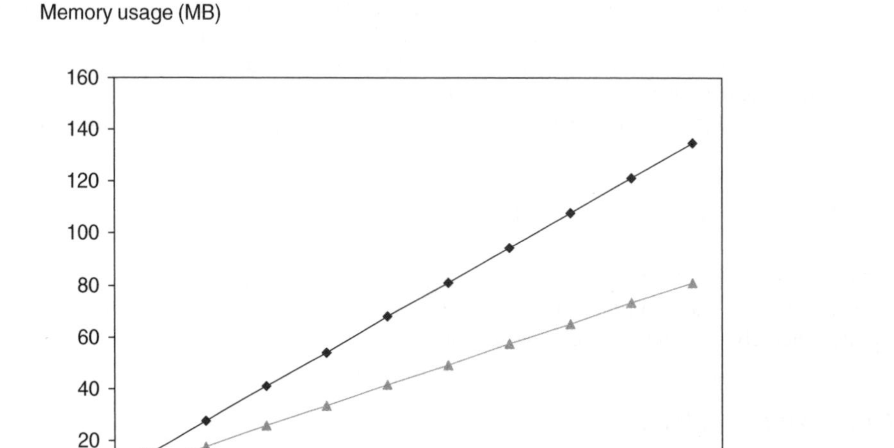


**Figure 9.3** Comparison of memory consumption with an increasing number of DF descriptions. One description consists of one service description with no custom parameters

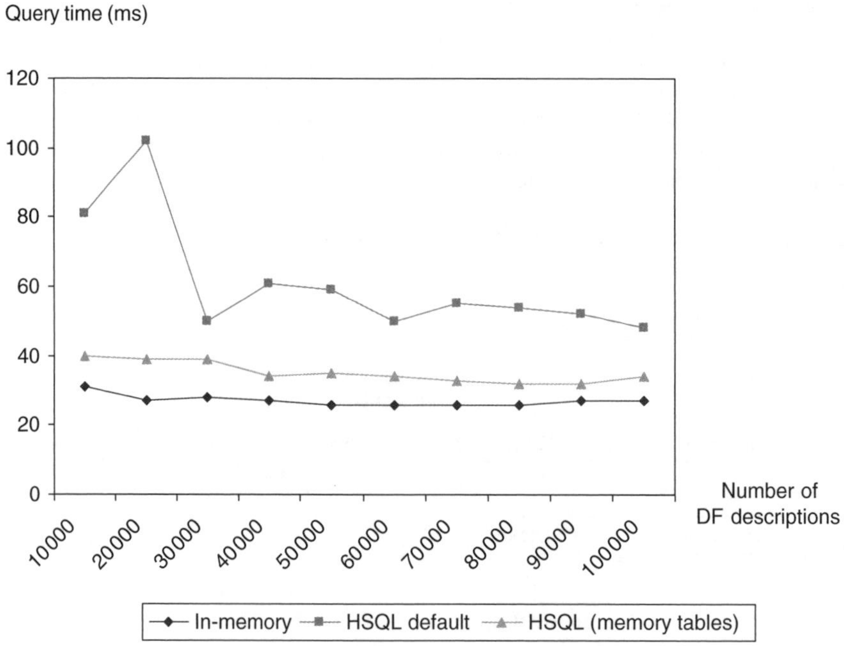

**Figure 9.4** Comparison of query times for a simple query by service name with no other search constraints and with an increasing number of DF descriptions

### *9.2.3 COMPARISON OF PERFORMANCE AND SCALABILITY*

It is useful to observe the scalability and performance differences between the three variants: default in-memory management, HSQL default implementation and a custom implementation which configures HSQL to hold all the data in memory. These are compared in Figure 9.3 which shows the memory consumption of the three variants. While the in-memory management causes a linear increase of the memory consumption the default implementation for HSQL levels off at about 2 MB. The custom persistence example is a weaker variant of the default implementation but still shows a linear increase.

Figure 9.4 then compares the average query times for a query by service name without any other search constraints. The test results show that the default HSQL implementation allows comparable query times at nearly constant memory consumption. This implies it is the best choice for scenarios where a compromise between high performance and good scalability is required.

# 10

# The JADE Web Services Integration Gateway

Many people responsible for making choices regarding technology provisioning are rapidly beginning to address the tangible prospects offered by Web service technology. Web services are simple in concept, yet effective in use, and represent a step toward widespread adoption of service-oriented computing. They are rapidly becoming the dominant means of building semi-automated computational networks using well-established Internet protocols and commonly used machine-readable representations. For this reason it seems prudent and useful to provide JADE platforms with a means of bidirectional interconnectivity between agents and Web services.

The Web Services Integration Gateway (WSIG) offers this interconnectivity. It is a JADE plug-in devised specifically to seamlessly and transparently connect the Web services domain with JADE platforms by offering bidirectional discovery and remote invocation of Web services by JADE agents, and of JADE agent services by Web services.

The WSIG v0.7 is available as an add-on within the main JADE distribution.

## 10.1 WEB SERVICE TECHNOLOGY

Web services are essentially captured application or business logic that is programmatically encoded to execute on a Web server by exposing its functional capabilities as methods available to clients over HTTP. Each method can therefore be published as a URL, can accept parameters and return data that is typically encoded in XML. The W3C defines a Web service as:

> ... a software system designed to support interoperable machine-to-machine interaction over a network. It has an interface described in a machine-processable format (specifically WSDL). Other systems interact with the Web service in a manner prescribed by its description using SOAP messages, typically conveyed using HTTP with an XML serialization in conjunction with other Web-related standards. (W3C, 2004)

The encoding and protocol stack typically used for creating and publishing Web services consists of WSDL, SOAP and UDDI. WSDL (Web Service Description Language) is the standard means for expressing Web service descriptions. SOAP (Simple Object Access Protocol) is a protocol defining the exchange of messages containing Web service requests and responses. UDDI (Universal Description, Discovery and Integration) is the directory services schema commonly used to register and discover Web services.

*Developing Multi-Agent Systems with JADE* Fabio Bellifemine, Giovanni Caire, Dominic Greenwood

Web services are fast approaching the point of critical mass beyond which they will become a mainstream technology. Tracing this path through the forthcoming years we can estimate that emerging technologies, such as the Semantic Web (W3CSem, 2004) will substantiate these trends to take the basis of Web services towards powerful modalities that employ semantic expressivity, autonomy and knowledge processing, as highlighted by Davies *et al.* (2004). Although not discussed in this chapter, an enhancement to the WSIG is planned to support Web services described using WSDL-S, the semantic extension to standard WSDL. This extension will allow far richer choreography of agents and Web services through more powerful expression of capabilities.

## 10.2 THE UTILITY OF AGENT AND WEB SERVICE INTEGRATION

Integrating Web services and software agents brings about an obvious benefit: connecting application domains by enabling a Web service to invoke an agent service and vice versa. However, this interconnection is more than simply cross-domain discovery and invocation; it will also allow complex compositions of agent services and Web services to be created, managed and administered by controller agents. Several arguments have been established to support this case, including those made in Richards *et al.* (2003), Laukkanen and Helin (2003) and W3C (2004), all of which are supported by the clear statement made in the Web Services Architecture specification of the (W3C) which clearly expresses the notion that '... software agents are the running programs that drive Web services – both to implement them and to access them as computational resources that act on behalf of a person or organization'.

Providing a means of interoperation between JADE agents and Web services is the objective of the WSIG, which is designed to encapsulate the functionality required to connect the two domains, while ensuring minimal human intervention and service interruption. From the perspective of agents, Web services are computational entities that can be called upon to perform one or more advertised, or discovered, reactive operations.

To the users of Web services, whether human or computational, agents can be a powerful means of indirection by masking the Web service for purposes of, for example, redirection, aggregation, integration and administration. *Redirection* describes the case where a Web service may no longer be available for some reason, or the owner of the Web service wishes to temporarily redirect invocations to another Web service without removing the original implementation. *Aggregation* allows several Web services to be composed into logically interconnected clusters, providing patterned abstractions of behaviour that can be invoked through a single service interface. *Integration* describes the means of simply making Web services available to consumers already using, or planning to use, agent platforms for their business applications, and *administration* covers aspects of automated Web service management where the agent autonomously administers one or more Web services without necessary intervention from a human user.

## 10.3 THE WSIG ARCHITECTURE

A schematic view of the WSIG architecture is shown in Figure 10.1. As illustrated, the WSIG consists of several components the specific operation of which is discussed in Section 10.6. A brief description of each component follows.

### *10.3.1 JADE DF*

This is the standard JADE DF provided with the JADE v3.4 distribution or greater. As of JADE v3.4 the JADE DF has been slightly modified to detect when an agent service registration is made to the DF by an agent that intends it to be published as a Web service. Such requests are registered as normal and then automatically forwarded to the WSIG Gateway Agent. (Note: This is also true for deregistration and modification requests.)

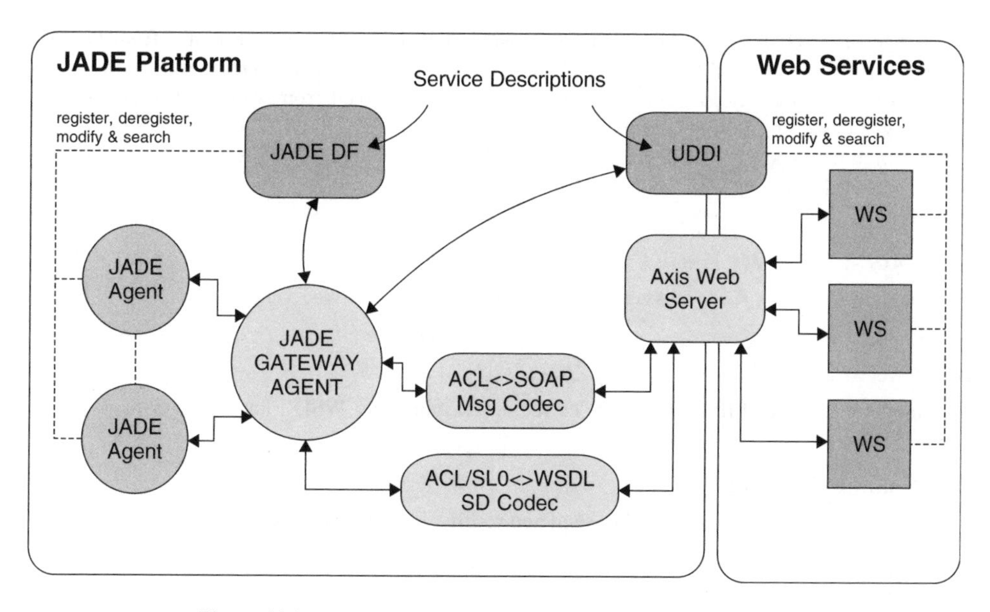


**Figure 10.1** Architecture of the Web Services Integration Gateway

### *10.3.2 UDDI*

This a standard UDDI directory, such as that provided by jUDDI as described in Section 10.4. Any registration, deregistration of modification of a WSDL Web service description from an external service is monitored by the Gateway Agent which transparently creates or updates corresponding FIPA DF service description.

### *10.3.3 JADE AGENTS*

Any agent service description published by a JADE agent as a Web service is registered, deregistered, modified and discovered as normal with the JADE DF, except that the *type* or service is special (see Section 10.6.3). Any Web service description that has been registered with the JADE DF by the Gateway Agent can be discovered and invoked by any JADE agent, but can only be deregistered, or modified, by the Gateway Agent.

To invoke a Web service, a JADE agent first seeks its service description in the JADE DF (or otherwise, be provided the description), and then sends an invocation request to the Gateway Agent.

### *10.3.4 JADE AGENT GATEWAY*

This is a specialized JADE agent that manages the entire WSIG system. Its operations are as follows:

- Receive and translate agent service registrations from the JADE DF into corresponding WSDL descriptions and register these with the UDDI repository as tModels. This also applies to deregistrations and modifications.
- Receive and translate Web service operation registrations from the UDDI repository into corresponding ACL descriptions and register these with the JADE DF. This also applies to deregistrations and modifications.
- Receive and process Web service invocation requests received from JADE agents. Processing includes retrieving the appropriate tModel from the UDDI repository, translating the invocation

message into SOAP and sending it to the specified Web service. Any response from the Web service will be translated back into ACL and sent to the originating JADE agent.
- Receive and process agent service invocation requests received from Web service clients. Processing includes retrieving the appropriate tModel from the UDDI repository, translating the invocation message into ACL and sending it to the specified agent. Any response from the agent will be translated back into SOAP and sent to the originating Web service.

### *10.3.5 ACL<>SOAP MESSAGE CODEC*

This module translates ACL messages into SOAP messages and vice versa.

### *10.3.6 ACL/SL0<>WSDL MESSAGE CODEC*

This module translates FIPA-ACL/SL0 service descriptions into WSDL and vice versa.

### *10.3.7 AXIS WEB SERVER*

This is a third-party technology used to send and receive SOAP messages to and from Web services.

### *10.3.8 WEB SERVICES*

Any Web service description published by a Web service is registered, deregistered, modified and discovered as normal with the WSIG UDDI. Any agent service description that has been registered with the UDDI by the Gateway Agent can be discovered and invoked by any Web service, but can only be deregistered, or modified by the Gateway Agent.

To invoke a JADE agent service a Web service first seeks its service description in the UDDI (or otherwise, be provided the description), and then sends an invocation request to the Gateway Agent via the Axis Web server.

## 10.4 INSTALLATION REQUIREMENTS

The WSIG requires several third-party libraries to be installed before operation. These are described below in terms of the versions of those libraries known to work with WSIG release v0.7.

- *At least Java v1.4, JADE v3.3 and Apache ANT.*Java v1.5 is recommended.
- *Apache Forrest.* As of WSIG v0.7, documentation is managed by Apache Forrest. An installation is provided with the WSIG package.
- *Apache Axis v1.4.* Apache Axis is a Java platform for creating and deploying Web service applications. The SOAP manipulation interface of Axis is used by the WSIG to handle SOAP's messages. Axis v1.4 is currently used by WSIG v0.7 with the Axis Jar files located in the wsig\lib directory.
- *Xerces XML parser v2.8.1.* The Xerces XML parser is required for use with Apache Axis. Xerces Jar files are located in the wsig\lib directory.
- *WSDL4j v1.6.2.* WSDL4j v1.6.2 is used to manipulate WSDL files directly from Java code. The Jar files are located in the wsig\lib directory.
- *Log4j v1.2.14.* Logging within the WSIG code is managed by the log4j project. The jar is located in the wsig\lib directory.
- *UDDI4j v2.0.5.* UDDI4j v2.0.5 is used to access a UDDI repository from Java code. The Jar files are located in the wsig\lib directory.
- *jUDDI v0.9rc4.* Access to a UDDI v2.0 compliant registry is required by the WSIG as the location to store service descriptions relating to JADE agent services registered with a platform

DF. External Web service clients will use this UDDI registry to search for JADE agent services that are accessible via SOAP and WSDL.

jUDDI v0.9rc4 is used in conjunction with a MySQL database to hold the registry records. Installation of jUDDI is non-trivial, thus an instruction document (jUUDI install.txt) is provided with the WSIG v0.7 release.

## 10.5 WSIG INSTALLATION PROCEDURE

The WSIG should be extracted from its distribution archive and placed in the JADE add-on directory. The package is provided with a precompiled Jar library in wsig\dist, although a fresh recompilation can be initiated by running Ant in the WSIG's base directory. Documentation can be recompiled if necessary by typing *forrest* in the wsig\doc directory; the compiled result is placed in the wsig\doc\build directory.

### *10.5.1 CONFIGURATION OF THE WSIG*

The WSIG is configured using the *wsig.properties* file that must be located in the directory where the WSIG runs. A sample version of the file is available in the wsig\misc directory. The configurable attributes are as follows:

- `wsig.host.name` The host name of the machine where the WSIG runs. For example,

  `wsig.host.name = pc7`

- `wsig.host.port` The port number where a gateway will be accessible. For example,

  `wsig.host.port = 2222`

- `wsig.agent_id` The agent ID (AID) of the WSIG Gateway Agent. The host name component must match the host machine name. Avoid using `localhost` as this can result in identification problems. For example,

  `wsig.agent_id = wsigs@pc7\:1099/JADE`

- `uddi.wsig_businessKey` This is the business key of a business entity registered in a UDDI repository. This key can either be created ad hoc or match a published business type; for many business types there is an associated business key. Lists are available on the Internet. For example,

  `uddi.wsig_businessKey = 8C983E50-E09B-11D8-BE50-DA8FBF3BDC61`

- `uddi.userName, uddi.userPassword` A user name and a password pair that will be provided to the UDDI repository to regulate access control. For example,

  `uddi.userName = bob` and `uddi.userPassword = somepassword`

- `uddi.lifeCycleManagerURL` This is the publish access point into the UDDI repository. For example,

  `uddi.lifeCycleManagerURL=http\://localhost\:8080/jUDDI/publish`

- `uddi.queryManagerURL` This is the inquiry access point into the UDDI repository. For example,

```
uddi.queryManagerURL = http\://localhost\:8080/jUDDI/inquiry
```

Certain start-up files must also be configured when the WSIG is running on a machine other than `localhost`. The host name must be changed in wsig\misc\wsig.bat, wsig\misc\run_TestAgent Client.bat and \wsig\misc\run_TestAgentServer.bat. Note that if run under *nix the scripts with the extension. sh should be changed.

The level of debug messages is set in the `log4j.properties` file. The misc/log4j.properties.txt must be copied into the running directory and edited before use, removing the .txt extension.

### *10.5.2 RUNNING THE WSIG*

The WSIG is initiated by running the following command-line scripts in order (note that the. sh versions should be used under *nix):

1. Start JADE with run_jade_main.bat.
2. Start the WSIG with wsig.bat.

To start the WSIG GUI edit the wsig.bat (or wsig.sh) script and append (-gui) to the start-up command as shown:

```
..\misc\runjade.bat -container -host localhost wsig:com.whitestein.
   wsig.fipa.GatewayAgent (-gui)
```

### *10.5.3 WSIG EXAMPLES*

Four examples are provided for testing purposes and are located in the wsig\misc directory. The first example, detailed in Section 10.7, starts an agent with some application services and a Web service client which will invoke the agent service via the WSIG gateway. This example is initiated by running the following two scripts in order:

1. run_TestAgentServer.bat
2. run_TestSOAPClient.bat

The second example, detailed in Section 10.8, starts a SOAP server and an agent client that will invoke a Web service registered on the SOAP server. This example is initiated by running the following two scripts in order:

1. run_TestSOAPServer.bat
2. run_TestAgentClient.bat

## 10.6 WSIG OPERATION

This section describes the way in which the WSIG operates. In effect, its operation is relatively transparent, with the user's main task being to code agent behaviours required to interact with Web services in a manner specific to the way the Web service is invoked. This information is, as expected, available in the DF description of the Web service. One important point to recognize is that all interactions between the WSIG Gateway Agent and other JADE agents use FIPA-ACL Request and FIPA-ACL Inform performatives. This simplifies the operation of the WSIG and is essentially all that is needed to invoke basic request–response Web services.

### *10.6.1 INITIATION OF THE WSIG*

When the command indicated in Section 10.5.2 is executed, the WSIG is launched and a new agent is created in the specified container. This agent is called the WSIG Gateway Agent and will handle all processes relating to the publication of agent services as Web services and Web services as agent services. The WSIG automatically registers itself with the platform DF at start-up.

### *10.6.2 DF AND UDDI REPOSITORIES*

The WSIG uses two repositories, the DF of the hosting JADE agent platform and a UDDI directory explicitly created for use with the WSIG.

The typical actions associated with both these repositories are register, deregister, modify and search. All of these are supported. In the DF case as its visibility is restricted to the JADE platform, these actions are available to JADE agents only. The UDDI is more or less the mirror case of this, where these actions are visible to external (to the JADE platform) Web services and Web service clients. The exception in the UDDI case is that it is also visible to the WSIG Gateway Agent, but not directly to any other JADE agent.

All register, deregister and modify actions performed on either the platform DF or WSIG UDDI repositories are automatically echoed onto the counterpart repository. For example, if an agent service is deregistered from the platform DF, the WSIG Gateway Agent will ensure that the corresponding tModel is removed from the UDDI repository. This ensures that the two repositories remain consistent.

### *10.6.3 PUBLISHING JADE AGENT SERVICES AS WEB SERVICES*

The WSIG allows an agent to publish a service description as a Web service endpoint. This process is described by Figure 10.2.

When a JADE agent wishes to make an application service it provides accessible to external (to the JADE platform) Web services, it need only ensure that the *type* slot of the agent service description is set to `web-service`:

```
sd.setType("web-service")
```

Although use of the type slot is the default case, if the agent prefers to reserve the type slot for another purpose, a property slot may be used instead.

Once this is done the agent simply sends the registration to the platform DF as it would normally when publishing a service for consumption by other agents. Any service labelled with a type or property slot as `web-service` will be registered as normal with the DF. However, the registration request will also be automatically forwarded to the WSIG Gateway Agent that will handle translation of the ACL/SL0 service description into the WSDL tModel, register it with the WSIG UDDI and create an endpoint for external Web service clients to directly invoke the agent service.

When the Gateway Agent receives a registration request it creates a unique operation name for the service by which it will be known when published as a Web service. A WSDL structure is then created (see Section 10.7) and recorded in the WSIG UDDI repository.

### *10.6.4 PUBLISHING WEB SERVICE OPERATIONS AS AGENT SERVICES*

The WSIG allows an external Web service to publish a service description as a JADE agent service description with the JADE platform DF. This process is described by Figure 10.3.

The Web service WSDL description is registered with the WSIG UDDI as it would be with any other UDDI – nothing special is required here for WSIG operation. When a new registration is detected the Gateway Agent translates it into an equivalent FIPA directory entry and sends this as a registration message to the platform DF.

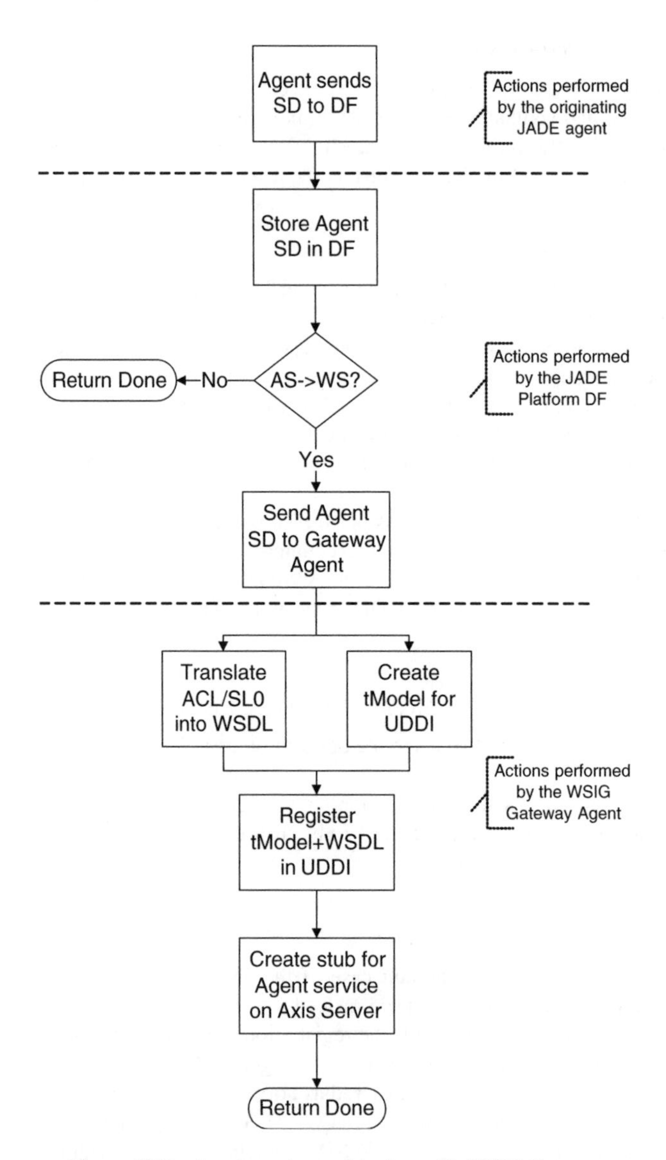


**Figure 10.2** Agent service registration with WSIG Gateway

With WSDL the capabilities, or services, that a Web service can perform are called *operations* (analogous to agent services). It is these operations that the WSIG will expose as JADE agent services by registering them in standard FIPA-ACL format in the platform DF. To achieve this, the WSIG has a UDDI interface through which, in similar form to agent service registration with the DF, the UDDI repository will inform the WSIG Gateway Agent whenever a new Web service has been registered. If a Web service WSDL contains several operations, these will be treated sequentially by the Gateway Agent, with each being translated into separate agent service descriptions for the platform DF.

Because of the obvious mismatch between WSDL and FIPA-ACL/SL0 service descriptions, certain attributes of the UDDI WSDL entry will be translated and inserted into properties attributes of the DF agent service description.

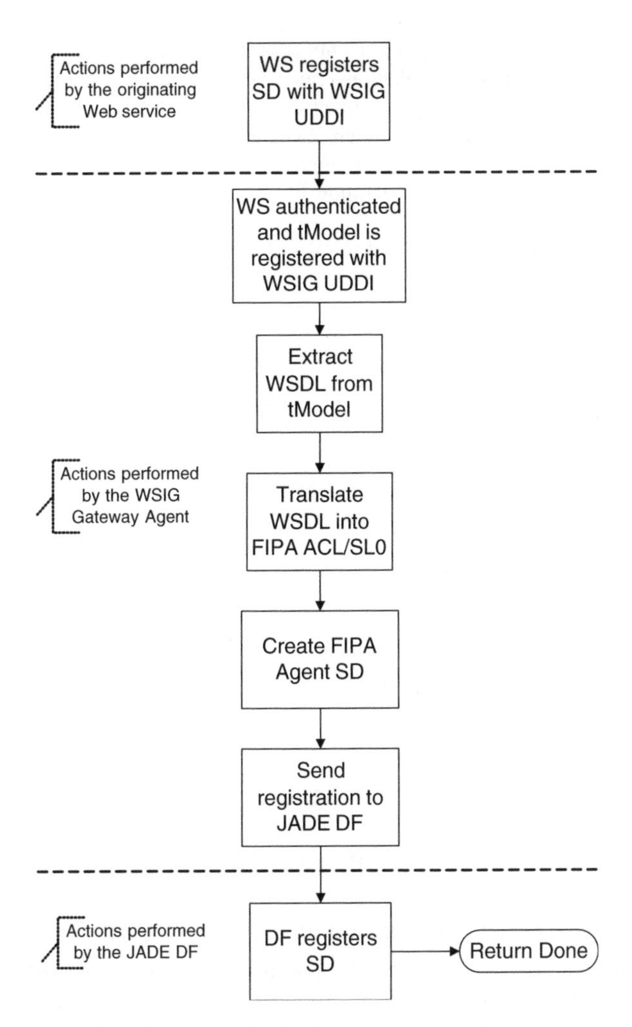


**Figure 10.3** Web service registration with WSIG Gateway

### *10.6.5 MESSAGE TRANSLATION*

The Message Translation codecs contain the parsing functionality for transforming SOAP-encoded messages into ACL-encoded messages and vice versa. In both directions parse-trees are constructed from the incoming message, which can then be traversed with each leaf node being translated into the corresponding encoding. This requires a schema that defines the SOAP equivalents of ACL slots and attributes, and vice versa.

The WSIG is restricted to treating ACL messages containing SL0 S-expressions. Each expression sequence is divided by the parser into the first field and the remaining. The first field provides the name of the data structure or function and is assigned as an XML tag name in an xml element. The remainder of the expression is translated recursively and placed into the content of the XML element.

SOAP messages consist of nested XML elements conforming to the SOAP schema. Every XML element is translated as follows: the name of each element is taken as the first field of an S-expression, the content of the element is recursively translated and structures are appended as the remainder of the S-expression.

The ACL-SL0 format allows named arguments (slots) in function calls. The translation process is thus slightly modified; each slot is taken as one structure instead of being separated into name and content. It is treated as a new XML element that encapsulates the slot content. The name of the element is the slot's name, indicated by an XML attribute called 'fipa-attribute' added to the element with value 'true'. The slot content is then recursively translated as an S-expression into XML. An example of slot translation is as follows:

```
(store :where box2 :value 20)

<store>
  <where fipa-atribute="true">
    <BO_String>box2</BO_String>
  </where>
  <value fipa-atribute="true">
    <BO_Integer>20</BO_Integer>
  </value>
</store>
```

If there are no element attributes then the translation is as above. If attributes do exist, one level of an S-expression is added and the original content is stored under the slot `:xml-element`. Attributes are stored as a set of FIPA management properties in the `:xml-attributes` slot. The name of the original content is prepended with `xml-tag-` and used as the first field in the added S-expression. An example of a XML attribute translation is as follows:

```
<q id="t12">Foo</q>

(xml-tag-q
   :xml-element (q Foo)
   :xml-attributes (set ( property :name id :value t12 )))
```

### *10.6.6 AGENT SERVICE INVOCATION BY A WEB SERVICE CLIENT*

The process of invoking an agent service with a Web service client is described by Figure 10.4. The agent service must have been previously registered with the platform DF.

A JADE agent would first seek a DF service description corresponding to a service type that it needs to perform some activity. It may or may not care whether one or more of the returned service descriptions represents a Web service rather than a typical agent service. If it does select a service description corresponding to a Web service, it must then construct a FIPA-ACL Request message specifying the service name and send it to the Gateway Agent. An example of the formulation of this message can be seen in Section 10.7.

A Web service client must first seek a WSDL tModel entry in the WSIG UDDI repository for suitable operation to perform some activity. This is identical to the process it would follow if discovering a 'normal' Web service. The UDDI tModel of any matching entries will indicate the access endpoint exposed by the Axis Web server for the given service. The Web service client constructs a SOAP request message and sends it to the exposed endpoint. The content of this SOAP message must contain the identity of the service to be invoked and any necessary parameters.

When received by the Gateway Agent, the SOAP request is parsed and the operation name translated into the corresponding DF service description name. A FIPA Request message is then created to invoke the agent service and populated with parameters from the received SOAP message. This message is sent to the appropriate agent and if a response is expected, the Gateway Agent will expect to receive a FIPA Inform from the target agent. Once a response is received it is parsed into a SOAP message and returned to the invoking Web service client.

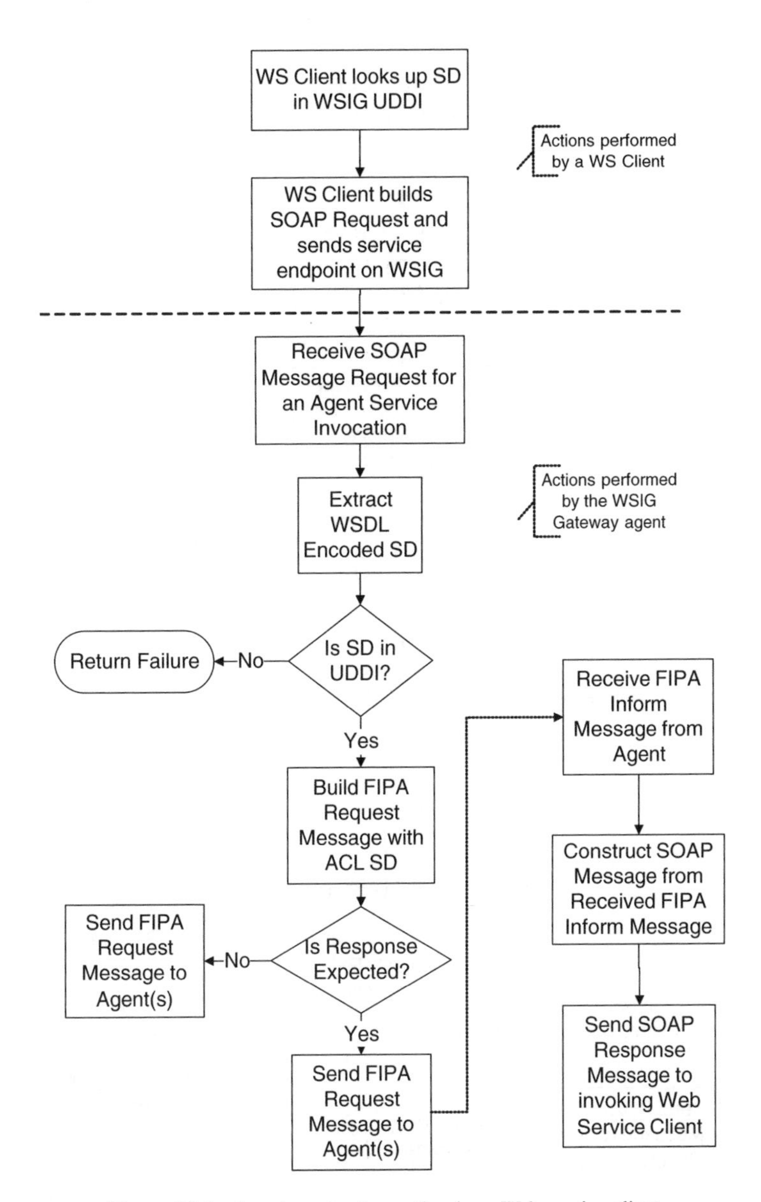


**Figure 10.4** Agent service invocation by a Web service client

### *10.6.7 WEB SERVICE INVOCATION BY AN AGENT*

The process of invoking a Web service with an agent is described by Figure 10.5. The Web service must have been previously registered with the WSIG UDDI repository,

A JADE agent must first seek a DF service description corresponding to a service type that it needs to perform some activity. It may or may not care whether one or more of the returned service descriptions represents a Web service rather than a typical agent service. If it does select a service description corresponding to a Web service, it must then construct a FIPA-ACL Request message specifying the service name and send it to the Gateway Agent. An example of the formulation of this message can be seen in Section 10.8.

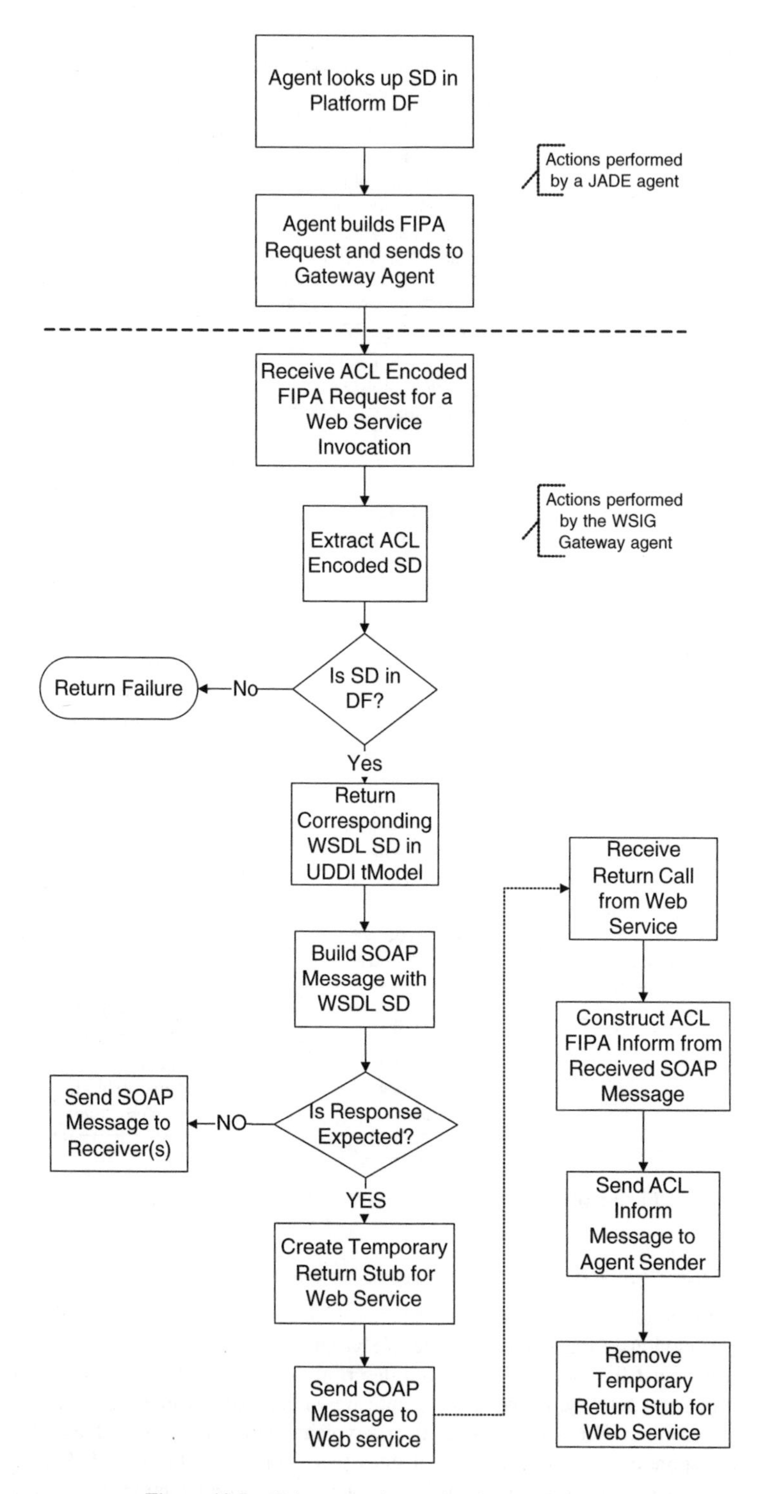


**Figure 10.5** Web service invocation by a JADE agent

Note that the only message performatives currently handled by the Gateway Agent are FIPA Request and FIPA Inform. This is because in the vast majority of cases interaction with Web services will only take a request–response form.

When received by the Gateway Agent, the FIPA Request is parsed and the service name translated into an operation identity and an access point. If the presence of an entry in the UDDI repository corresponding to the requested service is verified, the ACL message is translated into a SOAP message with the name of the operation and populated with parameters from the ACL message. This SOAP message is then sent to the access point of the Axis Web server.

If a response is received from the Web server it is translated back into an FIPA-ACL Inform message, which is sent by the Gateway Agent to the agent that originally made the Web service invocation.

## 10.7 EXAMPLE 1: WEB SERVICE CLIENT INVOKES AN AGENT SERVICE

This first example illustrates WSIG operation when an agent service has been registered with the JADE DF, automatically translated into a UDDI tModel and registered with the UDDI and then invoked by an external Web service client.

This example can be run by running the \wsig\misc\run_TestAgentServer and \wsig\misc \run_TestSOAPClient scripts in sequence.

The first step is to create an agent that provides a service. In the example below a service known as 'plus' is implemented which simply takes some numbers and performs the mathematical operation plus. A received FIPA Request message will trigger the service by invoking the `doFIPARequest` method.

```
...
// Basic initializations
private String UNNAMED = "_JADE.UNNAMED";
private Logger log =
   Logger.getLogger(TestAgentServer.class.getName());
private SLCodec codec = new SLCodec(0);
public static final String SERVICE_PLUS = "plus";
private int convId = 0;

// Setup behaviour of the "plus" agent
protected void setup() {
  log.info("TestAgentServer is starting.");

  // Add a cyclic behaviour to the Agent to receive REQUEST messages
  this.addBehaviour( new CyclicBehaviour( this ) {
    public void action() {
      ACLMessage msg = myAgent.receive();
      if ( msg != null ) {
        switch ( msg.getPerformative() ) {
          case ACLMessage.REQUEST:
            doFIPARequest( msg );
            break;
          default:
            // other messages are ignored
            break;
        }
        // Log debug information
```

```
        try {
          log.debug("TestAgentServer receives: "
              + SL0Helper.toString(msg) );
        } catch ( Exception e ) {
          log.error(e);
        }

      } else {
        block();
      }
    }
  } );
...

// Serve a received FIPA Request message
private void doFIPARequest( ACLMessage acl ) {
  ACLMessage resp = acl.createReply();
  AbsContentElement ac = null;
  AbsObject ao, ao2;
  long sum = 0;
  String str = "";

  // Decode the request
  try {
    ac = codec.decode(BasicOntology.getInstance(),acl.getContent());
  } catch ( CodecException ce ) {
    // Return a NotUnderstood message if received message
    // could not be decoded
    str = "(error CodecException ( " + ce + " ))";
    SL0Helper.fillAsNotUnderstood( acl, resp, str );
    send(resp);
    return;
  }

  // Return a NotUnderstood message if received message action is
  // erroneous
  if( null == ac ) {
    str = "(error action null)";
    SL0Helper.fillAsNotUnderstood( acl, resp, str );
    send(resp);
    return;
  }

  // Verify the action type and return NotUnderstood message if it
     is
  // an unknown type.
  if ( ! SL0Vocabulary.ACTION.equalsIgnoreCase( ac.getTypeName()) )
     {
    str = "(unknown action_format)";
    SL0Helper.fillAsNotUnderstood( acl, resp, str );
    send(resp);
    return;
```

```
    }else{
      // Parse the action
      ao = FIPASL0ToSOAP.getActionSlot( ac );
      if ( null == ao ) {
        str = "(unknown action_slot_format)";
        SL0Helper.fillAsNotUnderstood( acl, resp, str );
        send(resp);
        return;
      }

      // Check the service name
      String opName = ao.getTypeName();
      if ( SERVICE_PLUS.equalsIgnoreCase(opName) ) {
        // unnamed parameters are expected
        if ( ! FIPASL0ToSOAP.isWithUnnamed(ao) ) {
          str = "(unknown (format " + opName + " ))";
          SL0Helper.fillAsNotUnderstood( acl, resp, str );
          send(resp);
          return;
        }

        // Perform the 'plus' operation on the supplied arguments
        String[] name = ao.getNames();
        for(int i = 0; i < ao.getCount(); i ++ ) {
          // get unnamed slot
          ao2 = ao.getAbsObject( UNNAMED+i );
          try{
            sum += ((AbsPrimitive)ao2).getLong();
          }catch(java.lang.ClassCastException cce) {
            str = "(error (argument_format "+ opName+ " at " + i
               +"))";
            SL0Helper.fillAsNotUnderstood( acl, resp, str );
            send(resp);
            return;
          }
        }
        // Create the INFORM message to return the result
        resp = SL0Helper.createInformResult( acl, "" + sum );
      }else{
        str = "(unknown (service " + opName + " ))";
        SL0Helper.fillAsNotUnderstood( acl, resp, str );
      }
    }
    // Send the result to the invoker
    send(resp);
  }
  ...
```

The agent must now register the service in the JADE DF with the type slot of the registration request set to 'web-service'. An alternative to using the type slot is to specify the service type as a property (as shown).

```
...
private DFAgentDescription dfad = new DFAgentDescription();

protected void setup() {
...
  // Register the agent with the DF

  // prepare the registration message
  ACLMessage msg = new ACLMessage( ACLMessage.REQUEST );
  AID  dfAID = new AID( "df", AID.ISLOCALNAME );
  msg.addReceiver( dfAID );
  msg.setSender( this.getAID());
  msg.setConversationId( "conv_" + convId ++ );
  msg.setLanguage(FIPANames.ContentLanguage.FIPA_SL0);
  msg.setOntology(FIPAManagementVocabulary.NAME);

   // prepare the DFAgentDescription
  dfad.setName( this.getAID());
  dfad.addLanguages( FIPANames.ContentLanguage.FIPA_SL0 );
  dfad.addProtocols( FIPANames.InteractionProtocol.FIPA_REQUEST );
  ServiceDescription sd;
  sd = new ServiceDescription();
  sd.setName( SERVICE_PLUS ); // here is the service name
  sd.addLanguages( FIPANames.ContentLanguage.FIPA_SL0 );
  sd.addProtocols( FIPANames.InteractionProtocol.FIPA_REQUEST );
  sd.setType("web-service");

  // A property slot can also be used to express the service
    'type'.
  Property p = new Property("type","(set web-service)");
  sd.addProperties( p );
  dfad.addServices(sd);

  //Set the register argument
  Register reg = new Register();
  reg.setDescription(dfad);

  // Create the registration action
  Action action = new Action( this.getAID(), reg );

  // Send the request for registration to the DF
  try {
    getContentManager().registerLanguage( codec );
    getContentMan-
       ager().registerOntology(FIPAManagementOntology.getInstance());
    getContentManager().fillContent(msg, action);
    send(msg);
  }catch (Exception e) {
    // something is wrong
    e.printStackTrace();
  }
 ...
```

This example is available in `com.whitestein.wsig.test.TestAgentServer` and is precompiled with the WSIG by default. The wsig\misc\run_TestAgentServer script runs the example. Ensure that JADE and the WSIG are already running before initiating the example.

We now look at how to construct a Web service that will invoke the 'plus' agent service. The first step is that the client must perform a search for an operation matching its requirements on the WSIG UDDI repository. The correct access point, operation name and WSDL structure is returned by the UDDI from which the client constructs a SOAP request message to invoke the agent service. This SOAP message is then sent to the access point published by the WSIG Axis server. If the invocation succeeds, as answer will be received by the client from WSIG containing the result from the plus operation.

The codified WS client is available in `com.whitestein.wsig.test.TestSOAPClient`. The wsig\misc\run_TestSOAPClient script runs the example.

```
...
// Some initializations
private final static String fipaServiceName = "plus";
private static Logger log =
  Logger.getLogger( TestSOAPClient.class.getName());
private UDDIProxy uddiProxy;

// Set up the uddi4j UDDI repository
private void setupUDDI4j() {
  Configuration c = Configuration.getInstance();
  synchronized ( c ) {
    System.setProperty( Configuration.KEY_UDDI4J_LOG_ENABLED,
      c.getUDDI4jLogEnabled());
    System.setProperty( Configuration.KEY_UDDI4J_TRANSPORT_CLASS,
      c.getUDDI4jTransportClass());
    uddiProxy = new UDDIProxy();

    // Select the desired UDDI server node
    try {
      // contact a back end UDDI repository
      uddiProxy.setInquiryURL(c.getQueryManagerURL());
      uddiProxy.setPublishURL(c.getLifeCycleManagerURL());
    }catch( Exception e ) {
      log.error(e);
    }
  }
}

//Discover available services
private ServiceList findServices() {
  ServiceList sl = new ServiceList(); // default is an empty list
  try {
    String businessKey = "";
    Vector names = new Vector(1);
    names.add( new Name("%WSIG%") );  // substring is WSIG

    CategoryBag cb = new CategoryBag();
    KeyedReference kr = new KeyedReference();
    kr.setTModelKey("uuid:A035A07C-F362-44dd-8F95-E2B134BF43B4");
```

```
        kr.setKeyName("fipaServiceName");
        kr.setKeyValue( fipaServiceName );
        cb.add( kr );
        TModelBag tmb = new TModelBag();
        FindQualifiers fq = new FindQualifiers();

        // Construct request
        sl = uddiProxy.find_service(
            businessKey,
            names,
            cb,
            tmb,
            fq,
            10 );
    } catch ( UDDIException ue ) {
        log.debug( ue );
    } catch ( TransportException te ) {
        log.debug( te );
    }
    return sl;
}

// Write the list of discovered services into the log
private void writeToLog( ServiceList list ) {
    ServiceInfo info;
    ServiceInfos infos = list.getServiceInfos();
    String s;
    int k;
    for ( k = 0; k < infos.size(); k ++ ) {
        info = infos.get( k );
        s = info.getDefaultNameString();
        log.debug(" a service found: " + s );
    }
}

// Test to call the UDDI discovery process
private void test(){
    setupUDDI4j();

    // find Services
    ServiceList sList = findServices();
    if ( log.isDebugEnabled() ) {
        writeToLog( sList );
    }
...
```

An AccessPoint, an operation name and a WSDL name space is required to call the operation. UDDI records related to service must be traversed to obtain this information.

```
...
    ServiceInfo info;
    ServiceInfos infos = sList.getServiceInfos();
```

```
// Log if no service is available
if ( infos.size() < 1 ) {
  log.info(" No service is available.");
  return;
}

// Attempt to retrieve service details from the UDDI
info = infos.get( 0 );
ServiceDetail sd = null;
try {
  sd = uddiProxy.get_serviceDetail( info.getServiceKey() );
}catch ( UDDIException ue ) {
  log.debug( ue );
}catch ( TransportException te ) {
  log.debug( te );
}
if ( null == sd ) {
  log.info(" No service detail is available.");
  return;
}

Vector sv = sd.getBusinessServiceVector();
if ( sv.size() < 1 ) {
  log.info(" No service detail is available.");
  return;
}

// Take the first service available
BusinessService bs = (BusinessService) sv.elementAt( 0 );

// Obtain an accessPoint
BindingTemplates bts = bs.getBindingTemplates();
if ( bts.size() < 1 ) {
  log.info(" No bindingTemplate is available. ");
  return;
}
BindingTemplate bt = bts.get(0);
AccessPoint aPoint = bt.getAccessPoint();
URL ap = null;
try {
  log.info(" An accessPoint is " + aPoint.getText()
    + " and type is " + aPoint.getURLType() );
  ap = new URL( aPoint.getText() );
}catch (MalformedURLException mfe) {
  log.error( mfe );
  return;
}

// Get the tModel - only one is expected
TModelInstanceDetails tmids = bt.getTModelInstanceDetails();
if ( tmids.size() < 1 ) {
  log.info(" No TModelInstanceInfo is available. ");
```

```
  return;
}
TModelInstanceInfo tmii = tmids.get(0);
String tmk = tmii.getTModelKey();
TModelDetail tmd = null;
try {
  tmd = uddiProxy.get_tModelDetail( tmk );
}catch ( UDDIException ue ) {
  log.debug( ue );
}catch ( TransportException te ) {
  log.debug( te );
}

if ( null == tmd ) {
  log.info(" No TModelDetail is available.");
  return;
}

Vector tmdv = tmd.getTModelVector();

if ( tmdv.size() < 1 ) {
  log.info(" No TModel is available.");
  return;
}
TModel tm = (TModel) tmdv.get(0);

// Extract the WSDL URL from the tModel - only one is expected
OverviewDoc ovd = tm.getOverviewDoc();
if ( null == ovd ) {
  log.info(" No OverviewDoc is available in TModel.");
  return;
}
String wsdlURL = ovd.getOverviewURLString();
if ( null == wsdlURL ) {
  log.info(" OverviewDoc's URL is null.");
  return;
}
log.info(" TModel refers to wsdl: " + wsdlURL );

// Get an operation for fipaServiceName
CategoryBag cb = bs.getCategoryBag();
KeyedReference kr;
int k;
for ( k = 0; k < cb.size(); k ++ ) {
  kr = cb.get( k );

  if ( fipaServiceName.equalsIgnoreCase( kr.getKeyName() )) {

    // Call the located fipeService
    callOperation( ap, kr.getKeyValue(), wsdlURL );
    return;
  }
```

```
    }
}
...
```

A SOAP request is now constructed in the following `generatePlus()` method and a invocation call onto the accessPoint then made.

```
...
// Calls the operation
// accessPoint is the access point
// opName is the operation name
// wsdlNS is the WSDL name space
private void callOperation( URL accessPoint, String opName, String
    wsdlNS ){
  URL serverURL = accessPoint;
  HttpURLConnection c = null;

  SOAPMessage retSOAP = null;

  // Generate a test message
  String str;
  int[] values = { 3, 5, 7 };
  str = generatePlus( opName, values, wsdlNS );

  SOAPMessage soap;
  soap = new Message(str, false,
                    "application/soap+xml; charset=\"utf-8\"", "" );

  // Record some debug information
  ByteArrayOutputStream baos;
  try {
    baos = new ByteArrayOutputStream();
    soap.writeTo(baos);
    log.info("A SOAP sent: \n   " + baos.toString());
  } catch (SOAPException e) {
    log.error(e);
  } catch (IOException ioe) {
    log.error(ioe);
  }

  try {
    // Send the invocation request
    c = WSEndPoint.sendHTTPRequest( serverURL, soap );

    // Read the invocation response
    retSOAP = WSEndPoint.receiveHTTPResponse( c );

    // Catch and log information if the response is null
    if ( retSOAP != null ) {
      try {
        baos = new ByteArrayOutputStream();
        retSOAP.writeTo(baos);
```

```
        log.info("A SOAP received: \n  " + baos.toString());
      } catch (SOAPException e) {
        log.error(e);
      } catch (IOException ioe) {
        log.error(ioe);
      }
    }else {
      log.info("SOAP received: null.");
    }

    // Release resources
    c.disconnect();

  }catch (SOAPException se) {
    log.error(se);
  }catch (IOException ioe) {
    log.error(ioe);
  }finally{
    if (c != null) {
      c.disconnect();
    }
    isRunning = false;
    //return;
  }
}

// Method to generates a SOAP message.
// op_name is the operation name
// nums is an array of integers as arguments to the plus operation
// wsdlNS is the WSDL namespace
public static String generatePlus(String op_name, int[] nums, String
   wsdlNS) {
  String str =
      "<?xml version=\"1.0\" encoding=\"UTF-8\"?> " +
      "<soapenv:Envelope
        xmlns:soapenv=\"http://schemas.xmlsoap.org/soap/envelope/\"
        xmlns:xsd=\"http://www.w3.org/2001/XMLSchema\"
        xmlns:xsi=\"http://www.w3.org/2001/XMLSchema-instance\">  " +
      "<soapenv:Body>\r\n" +
      "<tns:" + op_name + " xmlns:tns=\"" + wsdlNS + "\" >   ";

  for ( int k = 0; k < nums.length; k ++ ) {
    str += "<tns:BO_Integer>" + nums[k] + "</tns:BO_Integer>\r\n";
  }

  str +=
      "   </tns:" + op_name + ">  " +
      "  </soapenv:Body>   " +
      " </soapenv:Envelope>\r\n";
  return str;
}
...
```

## 10.8 EXAMPLE 2: AGENT SERVICE INVOKES A WEB SERVICE

This second example illustrates WSIG operation when a Web service has been registered with the WSIG UDDI, automatically translated into an ACL service description and registered with the JADE DF and then invoked by a JADE agent.

This example can be run by running the \wsig\misc\run_TestSOAPServer and \wsig\misc \run_TestAgentClient scripts in sequence.

The initial step shown below is the setting up the Web service to receive an incoming invocation request for a simple 'echo' operation which will return an echo string. It is implemented in the `doRequest()` method of `com.whitestein.wsig.test.TestSOAP ServerConnection`.

```
...
// Name the operation
public static final String OP_1 = "echo";
...
private void doRequest(WSMessage wsMsg, OutputStream os)
            throws IOException {

  String answer = "";
  String opName =
         wsMsg.getTheFirstUDDIOperationId().getWSDLOperation();
  if ( OP_1.equals( opName )) {
    answer =
      "<?xml version=\"1.0\" encoding=\"UTF-8\"?> " +
      "<soapenv:Envelope
        xmlns:soapenv=\"http://schemas.xmlsoap.org/soap/envelope/\"
        xmlns:xsd=\"http://www.w3.org/2001/XMLSchema\"
        xmlns:xsi=\"http://www.w3.org/2001/XMLSchema-instance\">  " +
      "<soapenv:Body> " +
      "<tns:String xmlns:tns=\""+ wsdlTargetNamespace +"\" >" +
      "Echo string" +
      "</tns:String>   " +
      "</soapenv:Body>   " +
      "</soapenv:Envelope>\r\n";
  }else {
    answer =
      "<?xml version=\"1.0\" encoding=\"UTF-8\"?> " +
      "<soapenv:Envelope
        xmlns:soapenv=\"http://schemas.xmlsoap.org/soap/envelope/\"
        xmlns:xsd=\"http://www.w3.org/2001/XMLSchema\"
        xmlns:xsi=\"http://www.w3.org/2001/XMLSchema-instance\">  " +
      "<soapenv:Body> " +
      "<soapenv:Fault>" +
      "<soapenv:faultcode>soapenv:Client</soapenv:faultcode>" +
      "<soapenv:faultstring>Unknown
         operation.</soapenv:faultstring>"
      +
      "<soapenv:faultactor></soapenv:faultactor>" +
      "</soapenv:Fault>" +
      "</soapenv:Body>   " +
      "</soapenv:Envelope>\r\n";
  }
```

```
  byte[] content = answer.getBytes("UTF-8");
  Connection.sendBackSOAPContent( content, os );
}
...
```

The Web service must register itself with the WSIG UDDI once it is initiated. The WSIG provides a UDDI registration interface as a valid UDDI repository and all requests for the WSIG are sent directly to the UDDI repository. From these requests relevant information is extracted and used by the WSIG. Code exemplifying the registration phase is available in the `register()` method of `com.whitestein.wsig.test.TestSOAPServer`.

A JADE agent invoking the operation must know the service name 'echo', which can be discovered in the JADE DF. For the time being an agent must know the service name in advance.

```
...
// An agent searches the DF for a suitable service
private void doSearch() {
   DFAgentDescription template = new DFAgentDescription();
   ServiceDescription sd = new ServiceDescription();
   Property p = new Property(
     Configuration.WEB_SERVICE + ".operation",
     wsdlOperation );
   sd.addProperties( p );
   template.addServices( sd );
   try {
     DFAgentDescription[] res = DFService.search( this, template);
...
```

If the discovery is successful the agent sends an invocation request to the service host and a behaviour is added to await the expected response.

```
...
serviceName = findServiceName( res[0] );
if ( null == serviceName ) {
  log.info( "No service is found." );
  doDelete();
  return;
}

// Add a behaviour to await the expected response message
this.addBehaviour( new CyclicBehaviour( this ) {
  public void action() {
    ACLMessage msg = myAgent.receive();
    if ( msg != null ) {
      processResponse( msg );
    }else {
      block();
    }
  }
});

final ACLMessage m = createRequest( aid, serviceName );
this.addBehaviour( new OneShotBehaviour( this ) {
  public void action() {
```

```
    send( m );
  }
...
```

The request for the operation is created in the ACL-SL0 format. The message content is an action where the Web service operation forms the action body.

```
...
private synchronized ACLMessage createRequest(AID aid,
                                                String service ) {
  ACLMessage msg = new ACLMessage( ACLMessage.REQUEST );
  msg.addReceiver( aid );
  msg.setSender( this.getAID());
  msg.setProtocol(FIPANames.InteractionProtocol.FIPA_REQUEST);
  msg.setConversationId( "conv_" + convId ++ );
  msg.setLanguage(FIPANames.ContentLanguage.FIPA_SL0);
  msg.setOntology("AnOntology");
  msg.setContent(
    "((action\n" +
    "    (agent-identifier\n" +
    "      :name "+Configuration.getInstance().getGatewayAID()+"
      )\n"+
    "    (" + service + "\n" + "     ) ))");
  return msg;
}
...
```

# 11

# Agent-Society Configuration Manager and Launcher

The Agent-Society Configuration Manager and Launcher (ASCML) is a tool to create, configure, launch and monitor agent societies running on several hosts across distributed environments. In this context, an agent society is defined as a set of collaborating agents forming an application or a service. For example, the buyer and seller agents of the book-trading application, introduced in Chapter 4, form a society.

Starting an application consisting of multiple agents usually requires the manual configuration and launch of each agent in turn on their assigned platforms. This can be difficult, complicated and error-prone, not only when agents are distributed over several hosts, but also when running in a single location. It is also more demanding if the agents are in some way dependent on one another. The ASCML assists in creating different application configurations and manages the launching and monitoring of an application. Additionally, the ASCML supervises the distribution of agents over the network and resolves possible dependencies among agents, societies and services. While the application is running, the ASCML monitors the run-time states of the agents to check whether an application is operating correctly. If any functional dependencies or constraints are violated, the ASCML may actively attempt to restart non-functional components.

In order to configure and launch an application, both society and agent descriptions must first be created. These descriptions are based on XML meta-models which facilitate the reuse of descriptions later on. The ASCML's GUI supports the creation or modification of society and agent descriptions, but these can, if so desired, be written directly in XML. The GUI also allows control over starting, stopping and monitoring societies as well as managing a collection of active projects.

## 11.1 BASIC TERMS AND CONCEPTS

As mentioned, the ASCML is a deployment tool for multi-agent applications. *Deployment*, as defined by the Object Management Group (OMG), is the process between acquisition and execution of software. Different phases of the deployment process can be identified. The first phase is to gather all that the application needs (such as the software for the agent platform itself, the agent application and any additional software packages) and then set application-specific properties to produce a specific application configuration. Next the target environment is taken into account and a plan for distributing the software is developed. Using this plan, the placement of the software components on the machines hosting the application takes place and finally the application is launched.

*Developing Multi-Agent Systems with JADE* Fabio Bellifemine, Giovanni Caire, Dominic Greenwood

The ASCML defines agent types, agent instances, society types and society instances. Societies are defined simply as a group of agents and subsocieties as recorded in a societies description file. This description is processed by an ASCML resulting in an internal hierarchical model unique to the ASCML. On top of this hierarchical model there is only one ASCML responsible for managing the whole agent society. If a society contains remote subsocieties these are managed by dependent ASCMLs and represented in the hierarchical model accordingly.

Figure 11.1 depicts two agent platforms (called A and B). Both platforms host ASCML agents, which are able to communicate with one another. The ASCML running on platform A defines a society (called Society I), which consists of two agents (Agent 1 and Agent 2), that are started locally, and a remote subsociety (Society II), which is launched on platform B. Only the ASCML on platform B knows that this society consists of Agent 3 and Agent 4. Once Society I is started, the ASCML on platform A requests its local platform to start Agent 1 as well as Agent 2 and also sends a request message to the other ASCML to initiate the start-up of Society II. When all four agents are running their status is continuously monitored by their designated ASCMLs. When the status of an agent changes (e.g. from running to erroneous) the responsible ASCML informs other interested ASCMLs and finally reports the change to the user.

When discussing a society, we always distinguish between a society type and a society instance. A society type may be thought of as an application and a society instance as a special configuration or setting of this application. A society type specifies the static properties, such as the application's name and a human-readable description. Additionally it contains a set of society instances. A society instance specifies details such as the number of agents and subsocieties that are to be launched, the ASCML agent that is responsible for launching them on remote machines and whether there are dependencies between agents, societies or services that dictate a particular start-up sequence. Having a look at the book-trading example, we can observe that it is a society type and that one or more society instances may be declared by specifying how many agents of a specific agent type belong to the society instance. Two society instances may be created, for example, the first instance defining – for testing purposes – only one buyer and one seller agent, and the other instance defines a real-life scenario with 1000 buyer and 100 seller agents.

The difference between agent type and agent instance is similar to the difference between society type and society instance. An agent type is a description of an agent that could be used by referencing the agent type's name. It defines the type name, the main Java class name, the target agent platform which the agent is designed to run on (currently JADE and Jadex are supported), and a set of parameters that are passed to the agent's main class at start-up. Roughly speaking, an agent type

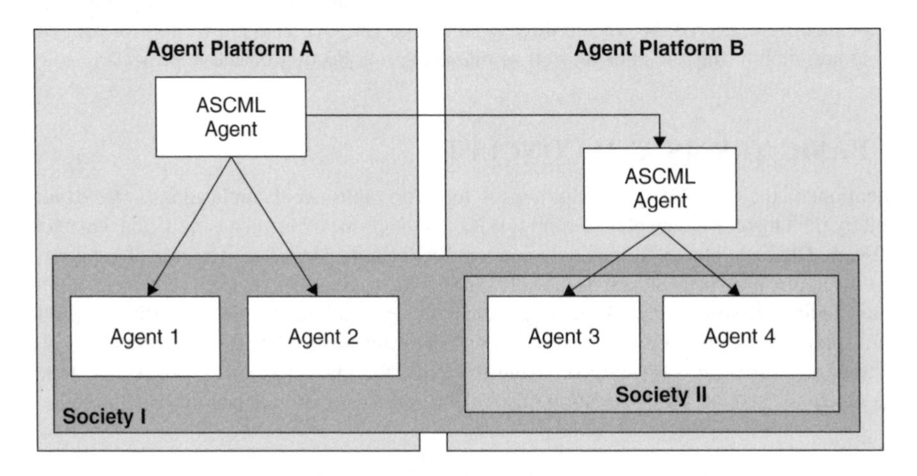


**Figure 11.1** Example agent system with ASCML agents

encapsulates the knowledge where an agent's class file is located, which parameters are supported by the agent stored in that class file, the valid values for each parameter, and finally, a human-readable description of all this information. This enables and encourages a developer to provide all his additional knowledge about the agents to the agents' user. Furthermore, it saves the user the necessity of discovering it through trial and error, or in the worst case, to read the agents' source code.

An agent instance is a special setting for an agent type which adds some additional information to the type definition and is therefore more specific. Additional parameters may be provided, default parameter values overwritten which were predefined by the agent type, dependencies specified between the agent instance and other agents, societies or services and much more. With respect to a book-trading example consisting of one seller and two buyer agents, although two agent types (buyer and seller) are involved, three instances are actually present.

An agent instance in this context does not necessarily correspond to a process (or thread) running on an agent platform. It is in fact simply a name for the model. However, to prevent confusion the term 'run-time instances' is used when referring to agent processes running on an agent platform. Thus an agent instance can be called 10 times resulting in 10 run-time instances, which are named by the ASCML according to a user-specified naming scheme.

The next notion is that of dependency, a key feature of ASCML that differentiates it from other deployment techniques. The ASCML distinguishes six different types of dependencies: agent type, agent instance, society type, society instance, delay and service dependencies. A dependency expresses a relationship between elements. If one element declares itself to be dependent on another the declaring element cannot be started until the referenced element is available. Dependencies may also be used to describe whether an application is functional or not. Once a dependency does not hold, the ASCML can be told to actively engage in fulfilling the dependency by starting the dependent part. In this manner monitoring and automatic (re)start of a society's non-functional parts can be realized.

## 11.2 BOOK-TRADING EXAMPLE

This section provides a hands-on tutorial on how to use the ASCML using the well-known book-trading agents discussed throughout this book. Firstly, a copy of the run-time version of the ASCML must be obtained. As the ASCML is a work in progress the development snapshots contain the newest fixes and improvements. Therefore, it is recommended to download the Jar file from

**http://intensivstation.informatik.rwth-aachen.de/ASCML/snapshot/ASCML-svn_snapshot-bin.jar**

To fully use the capabilities of the ASCML an instance of it is required on every platform that will participate in the scenario. In this example, we will only use one platform on a local machine. Starting the ASCML is as simple as starting any JADE agent; just ensure that the ASCML Jar file (e.g. ASCML-svn_snapshot-bin.jar) as well as [JiBX] and JADE are in the class path

```
java jade.Boot ascml;jade.tools.ascml.ASCML
```

or

```
java -cp jade.jar;http.jar;jadeTools.jar;iiop.jar;jibx-run.jar;
ASCML-bin-0.9.jar jade.Boot ascml;jade.tools.ascml.ASCML
```

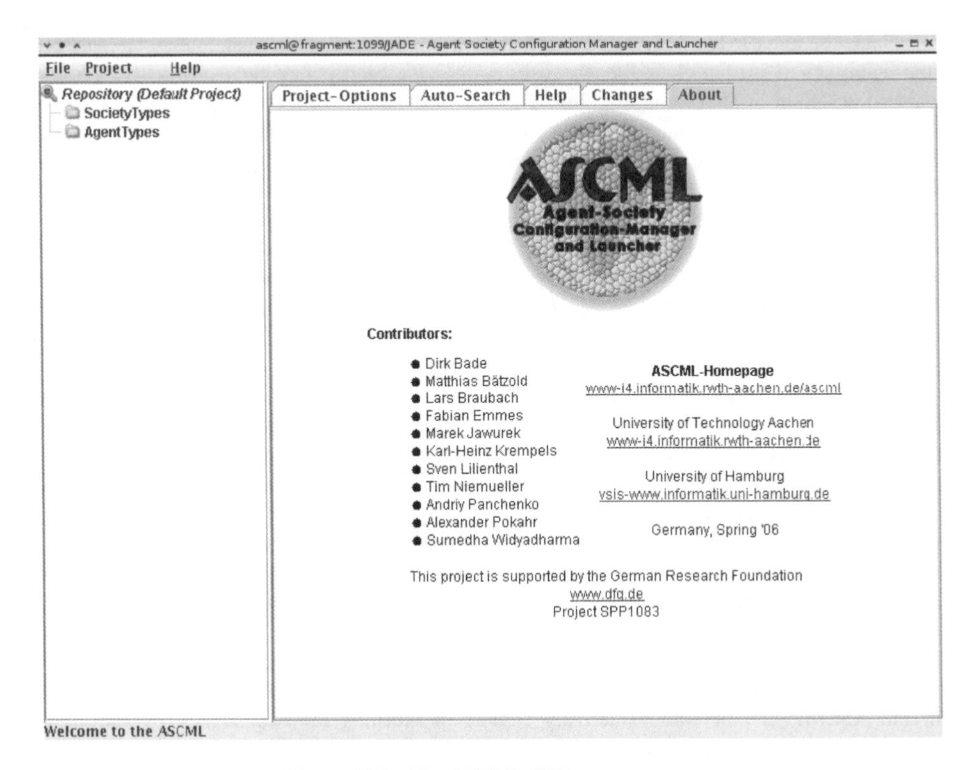


**Figure 11.2** The ASCML GUI start-up screen

If the ASCML cannot find a valid configuration file it will ask for a filename to be specified or a new file created in the current working directory. This file is called *ascml.repository.properties.xml*. When ASCML is terminated the current project will be saved into this file.

Once a configuration file has been loaded or created the GUI appears, as shown in Figure 11.2. On the left-hand side is a tree-structured repository containing every model which has been loaded by the ASCML. On the right-hand side context-specific information is provided according to the model selected in the repository tree.

The first thing to do is to create models for the two types of agents to be used in the example. This is achieved by selecting 'Create new **Agent Type**' from the 'Project' menu which will cause a new agent type to be inserted and an option panel displayed presenting an empty template for a new agent. This template is shown in Figure 11.3 with attribute fields completed for a book buyer agent.

These fields are defined as follows:

- Source-Path: the destination directory for the description file, preferably with the classes or the jar.
- Source-File: the filename for the description. This will be auto-generated from the type name.
- Type-Name: a descriptive name for this agent type. This does not need to be equal to the class name.
- Type-Package: the Java package name the agent class resides in.
- Type-Class: the name of the agent's Java class excluding the package name.
- Platform: the specific agent platform. Currently only JADE is available.
- Description: a description of the agent. Its parameters can have separate descriptions.

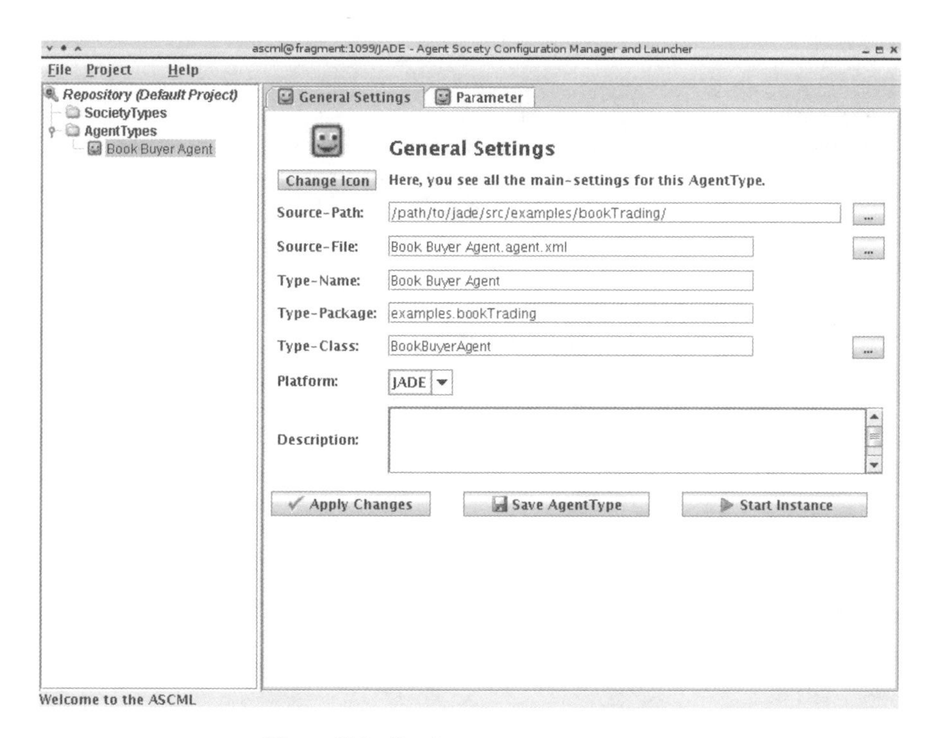


**Figure 11.3** Panel to create a new agent type

Because the buyer agent in the example requires a parameter to know which book he is supposed to buy, a new parameter must be created by selecting the 'Parameter' tab. This particular type of agent requires an unnamed parameter, thus the parameter name can be left empty with only a new value added by entering a book name in the 'Add new Value' field. This has now specified everything the buyer agent requires.

The same procedure is followed for the seller agent with the exception that the seller does not require any parameters.

To test the application, an instance of the seller agent type is initiated first by selecting the seller agent in the repository tree and pressing the 'Start Instance' button. The following dialogue will then initiate a run-time instance of the seller agent once a descriptive name is provided. The seller agent is now running and will ask for the name and price of the book to be sold. For the purposes of this example, this name must match the name of the book the buyer will look for.

The buyer agent is now started. In fact several buyer agents can be started, each looking for different books. The GUI will now show the various buyers contacting the seller and if possible, buying their desired book. Once a successful acquisition is made the buyer transitions into a stopped state. If the agents function correctly the next step is to create an agent society.

This procedure is similar to creating an agent type and initiated by selecting 'Create new **Society Type**' from the 'Project' menu. The template for an agent society is shown in Figure 11.4 with the attribute fields completed for a new society.

Once the society has been saved, the ASCML will check the model for any inconsistencies or missing elements. If the society is incomplete or erroneous an error icon is displayed in the repository tree instead of the society's normal icon. At least one society instance for our society type must be specified which is achieved by pressing the 'Create **Society Instance**' button and entering a name and description for the new instance (see Figure 11.5). All other parameters can be left as they are.

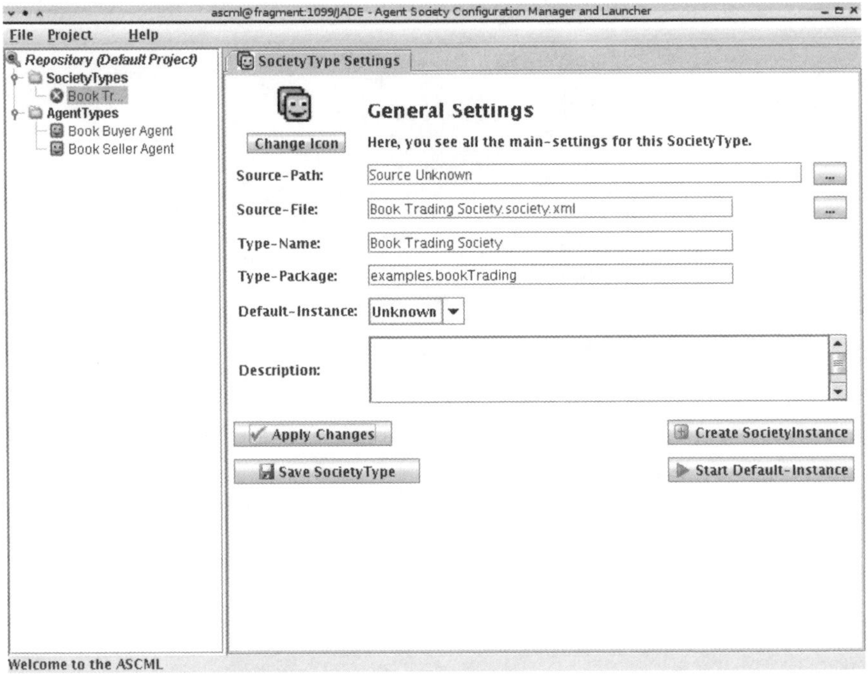

**Figure 11.4** Panel to create a new agent society type

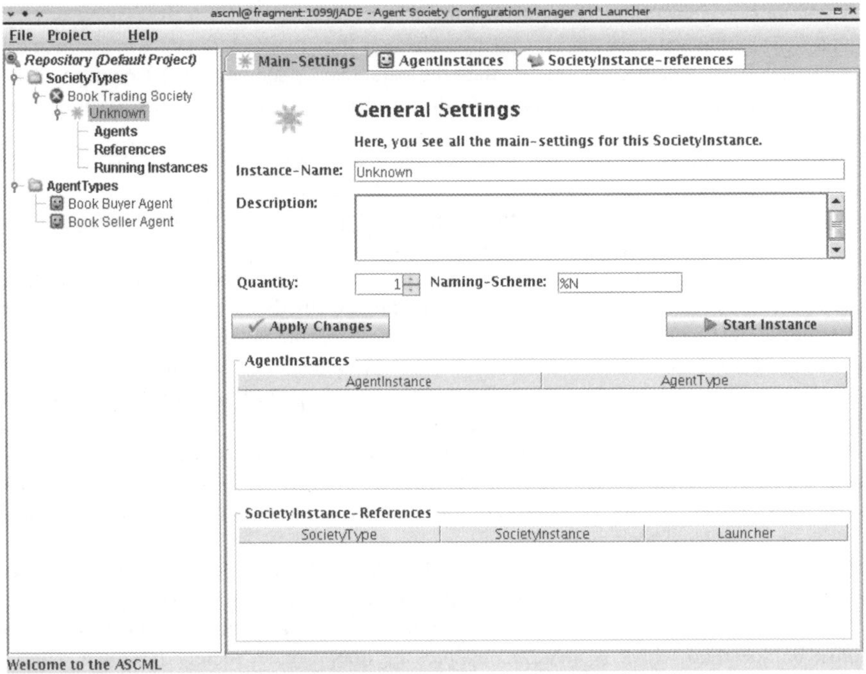

**Figure 11.5** Creating a society instance

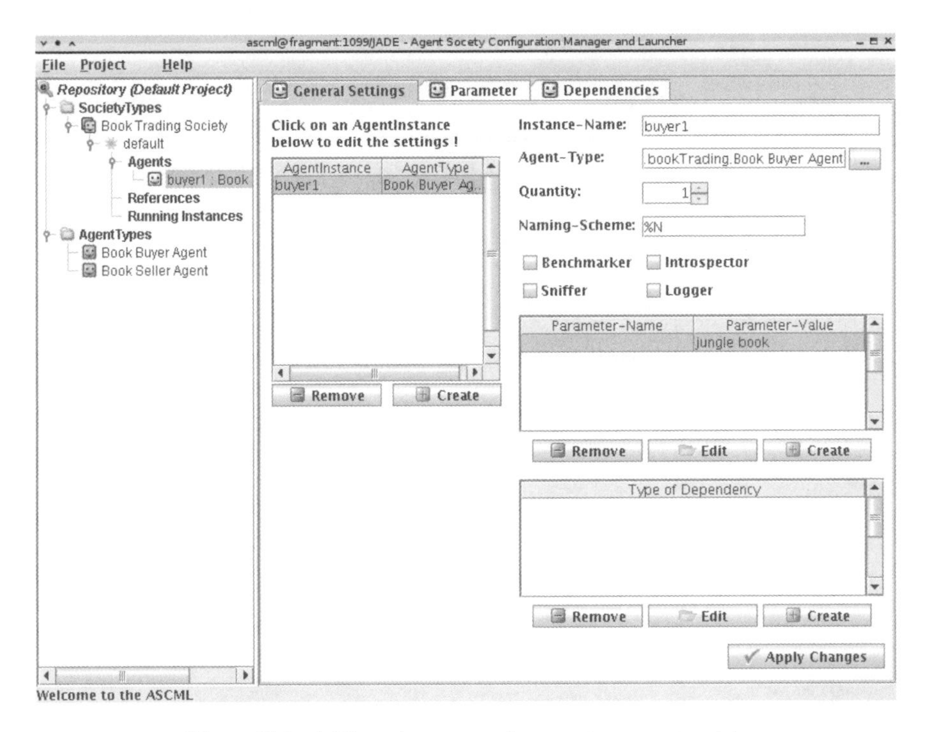


**Figure 11.6** Adding a buyer agent instance to an agent society

Selecting the 'Agent Instances' tab will now allow the addition of some agents to this society instance. The addition of a buyer agent is shown in Figure 11.6.

Next we need to create a parameter for this agent. Selecting 'Create' below the parameter field allows a new parameter to be created for this agent instance. As shown in Figure 11.6, the name field should be left blank and the value field set to the name of the book to buy. Use the 'Apply Changes' button to save both the parameter data and the parent society instance.

Selecting the '**Agent Instances**' tab of the society instance, shown in Figure 11.5, we can now also create a seller agent as a new agent instance. This should be done using the same procedure as for the buyer agent just described. The seller agent should be named 'seller1'.

We now have a functional instance of an agent society which can be launched by clicking on the society instance in the repository and selecting 'Start Instance'. It should be provided with a descriptive name, as asked for, and then the bookseller should be supplied with the book the buyer is expected to buy. All running agents can be listed by expanding the 'Running Instances' branch in the repository tree. As soon as the transaction has finished, stop the society instance by right-clicking its icon in 'Running Instances' and selecting 'Stop Instance'.

To create a more complex book-trading society with several buyers, we can create a separate dedicated buyer society in order to illustrate the usage of subsocieties. As shown in Figure 11.7, a new society should be created called 'Buyer Society' within which a new society instance can be created called, in this example, '3 buyers'. We can now add three buyer agents to the society, giving each a different name and book to buy.

A reference to this newly created society can now be created from within the 'Book Trading Society'. To achieve this a new society instance for the 'Book Trading Society' should be created and called, for example, 'seller + 3 buyers'. The agent instance for the seller agent should be added as described previously, after which the '**Society Instance**-references' tab should be selected to fill out the given empty template as shown in Figure 11.8. Here, Reference-Name is a descriptive

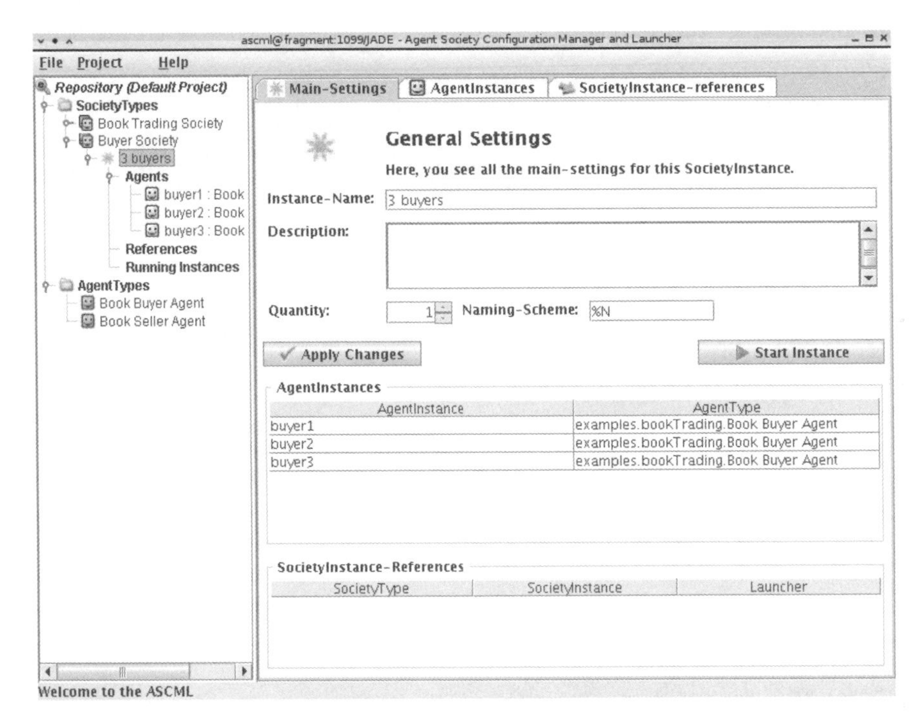


**Figure 11.7** Creating an agent subsociety

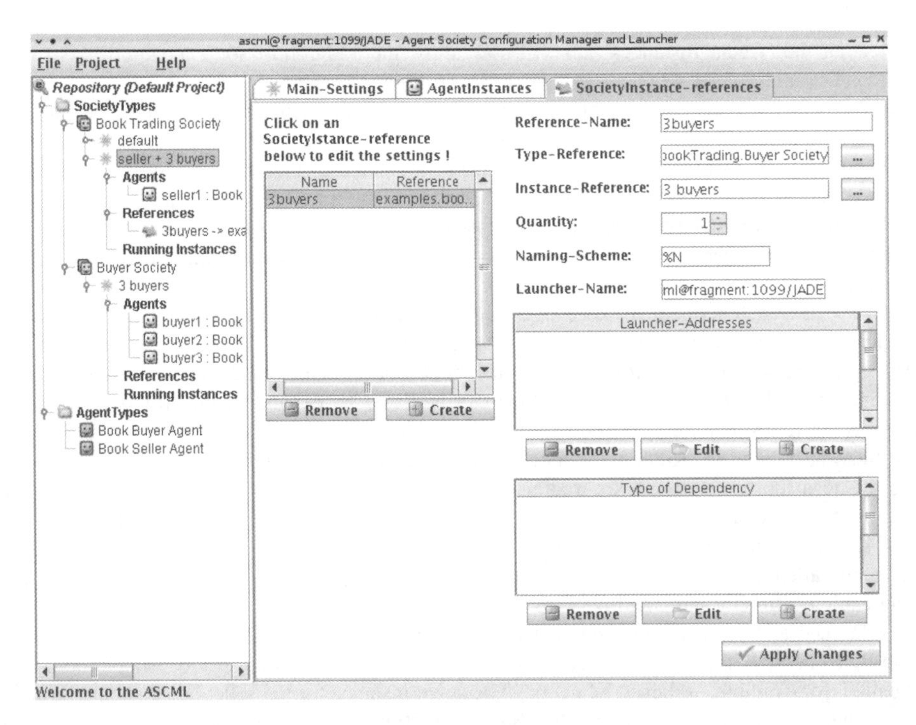


**Figure 11.8** Creating a reference to an agent society

name for the particular instance. Type-Reference selects which type of society to reference. Instance-Reference selects which instance of the type to use.

When this society is initiated, the contained seller agent will be started as well as the referenced subsociety, which contains the three buyer agents. Single agents, subsocieties or the whole society can be stopped by clicking on the respective leaves in the 'Running Instances' branch.

To learn more about how to create dependencies between elements, which is out of the scope of this chapter, please refer to the online ASCML tutorial on the third-party software area of the JADE website.

## 11.3 DISTRIBUTED DEPLOYMENT

To illustrate distributed deployment of the ASCML we will employ a simpler scenario than the book-trading system – the Ping example bundled with the ASCML download. This example contains two agents named Ping Agent and Pong Agent that simply bounce data to one another.

Two separate JADE platforms are required called, for the purposes of this example, platform1 and platform2. This can of course be altered if necessary to match an existing platform configuration. The easiest way to set this up is to have two platforms on the same physical machine on different ports. Make sure the jar containing the agents is in the class path.

The following steps need to be performed on both platforms. As shown in Figure 11.9, the 'Project-Options' tab should be selected and the examples.jar file added to the search path for agent/society descriptions.

Next, as shown in Figure 11.10, the 'Rebuild Index' button on the 'Auto-Search' tab should be used to allow the ASCML to search for agent and society descriptions. It will locate several within the examples.jar, from which the **Ping Agent** and **Pong Agent** (*PingAgent.agent.xml* and *PongAgent.agent.xml*) descriptions, and the society description (*PingPong.society.xml*) should be selected.

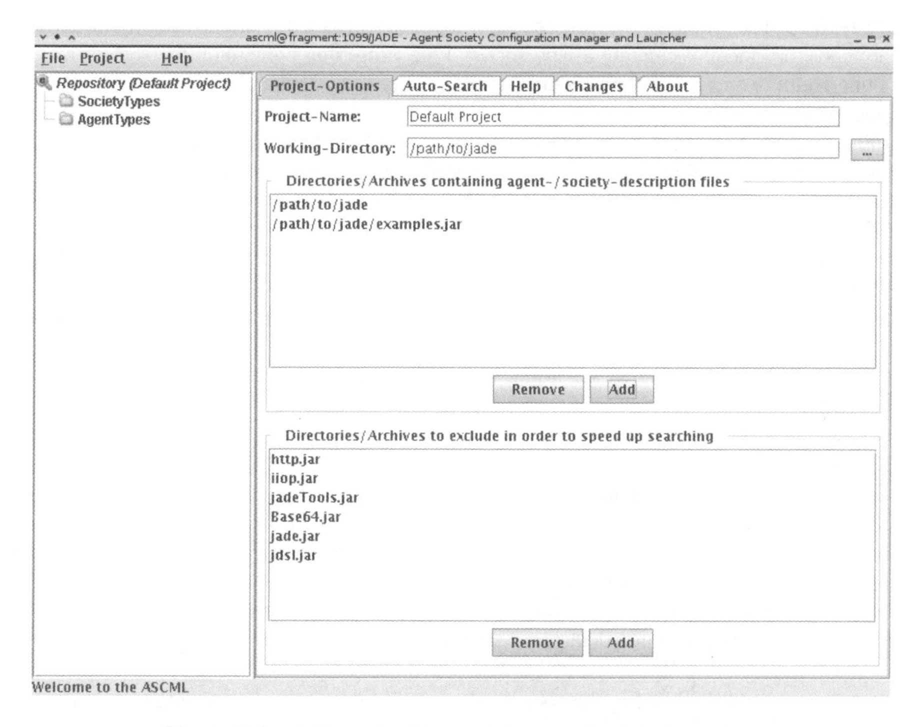


**Figure 11.9** Adding a Jar file containing agent/society descriptions

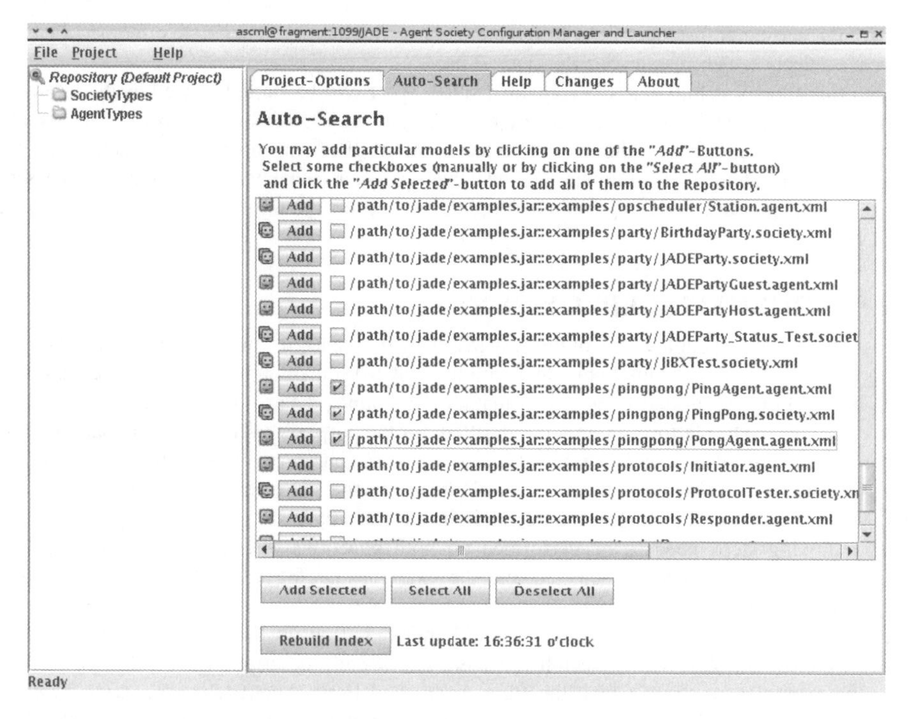


**Figure 11.10** Selecting agent and society models from a Jar file

Once each required model has been checked, the 'Add Selected' button loads the descriptions into the repository thereby creating a new society type **PingPong** containing the society instances **Ping Pong Remote** and **Pong Remote** among others. These are the society instances used for this example.

Turning to platform1, the initiating platform, the 'Society Instance-reference' of the **Ping Pong Remote** society must be altered to start the referenced instance on the correct remote machine. This is shown in Figure 11.11, where the *Launcher-Name* should be set to the address of the remote ASCML agent (e.g. `ascml@platform2:1099/JADE`). The *Launcher-Addresses* are the transport addresses used to contact the remote ASCML and should be set to e.g. `http://platform2:7778/acc`.

Now for platform2, the remote platform, the parameters of the **Pong Agent** in the **Pong Remote** society instance must be altered to allow it to locate the **Ping Agent**. This is shown in Figure 11.12. Platform names should be adapted according to system set-up with names that are resolvable from both computers respectively.

The following parameters must be set:

- *df* is the DF Agent that the Pong Agent will use to locate a Ping Agent. This will typically be of the form `df@platform1:1099/JADE`.
- *transport* is which transport to use to contact the DF Agent and Ping Agent. This will typically be of the form `http://platform1:7778/acc` if HTTP is used.
- *delay* is the delay milliseconds between replies. If blank a default value is selected.
- *verbose* turns on/off additional debugging output.

These changes should be saved to the society file to avoid the need to change them again.

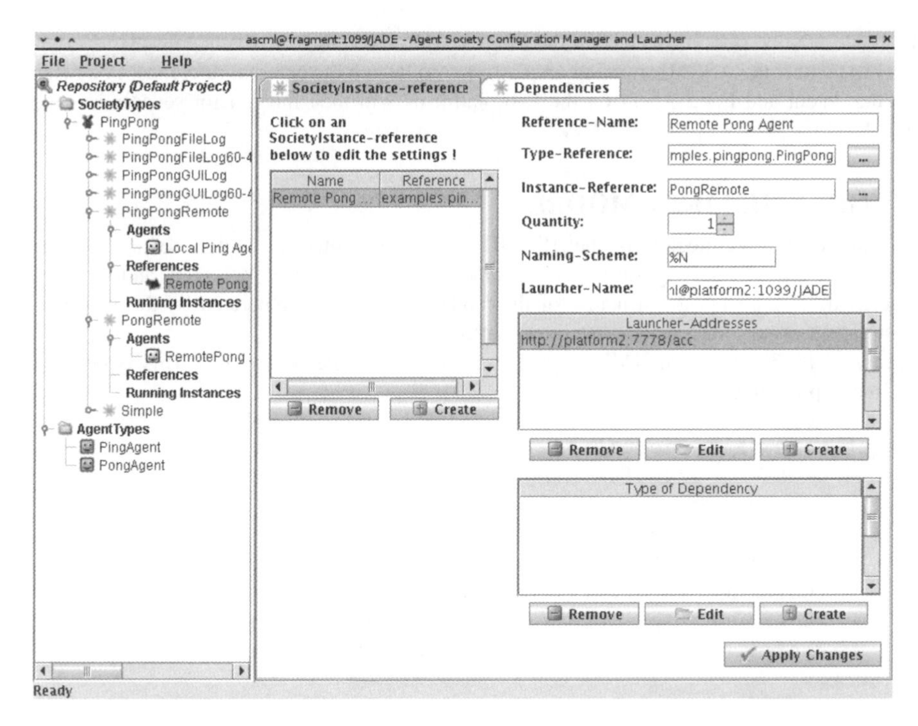


**Figure 11.11** Initiating a SocietyInstance-reference for the Remote Pong example

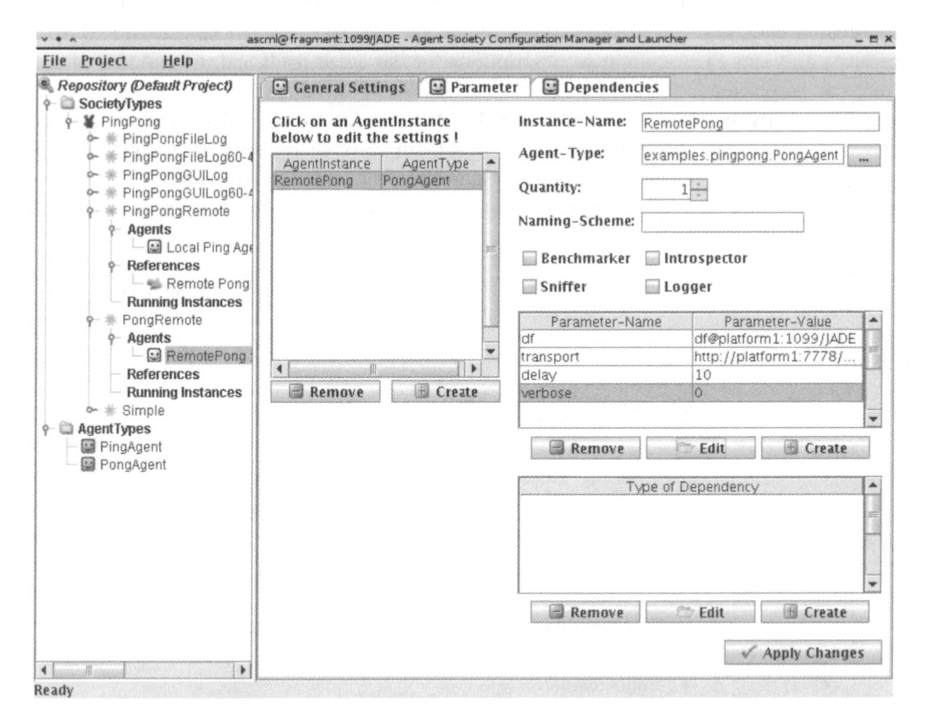


**Figure 11.12** The general settings for the RemotePong Agent

It should now be possible to launch the Ping Pong Remote society instance on platform1 which will try to contact the ASCML running on platform2 to launch the external reference: Pong Remote. The Ping Agent and Pong Agent on the two platforms will then interact for several seconds and terminate.

## 11.4 THE XML META-MODEL

Before the internal architecture of the ASCML agent is presented in the next section, the internal structure of the model description files (**.agent.xml* and **.society.xml*) for agent and society types is discussed. This covers the structure of the model files that are generated by the ASCML-GUI once an agent or society type is created and saved.

Figure 11.13 illustrates an XML snippet from an agent type description. This is described in XML markup as follows:

```
<agent name="PingAgent"
       package="examples.pingpong"
       type="jade"
       class="PingAgent"
       description="This Ping-agent sends ping-messages">
       <parameters> [...] </parameters>
       <agentdescriptions> [...] </agentdescriptions>
       <servicedescriptions> [...] </servicedescriptions>
</agent>
```

Figure 11.13 and the code snippet above depict the structure of an agent type specification. The root agent tag captures important properties of an agent such as the agent type's name, the agent's implementation class and the platform type. Within the `servicedescriptions` and `agent-descriptions` tags FIPA-compliant descriptions for the agent itself and the services offered by the agent can be declared.

The `parameter` tag encloses single-valued parameters and multi-valued parameter sets that are passed as *key=value*-arguments to the agent's main class at the time of creation. Parameters specified by the agent type can be declared as mandatory or optional. All parameters specified for an agent type are inherited by corresponding agent instances. The agent instances have to provide

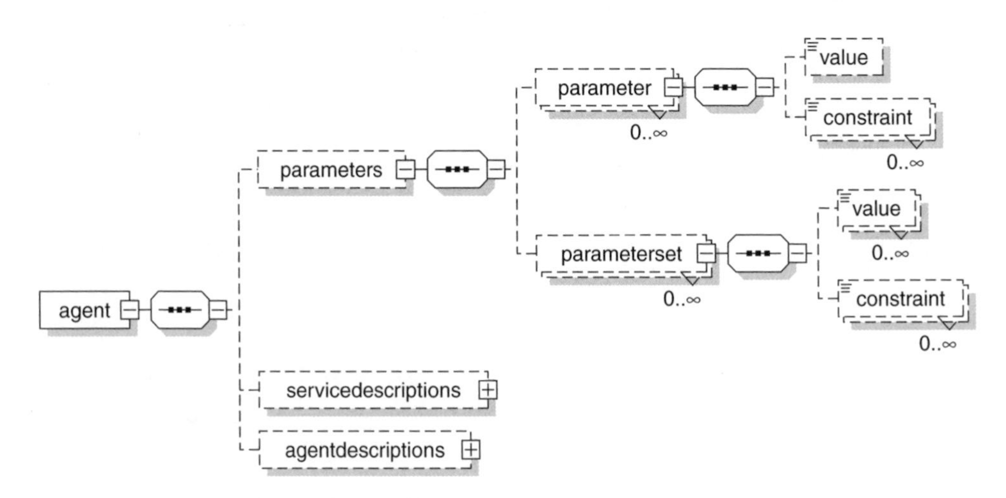


**Figure 11.13** An XML snippet from an agent type description

parameter values for all mandatory parameters for which the agent type provides no values. They may also overwrite the predefined values of the agent type with other values.

Below is part of an example XML description for the parameters of an agent type and an agent instance.

```
<!-- This is part of a definition for an agent type -->
<agent [...]>
   <parameters>
      <parameter name="pingCount" optional="false"
         description="Amount of pings to send"/>
      <parameter name="delay" optional="true"
                 description="Delay in msec after each ping">
      <parameterset name="receiver" optional="false"
                 description="One or more agentIDs">
           <value> ASCML@192.168.0.1:1099/JADE </value>
           <value> ASCML@134.100.11.68:1099/JADE </value>
      </parameterset>
   </parameters>
</agent>

<!-- This is an agent instance definition as part
                 of a society type definition -->
<agentinstance name="PingPing" type="examples.pingpong.PingAgent">
    <parameter name="pingCount">
        <value> 10 </value>
    </parameter>
</agentinstance>
```

When the parameters are passed to the agent's main class only the two mandatory parameters `pingCount` and `receiver` are passed as arguments, the optional parameter `delay` is omitted, because no value has been supplied by the agent instance.

In addition to the agent type meta-models we can also define a meta-model for society types. To facilitate reusability, the specifications for society types are stored separated from the agent type specification in files with the suffix **.society.xml*. Figure 11.14 depicts the XML root-element of a society type followed by a set of child tags. The imports, agenttypes and societytypes tags specify lists of referenced elements but may be omitted since they are optional and only used for model

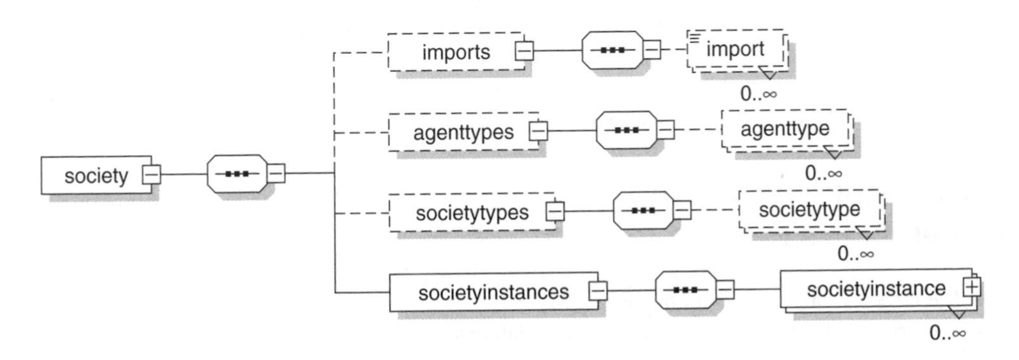


**Figure 11.14** XML root-element of a society type with child tags

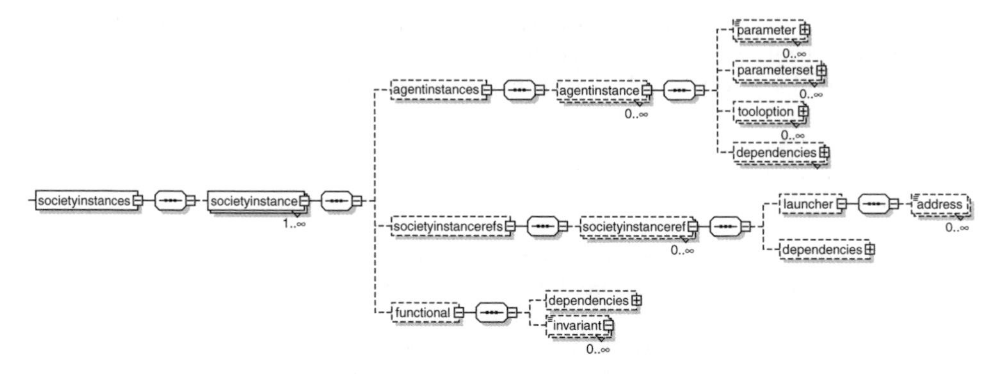


**Figure 11.15** The elements constituting a society instance

checking purposes. The societyinstances tag is the only mandatory tag within which one or more society instances representing different application settings may be listed.

The elements constituting a society instance are shown in Figure 11.15. A set of agent instances can be specified that will be created on the local platform once the society instance is initiated. Every agent instance can be provided, as mentioned above, with parameters and parameter sets. In addition, a set of platform-dependent tool options may be specified that are used to start special tool agents (e.g. sniffer or logger) together with the agent instances.

Besides agent instances, a society instance can also contain an arbitrary number of subsocieties that themselves can contain further subsocieties. This allows a recursive application definition and facilitates the creation of distributed multi-agent systems. Each referenced subsociety instance refers to a concrete society instance, which itself belongs to a declared society specification. For the purpose of starting a remote society, a launcher identifier can be declared which designates the remote ASCML agent responsible for starting the corresponding remote society instance.

It can be declared when a society instance is functional by specifying the set of dependencies and invariants that must be fulfilled. The ASCML supervises these dependencies and invariants during run-time and either reports any changes to the user, or in the case of the dependencies being marked as active, engages autonomously in restarting the non-functional parts.

The last type of model to illustrate is the meta-model for dependencies. As noted from the society instance schema in Figure 11.15, dependencies can be defined for agent instances, society instance-references and to describe functional state. Figure 11.16 shows the different types of these dependencies.

An agent type dependency can be used to wait for an arbitrary number of agents of a specified type to be running, while an agent instance dependency exactly refers to one designated agent, identified by its unique name. Both kinds of dependencies also exist for the society element. The delay dependency is used to wait a specified number of milliseconds before continuing. The last kind of dependency is the most abstract one and allows the definition of indirect relationships between elements as an element depends on a service to be available.

## 11.5 INSIDE THE ASCML

ASCML consists of four primary parts:

- The ***GUI*** for interacting with the user.
- The ***Repository*** which stores all information relating to available and running agents and societies.
- The ***Launcher*** which starts and stops local agents and societies and communicates with other ASCMLs.

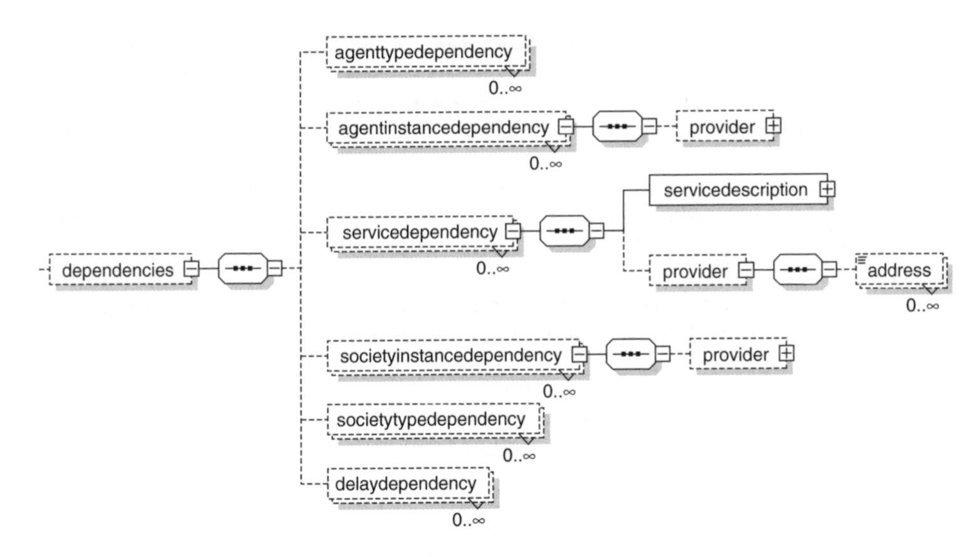


**Figure 11.16** Types of dependencies

- The ***DependencyManager*** which resolves all dependencies at start-up and monitors the functional state of all launched societies.

When the start button is pressed to launch a society or an agent, after asking the user for a name to identify the society or agent, the GUI tells the repository to launch the selected model, providing the name and all the attributes the user filled in. The repository now creates one or more so-called `RunnableInstance` models. These models can be seen as a copy of an agent instance or society instance provided with a unique name and they represent the run-time instances of the agents and societies. The repository will then forward the request to the launcher through the `LauncherInterface`, providing the newly created `RunnableInstance` models. The `LauncherInterface` subdivides these models into all included societies and then gives the control to the `DependencyManager`.

The `DependencyManager` keeps track of all started societies and agents through its library, where it stores the name and types of all running models. When starting a society, it first adds the society to the library and then tries to start it. All models are started through the `DependencyS-tateController`, which tries to resolve active dependencies and monitors passive ones. After all dependencies are resolved or if there have not been any, the models' status is set to Starting. The `RunnableStarter`, created by the `DependencyManager` at start time, implements the `ModelChangeListener` interface. If it receives the message that a model's status changed to *Starting*, it will try to actually start this model.

Agents are started by the `AgentLauncherThread`, which will be explained later. Societies are first added to the `FunctionalStateController`, which is responsible for setting the societies' status. The status is set to *Functional* when all required models are running, it is set to *NonFunctional* if not all are running and to *Dead* when all models are stopped after stopping the society. After adding the society to the `FunctionalStateController`, all agents included in this society are started; afterwards all remote societies are started as well. To start an agent an `AgentLauncherThread` is created, checking that all tools (e.g. *Sniffer*, *Logger*, *Benchmarker*) for which tool options have been provided, are running and the agent is registered with them. An agent is then launched and passed all the provided parameters. When all agents and remote societies are started and running, the `FunctionalStateController` will switch the societies' status to *Functional*.

Stopping a society is simpler. The DependencyManager sets the societies' status to *Stopping*, signalling the now slightly misnamed RunnableStarter that it should stop the specific model. For societies, all included agents and remote societies will be stopped. For every agent, an AgentKillThread will be started, requesting the AMS to kill that agent. If every agent is stopped, the societies' status will change to *Dead*.

Whenever an ASCML wants to start a remote agent or society, it has to send this request to another ASCML. The address for the remote ASCML is given by the launcher-item of the societyinstanceref tag or the provider tag in the XML-files. To start and stop models, a simple action request is performed. A more complex scenario is where one ASCML subscribes to another ASCML to receive updates, whenever the status of some specified model changes. To achieve this, the local ASCML creates a FIPA Confirm message containing an expression such as:

```
(iota ?x (=
      (AgentInstance :ModelStatus ?x
      :Name AgentTypeName.AgentInstanceName)
      ?x))
```

This message is then wrapped into a subscription message and sent to the remote ASCML. The message will be received through the remote ASCML's StatusSubscriptionManager, which implements the SubscriptionManager interface. This manager will notify all subscribed ASCMLs when the status of a model changes through a message such as:

```
((= (iota ?x
      (= (Agent Instance? :Model Status? ?x
            :Name Agent Type Name.Agent Instance Name?)
            x?)
      (Available)))
```

These two messages are also used for requesting the actual status of a model. The first message can be wrapped into a Query-Ref to receive the second message expressing the momentary status.

In terms of how the ASCML handles tool options, which is important if ASCML is to be used to handle a new tool, there are several things to do:

1. Insert an identifier into Tool Option and ensure the XML-parser recognizes the tool.
2. Create a subclass of the ToolRequester to work with the tool.
3. In the AgentLauncher create a ToolRequester object for the tool and add it to the toolMap.

The ASCML will now be able to start the tool together with the agents for which the specific tool options are set.

## 11.6 DISTRIBUTED MONITORING, LOGGING AND DEBUGGING

Distributed monitoring of societies and agents is supported by the ASCML in a limited way. For societies the functional state is monitored as it is defined in the agent society description. This means that all the dependencies defined have to be satisfied regardless of whether they depend on local or remote agents. This requires the forwarding of status messages relating to the considered agents from all the participating agent platforms to the one hosting the ASCML agent. On that platform the ASCML checks the dependencies and shows the final state of a society to the user.

Distributed logging is provided. All agents selected to be logged are registered with the Logger Agents from their corresponding agent platform. This means that the configured output streams of one of these agents will be written to a file on the respective agent platform. The resulting files are not collected automatically and merged for improved debugging visibility because the contained time stamps are not in a chronological order. This is because no guarantee of clock synchronization can be made in distributed systems. However, many third-party tools are available to merge the log files whereby time stamps can be recomputed and the drift caused by the unsynchronized clocks removed.

Debugging agent applications is often a very time-consuming task. Therefore, the ASCML assists with the set-up of a debugging environment. If the check buttons labelled with Sniffer and/or Introspector are selected for a society or an agent to be launched, the ASCML will automatically register all these agents with the chosen tools. This also works in distributed scenarios.

## 11.7 OUTLOOK

ASCML can be used to create online repositories of agents and agent societies that can be shared via the Internet. One can imagine simply subscribing to some of those repositories when designing a new agent-based application with an application design CASE-tool. As with the vision of reuse expounded by the Object Oriented Programming community, the idea of sharing agents, agent behaviours and agent societies could help to promote acceptance and widespread use of this fascinating and useful technology.

Now all we need is more applications. If the reader wishes to contribute, please remember to provide an agent description file for each agent and additionally a society description file with a few sample configurations to demonstrate the features of the application. The best place for agent description files is the directory containing an agent's class file. A society description file should be located at the root of the class package.

# 12

# JADE Semantics Framework

While the basic JADE framework allows developers to implement their own agent model, the JADE Semantics add-on (JSA) provides one particular BDI[1]-like agent abstraction, namely the one used to formally define the FIPA-ACL semantics (for theoretical details, see Louis and Martinez, 2005a and b). In other words, JSA-based agents automatically interpret the meaning of received and sent messages, in compliance with FIPA-ACL specifications, and behave accordingly. At least two motivations would lead to using this add-on. The first is to build agents that conform fully to the FIPA-ACL semantics. Such 'semantic agents' are intrinsically more flexible and open. For example, they do not need to strictly follow interaction protocols, as they infer how to react according to the genuine meaning of exchanged messages. The second motivation is to benefit from intrinsically handled advanced features for developing agents. For example, without any coding, a semantic agent is able to answer any query about a fact he has been informed of previously.

To illustrate the main differences between 'classical' and 'semantic' JADE programming, we will again use the book-trading example. The complete code of this example is included in the JSA releases from version 1.4. As shown in the code skeleton of the seller and buyer agents hereunder, the programming style is more or less the same. The first difference lies in the choice of the class to extend: for a semantic agent, it is the `SemanticAgentBase` instead of the `Agent` one. But of course, the former extends the latter.

```
public class BookSellerAgent extends SemanticAgentBase {
  class BookSellerCapabilities extends SemanticCapabilities {...}
  class UpdateBookListBehaviour extends TickerBehaviour {...}
  public BookSellerAgent() {...}
  public void setup() {...}
}
```

The `setup` method of the `BookSellerAgent` class initializes the list of books to be sold. Additionally, it installs the `UpdateBookListBehaviour` JADE behaviour to periodically reduce the price of the unsold books or to remove them when the selling delay has expired. The semantic agent-specific programming is actually located in the `BookSellerCapabilities` class,

[1] BDI is a well-known model for cognitive intelligent agents, which specifies agents through the concepts of beliefs, desires and intentions. Most BDI implementations are based on the formalization proposed by Rao and Georgeff in the early 1990s (e.g. Jadex, a JADE third-party add-on). The JSA framework implements a quite different BDI formalization, which was proposed by Sadek at the same time.

*Developing Multi-Agent Systems with JADE* Fabio Bellifemine, Giovanni Caire, Dominic Greenwood

which extends the SemanticCapabilities class of the JSA framework. This class defines the 'semantic capabilities' of the agent and is discussed later in this chapter.

```
public class BookBuyerAgent extends SemanticAgentBase {
  class BookBuyerCapabilities extends SemanticCapabilities {...}
  class AdjustBuyingPriceBehaviour extends TickerBehaviour {...}
  public BookBuyerAgent() {...}
  public void setup() {...}
}
```

Similarly, the setup method of the BookBuyerAgent class initializes the book to buy and the list of available sellers. It additionally installs the AdjustBuyingPriceBehaviour JADE behaviour, which periodically increases the buying price as long as the required book has not been bought. The semantic capabilities of the buyer agent are implemented by the BookBuyerCapabilities class.

## 12.1 FIPA-SL LANGUAGE

JSA-based semantic agents are capable of extracting the precise meaning of each received message, expressed as a FIPA-ACL communicative act, and reacting accordingly. As the FIPA specifications formally define the meaning of exchanged messages using the FIPA-SL logical language, the JSA framework makes heavy use of data structures representing FIPA-SL expressions. It is therefore important for developers to have a basic understanding of FIPA-SL statements to handle them efficiently within the JSA software framework.

### *12.1.1 UNDERSTANDING FIPA-SL EXPRESSIONS*

FIPA-SL is a logical language with a prefixed syntax (like the LISP language). It consists of a first-order predicate calculus language, extended with additional modal operators to represent cognitive attitudes of agents (beliefs, uncertainties, intentions) and the occurrence of actions.

FIPA-SL mainly allows two kinds of expressions. On the one hand, the *terms* represent concrete objects of the world. For example, "(plus 3 4)" represents the object resulting from applying the function named "plus" to the objects represented by the "3" and "4" terms. On the other hand, the *formulas* represent facts that can be true or false. For example, "(= (plus 3 4) 7)" states that the term "(plus 3 4)" represents the same object as the term "7".

#### 12.1.1.1 FIPA-SL Terms

The most simple term constructs in FIPA-SL represent the usual primitive data type constants with a quite intuitive syntax. Numbers are represented using whole numbers (e.g. "123") or floating-point numbers (e.g. "-45.6E1"). Strings are classically surrounded by quotation marks[2] (e.g. "prime numbers"). Dates are written with the following format: "<YYYYMMDD>T<HHMMSSmmm>z".

The compound term constructs "(set <elem1> <elem2> ... <elemN>)" and "(sequence <elem1> <elem2> ... <elemN>)" (*N* being positive or null) respectively represent a set and an ordered list of elements. The elements within a set are order-insensitive, without duplicate. The elements within a list are order-sensitive and may include duplicates. For example, "(set 2 2 1)" and "(set 1 2)" are equivalent, while "(sequence 2 2 1)", "(sequence 1 2)" represent distinct objects.

[2] Quotation marks may be omitted for single words starting with alphanumeric characters. Strings may also be specified as byte contents with the format "#<number_of_bytes>"<sequence_of_bytes>", which is very useful to represent binary data such as a picture or a sound.

Functional terms represent objects resulting from the application of a function to a list of objects of the world. These objects, called parameters, are recursively specified as terms. For example, "(plus 1 2)" may be interpreted as the result of adding the two numbers 1 and 2 together. Two variant syntaxes are allowed:

- Either the function symbol is followed by an order-sensitive list of unnamed parameters, which are directly terms. For example, "(minus 1 2)" or "(minus 2 1)". Note these functional terms are not equivalent because the order of parameters is relevant.
- Or the function symbol is followed by an order-insensitive list of named parameters, which are pairs of a parameter name (prefixed by a colon) and a parameter value given as a term. For example, "(minus :left 1 :right 2)". Note that, in this case, it is equivalent to "(minus :right 2 :left 1)". Such a syntax is often used to represent instances of an ontology, the function symbol denoting a class and the named parameters its slots: "(Person :name "john" :birthday 20000512T143000000z)".

Action expressions "(action <actor> <action_type>)" represent the performance of an action on the world. <actor> and <action_type> are terms that respectively represent the agent who performs the considered action and the performed action itself. The latter is generally expressed using a functional term with named parameters (see above). For example, FIPA-ACL communicative acts are handled as FIPA-SL actions: "(action john (Inform :content "(prime 1)"))" represents the performance of an Inform act by an agent identified as "john".

Last, Identifying Referential Expressions (or IREs) make it possible to represent a set of objects that satisfy particular constraints. Their general syntax is "(<quant> <term> <formula>)", where <quant> is an IRE quantifier (either "iota", "all", "any" or "some"), <term> is the pattern of the terms denoted by the IRE and <formula> is a regular FIPA-SL formula (see below) expressing the constraints the terms must satisfy. The pattern of terms includes variables (which are special terms prefixed by a question mark, e.g. "?x"), which are bound to their occurrences within the formula to generate the instantiated terms denoted by the IRE.

For example, "(all ?x (prime ?x))" represents the set of all prime numbers (assuming "prime" is a predicate identifying prime numbers). Similarly, "(all (square ?x) (prime ?x))" represents the set of all squared prime numbers (assuming "square" is the square function, which therefore returns a term). More precisely, the latter IRE represents a set of terms of the form "(square ?x)", where "?x" is replaced with all values satisfying the formula "(prime ?x)". "(iota ?x (prime (square ?x)))" represents the unique element ("?x") whose square is a prime number (the result of a "iota" construct is a single element, which is assumed to be unique). "(any (square ?x) true)" represents the square of any number (the result of an "any" construct is a single element). "(some ?x (prime ?x))" represents any set of prime numbers (the result of a "some" construct is a set).

#### 12.1.1.2 FIPA-SL Formulas

Atomic formulas are built from a predicate name directly followed by a list of ordered parameters, expressed as terms. For example "(prime 7)" means the predicate named "prime" is true for the value represented by the term "7". If the predicate expects no parameter, surrounding parentheses may be omitted: "idle" is equivalent to "(idle)". FIPA-SL provides some predefined predicates: "=", which is true if and only if its two parameters represent the same object (for example "(= (square 2) 4)" is usually true); "true" and "false", which are the two usual boolean constants.

Atomic formulas may be combined using classical logical connectors: "not" (unary), "or", "and", "implies", "equiv" (binary). For example "(or (prime 2) (not (prime 2)))" means that 2 is a prime number or is not a prime number. Atomic formulas may also be quantified, the allowed quantifiers being "exists" and "forall" (with the usual meaning) and the quantification variables being prefixed by a question mark (just like for IREs, see above). For example, "(forall ?x (or (prime ?x) (not (prime ?x))))" means that any number ("?x") is prime or not prime.

Finally, formulas may be nested at any level within modal operators. Modal operators make it possible to represent cognitive attitudes of agents, namely beliefs (operator "B"), uncertainties (operator "U") and intentions (operator "I"), and the occurrence of past (operator "done") and future (operator "feasible") actions, and thus temporal constraints. They all use the same general syntax: "(<modal_op> <param> <nested_formula>)". Cognitive operators expect as a parameter a term representing the agent owning the specified cognitive attitude, while action operators expect as a parameter a term representing the occurred action (in the past or the future). For example, "(B john (prime 4))" means that the agent identified as "john" believes that 4 is a prime number. "(done a1 (B john (prime 4)))" means that "john" believ*ed* 4 was a prime number (more precisely, he believed it just before the action described by "a1" occurred). "(feasible a2 (B john (not (prime 4))))" means that "john" *will* believe 4 is not a prime number (more precisely, he will believe it just after the occurrence of action "a2" if it is actually performed).

### *12.1.2 HANDLING FIPA-SL EXPRESSIONS*

To efficiently handle FIPA-SL expressions, the JSA framework relies on a hierarchy of Node classes, each of them representing a kind of node in the FIPA-SL language. The Node hierarchy is similar, to some extent, to the JADE Abs hierarchy (see Section 5.1.2). For example, all nodes representing FIPA-SL terms and formulas respectively inherit the Term and Formula classes. The main difference is that the Node hierarchy closely maps the FIPA-SL grammar and provides efficient operations to handle FIPA-SL expressions.

Actually, the main entry to handle FIPA-SL expressions is the SLPatternManip class. First of all, it provides static methods (named from*) to parse FIPA-SL expressions (as Java strings) and create the proper objects in the Node hierarchy. Conversely, any instance of the Node hierarchy may be unparsed into a FIPA-SL string using the toString method. Below is a typical code sample:

```
// Create a simple formula
Formula anIsbn = SLPatternManip.fromFormula("(isbn \"ISBN
   0439784549\")");
// Create formula patterns (see below)
Formula ISBN_FORMULA = SLPatternManip.fromFormula ("(isbn ??isbn)");
Term SELL_ACTION_TERM = SLPatternManip
     .fromTerm("(SELL_BOOK :buyer ??buyer :isbn ??isbn :price
        ??price)");
ActionExpression SELL_ACTION_EXPRESSION = (ActionExpression)
   SLPatternManip
     .fromTerm("(action ??actor "+ SELL_ACTION_TERM +")");
Formula I_DONE_SELL_ACTION = SLPatternManip
     .fromFormula("(I ??buyer (done "+ SELL_ACTION_EXPRESSION +"))");
```

Interestingly, the Node hierarchy implements an extension of the FIPA-SL language, which makes it possible to use meta-references within expressions. Syntactically, meta-references look like FIPA-SL variables, except they are prefixed by a double question mark (e.g. "??isbn") instead of a single one. Expressions including meta-references work as patterns, which may be matched against other expressions and instantiated into new expressions. The matching operation is useful in recognizing expressions of a given form, while the instantiation operation is useful in building expressions of a given form.

More precisely, a meta-reference may be replaced with virtually any FIPA-SL expression, the type of which (essentially term or formula) is consistent with the meta-reference type. If the same meta-reference occurs more than once in an expression, each occurrence must be substituted for exactly the same value. For example, in the formulas "`(isbn ??isbn)`" and "`(B john ??phi)`", the meta-references "`??isbn`" and "`??phi`" may respectively be replaced with any term and any formula. The former matches the instantiated formula "`(isbn "ISBN 0439784549")`" ("`??isbn`" being bound to "ISBN 0439784549"). In the formula "`(B john ??phi)`", it is also possible to bind "`??phi`" to any pattern of formula, as for example "`(isbn ??isbn)`". When instantiated, it yields the new pattern of formula "`(B john (isbn ??isbn))`".

The matching operation is performed by the method `match()` of the pattern to match (handled as a `Node` subclass). This method expects a FIPA-SL expression (as a `Node` object) as argument and returns either `null` if it does not match the pattern or a `MatchResult` object containing the required bindings for the matching. Bindings may be easily retrieved from a `MatchResult` object using the `get*` methods, which expect the name of the meta-reference (without the double question mark) as a parameter. For example, the following code sample displays the ISBN and the title of a particular book:

```
Formula BOOK_FORMULA = SLPatterManip.fromFormula("(and (isbn ??isbn)
   (title ??isbn ??title))");
Formula aHarryPotterFormula = SLPatterManip
  .fromFormula("(and (isbn \"ISBN 0439784549\")
                     (title \"ISBN 0439784549\" \"Harry Potter and
                        the Half-Blood Prince\"))");
MatchResult result = BOOK_FORMULA.match(aHarryPotterFormula);
if (result != null) {
  System.out.println("ISBN = " + result.getTerm("isbn"));
  System. out.println("Title = " + result.getTerm("title"));
}
```

The instantiation operation is provided by the method `instantiate` of the pattern to instantiate (i.e. a `Node` subclass). This method expects the name of the meta-reference to instantiate and the FIPA-SL expression to substitute as arguments. It returns an expression equal to the pattern, where each occurrence of the specified meta-reference has been replaced with the specified value. Alternatively, the static methods `instantiate` of the `SLPatternManip` class make it possible to instantiate one to four meta-references at a time within a pattern. They expect the pattern to instantiate as a first argument, and one to four couples of meta-reference names and values to substitute as second to ninth parameters. Note that all these methods work on a clone of the pattern to instantiate, which is therefore not affected and may be further reused. For example, the `setup` method of the `BookBuyerAgent` builds an IRE to query the price of a particular book and to subscribe to this piece of information:

```
// Get the book ISBN and the seller AID from the agent's arguments
isbn = new StringConstantNode((String)getArgument()[0]);
seller = Tools.AID2Term(new AID((String)getArgument()[1],
   AID.ISLOCALNAME));
// Create and instantiate the IRE pattern
purchase_ire = (Identifying_Expression)SLPatternManip
               .fromTerm("(some ?x (selling_price ??isbn ?x
                  ??seller)")
               .instantiate("isbn", isbn)
```

```
                .instantiate("seller", seller);
// Use convenient methods of SemanticCapabilities to send the
   messages
semanticCapabilities.queryRef(purchase_ire, seller);
semanticCapabilities.subscribe(purchase_ire, seller);
```

Additionally, some `Node` subclasses provide useful methods specific to particular kinds of FIPA-SL expressions. For example, the method `getSimplifiedFormula()` on `Formula` objects computes a conjunctive normal form of a formula, which is logically equivalent. For a comprehensive list of such methods, please refer to the documentation provided with the JSA framework.

## 12.2 INTERPRETATION ENGINE

It is not the goal of this chapter to precisely describe the internals of the JSA framework, but it is nevertheless important that the reader understands the basic principles and architecture. Figure 12.1 illustrates the main components of the JSA framework and the way in which they interact with one another and with JADE elements such as behaviours.

Whereas programming classical JADE agents consists mainly of writing behaviours that make the receipt and the analysis of FIPA-ACL messages explicit, programming a JSA-based semantic agent rather consists of extending the Semantic Interpreter Behaviour, which is implemented by the `SemanticInterpreterBehaviour` class. The main activity of this behaviour is drawing deductions from perceived events (such as a message receipt) and accordingly modifying the agent's beliefs and behaviours. Each inferred deduction is represented by a Semantic Representation (SR), which is in fact a FIPA-SL formula. Basically, the receipt of a message produces an initial SR, from which the Semantic Interpretation Behaviour then deduces new SRs by applying a set of rules called Semantic Interpretation Principles (SIPs). Each SIP reifies a particular interpretation principle, which can be generic or specific to an agent. Besides the new SRs produced, a SIP can also update the agent's belief base, by adding or removing facts, or adding behaviours to/from the agent. New behaviours can be directly created by the SIP or retrieved from action prototypes which implement actions according to the FIPA-ACL specification.

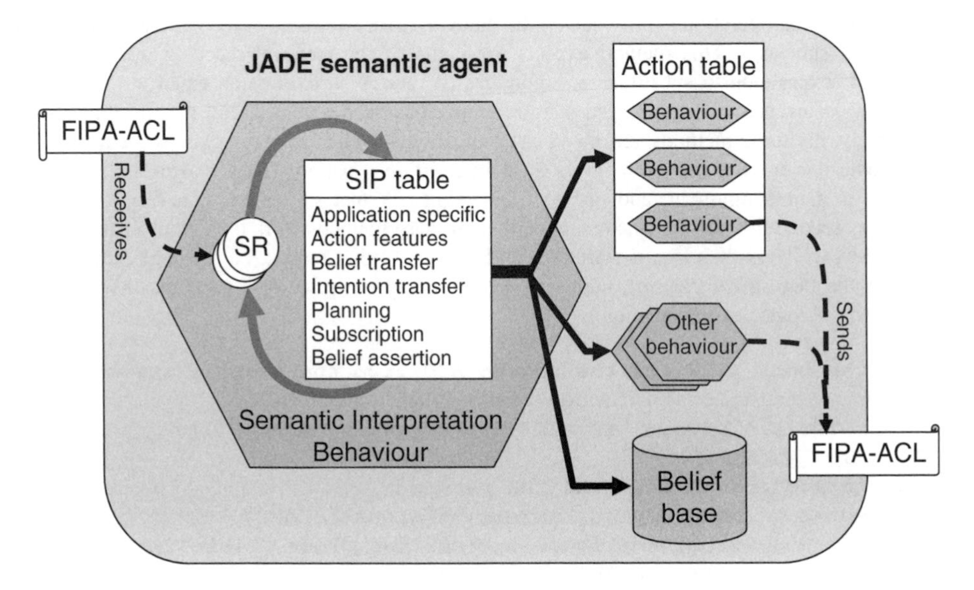


**Figure 12.1** The Semantic Interpretation Behaviour

The SIP table is divided into SIP classes. The main classes are as follows.

***Application-Specific***: These SIPs are really specific to an agent or an application. For example, they can be used to update a GUI, change a global property, and so on.

***Action Features***: Upon receipt of a message, these SIPs compute a set of SRs that represent the formal meaning of the message according to the FIPA-ACL specification (e.g. there is an SR for its effect, an SR for its precondition). The produced SRs are further interpreted by other SIPs.

***Belief Transfer***: These SIPs apply to SRs stating that an external agent intends the semantic agent (who is running the interpretation algorithm) to believe a particular fact. They consist of checking whether the semantic agent accepts the belief of this fact and producing the SRs with the corresponding meaning. These SIPs are useful to interpret incoming Inform messages.

***Intention Transfer***: These SIPs apply to SRs stating that an external agent intends a particular goal. Similarly, they consist of checking whether the semantic agent agrees to adopt this goal and producing the SRs with the corresponding meaning. These SIPs, which implement a form of cooperation, are useful to interpret all forms of Request messages.

***Planning***: These SIPs apply to SRs stating that the semantic agent intends to achieve a particular goal. They consist of adding a proper behaviour to the agent enabling it to reach the goal. Among the predefined SIPs, the *Action Performance* SIP and the *Rationality Principle* SIP perform a primitive action, selected from the agent's action table either by its name or by its effect. The developer has to complete this class of SIPs to handle more sophisticated planning capabilities.

***Subscription***: These SIPs apply to SRs stating an external agent intends to be notified of some event. They consist of installing a behaviour in the semantic agent that monitors this event and sends the expected notifications. In particular, such predefined SIPs implement interpretation of the Subscribe, Request-When and Request-Whenever acts.

***Belief Assertion***: Such a SIP applies to SRs stating the semantic agent believes a particular fact, and consequently asserts it into the belief base.

The JSA framework provides about 20 generic SIPs, which allow semantic agents to react to incoming messages according to the FIPA-ACL formal specifications. In order to specialize semantic agents for application needs, the JSA framework provides the developer with three extension mechanisms. The first consists of adapting the interpretation rules of the agent by adding new SIPs to its SIP table (see Section 12.4). The second mechanism consists of customizing the management of the agent's internal state by specializing its belief base. The third mechanism consists of increasing the knowledge of the agent by adding new action prototypes to its action table (see Section and 12.6).

## 12.3 BASIC SEMANTIC AGENT

Because all the basic SIPs are already part of the framework, the simplest JSA-based semantic agent is already able to perform some advanced semantic interpretation tasks. To make things more concrete, this 'simplest' agent can be run as follows:

```
java -cp ... jade.Boot -gui
   simplest:jade.semantics.interpreter.SemanticAgentBase()
```

Without any additional code, our 'simplest' agent is able to answer invalid messages with Not-Understood messages, ranging from syntactically incorrect messages to logically inconsistent messages. 'Simplest' also automatically asserts all facts it is informed of and is able to correctly answer queries about these facts. For example, if 'simplest' receives an Inform message with the following content, which gives the set of all books with their identifier, title and price,

```
((= (all (sequence ?isbn ?title ?price)
          (and (title ?isbn ?title) (selling_price ?isbn ?price)))
     (set (sequence "ISBN 0439784549" "Harry Potter and the
        Half-Blood Prince" 9.99)
           (sequence "ISBN 0618343997" "The Lord of the Rings"
              14.97))))
```

it automatically asserts the following facts into his belief base:

```
(title "ISBN 0439784549" "Harry Potter and the Half-Blood Prince")
(selling_price "ISBN 0439784549" 9.99)
(title "ISBN 0618343997" "The Lord of the Rings")
(selling_price "ISBN 0618343997" 14.97)
```

Then, if 'simplest' receives a Query-Ref message with the following content, which asks for all book titles,

```
((all (sequence ?x ?y) (title ?x ?y)))
```

it automatically answers with an Inform message the content of which would be:

```
((= (all (sequence ?x ?y) (title ?x ?y))
     (set (sequence "ISBN 0439784549" "Harry Potter and the
        Half-Blood       Prince")
           (sequence "ISBN 0618343997" "The Lord of the Rings"))))
```

Another example of the intrinsic sophistication of 'simplest' can be illustrated by these three messages, which express the same request with various formulations:

```
1) (REQUEST
      :sender    (agent-identifier :name other)
      :receiver (set (agent-identifier :name simplest))
      :content  "((action (agent-identifier :name simplest)
         (SELL_BOOK ...)))")
2) (INFORM
      :sender    (agent-identifier :name other)
      :receiver (set (agent-identifier :name simplest))
      :content  "((I (agent-identifier :name other)
                   (done (action (agent-identifier :name simplest)
                      (SELL_BOOK ...)))")
3) (INFORM
      :sender    (agent-identifier :name other)
      :receiver (set (agent-identifier :name simplest))
      :content  "((I (agent-identifier :name other) (book_sold
         ...)))")
```

'Simplest' is able to perform a requested action (e.g. to sell a book) whether (1) the requester (called 'other') directly requests him to do so, or (2) the requester informs 'simplest' about his intention for this action to be done (2), or (3) the requester informs 'simplest' about his intention for the effect of this action (i.e. a book is sold) to hold.

The last illustration of 'simplest''s capabilities is its ability to correctly handle some high-level communicative acts, such as Subscribe messages. For example, consider reception of a Request-Whenever message with the following requested action and notification condition:

```
(action (agent-identifier :name simplest)
        (INFORM-REF
             :sender   (agent-identifier :name simplest)
             :receiver (set (agent-identifier :name other))
             :content  \"((all (sequence ?x ?y) (title ?x ?y)))\"))
(exists ?z (B (agent-identifier :name simplest)
              (= (all ?x (exists ?y (title ?x ?y))) ?z)))
```

This will automatically cause the agent to inform the requester each time a new book title is asserted into his belief base.

Although 'simplest' intrinsically has several sophisticated interpretation capabilities provided by default with the JSA, real-world applications often require more specialized agents, e.g. agents able to drive particular devices, to deliver specific information, to deal with banks' accounts, and so on. To build such agents the JSA framework introduces the concept of semantic capabilities. Therefore to specialize a semantic agent the `SemanticCapabilities` class should be extended and an instance of it assigned to the semantic agent:

```
public class MySemanticAgent extends SemanticAgentBase {

  class MySemanticCapabilities extends SemanticCapabilities {
    protected SemanticInterpretationPrincipleTable
       setupSemanticInterpretationPrinciples() {
      SemanticInterpretationPrincipleTable t =
         super.setupSemanticInterpretationPrinciples();
       t.addSemanticInterpretationPrinciple(...);
       ...
       return t;
    }

    protected KBase setupKbase() {
       FilterKBase kb = (FilterKBase) super.setupKbase();
       kb.addKBAssertFilter(...);
       kb.addKBQueryFilter(...);
       ...
       return kb;
    }

    protected SemanticActionTable setupSemanticActions() {
       SemanticActionTable t = super. setupSemanticActions();
       t.addSemanticAction(...);
       ...
       return t;
    }
    ...
  }

   public MySemanticAgent() {
     this.semanticCapabilites = new MySemanticCapabilities();
  }
}
```

By default, a semantic agent holds a direct instance of the `SemanticCapabilites` class. Specializing this class mainly consists of redefining the methods described above: `setupSemanticInterpretationPrinciples()` to add semantic interpretation principles, `setupKbase()` to configure the agent's belief base, and `setupSemanticActions()` to add semantic actions the agent will be able to deal with. The redefinition of all these methods consists of obtaining the object to specialize from the super method (or creating a specific one if needed), configuring and returning it. The next subsections focus on these tasks.

Additionally, the `SemanticCapabilities` class provides the developer with some convenient methods. These include, for example, the ability to perform a particular communicative act (`inform()`, `request()`, etc. methods) or to run the interpretation algorithm with a particular SR (`interpret()` method). Here is an example of the `BookBuyerCapabilities` class.

```
class BookBuyerCapabilities extends SemanticCapabilities {

  class PriceProposalSIP extends ApplicationSpecificSIPAdapter
  {...}
  class SellActionDoneSIP extends ActionDoneSIPAdapter {...}
  protected SemanticInterpretationPrincipleTable
     setupSemanticInterpretationPrinciples() {
    SemanticInterpretationPrincipleTable t =
       super.setupSemanticInterpretationPrinciples();
     t.addSemanticInterpretationPrinciple(new
        PriceProposalSIP());
     t.addSemanticInterpretationPrinciple(new
        SellActionDoneSIP());
     return t;
  }
}
```

## 12.4 SPECIALIZING THE INTERPRETATION ACTIVITY

As discussed Section 12.2, the first way of customizing the behaviour of a semantic agent is to adapt its interpretation algorithm. This can be achieved by either adding application-specific SIPs or specializing the generic SIPs provided with the JSA framework. The usual place in the code for such operations is the `setupSemanticInterpretationPrinciples()` method of the `SemanticCapabilities` class (see the code template at the end of the previous section).

Note that directly removing or modifying generic SIPs should be avoided unless really necessary as it can often result in unexpected side effects.

### *12.4.1 ADDING APPLICATION-SPECIFIC SIPS*

Application-specific SIPs are useful in two main cases:

- *Reactive generation of application-specific SRs*. This consists of producing specific SRs while interpreting a given SR pattern. For example, from an SR stating an agent sells a book at an acceptable price, generate an SR stating the semantic agent intends to buy this book. The produced SR is expected to alter the observable behaviour of the agent when further interpreted by the interpretation algorithm.
- *Application-specific notification triggering*. This consists of executing a piece of specific code each time a given pattern of SR is interpreted, without modifying the interpretation process. For example, such SIPs make it possible to finely control GUIs.

Adding an application-specific SIP simply consists of creating an instance of the `ApplicationSpecificSIPAdapter` abstract class. The constructor needs a reference to the Semantic Capabilities of the interpreting agent as the first argument and a pattern of SR to which the applicative SIP applies as the second argument. This pattern of consumed SR is given as a `Formula` object and may of course include meta-references.

*Important note:* As SRs express meanings from the point of view of the interpreting agent, they should always be of the form "`(B ??myself...)`" or "`(I ??myself...)`", where the meta-reference "`myself`" is intended to match the AID of the interpreting semantic agent.

Each `ApplicationSpecificSIPAdapter` instance must implement the `doApply()` method, which specifies what the application of the SIP consists of. This method is called by the interpretation algorithm as soon as an SR that matches the specified pattern of consumed SR is produced during the interpretation process. The resulting `MatchResult` object is given as the first argument of the `doApply()` method. This method returns either `null` if the SIP is eventually considered to be not applicable to the selected input SR, or the list of the SRs eventually produced by the application of the SIP.

As special cases, if the returned list is empty (but not null), the SIP is said to be 'absorbent' with respect to the input SR. If this list contains exactly the input SR, the SIP is said to be 'neutral' (the input SR will be interpreted further as if this SIP had not been applied). This latter case is generally used for notification triggering as identified previously. To ease its implementation, the second argument of the `doApply()` method gives a list of SRs containing only the input SR, which can therefore be directly returned.

As an example of an application-specific SIP, the `PriceProposalSIP` is called each time the buyer is informed about the selling price of a book proposed by another agent. If the proposed book and price fit its needs, then the SIP makes the buyer agent adopt the intention to buy it.

```
// This behaviour regularly increases the buying price.
TickerBehaviour adjustPriceBehaviour = ...;
// This method returns true if the ISBN is the
// one of the book to buy and the price
// is acceptable.
boolean isAcceptable(Constant isbn, Constant price) {...}

class PriceProposalSIP extends ApplicationSpecificSIPAdapter {

  public PriceProposalSIP() {
    super(BookBuyerCapabilities.this,
          "(B ??myself (selling_price ??isbn ??price ??seller))");
  }

  protected ArrayList doApply(MatchResult applyResult, ArrayList
     result, SemanticRepresentation sr) {
    final Constant isbn = (Constant)applyResult.term("isbn");
    final Constant price = (Constant)applyResult.term("price");
    if (isAcceptable(isbn, price)) {
      try {
        result.add(new SemanticRepresentation
          ((Formula)SLPatternManip
                .instantiate(I_DONE_SELL_ACTION,
                             "actor", applyResult.term("seller"),
                             "isbn", isbn,
                             "price", price,
```

```
                                    "buyer", getAgentName()))),
        }
        catch(Exception e) {e.printStackTrace();}
        // After committing to buy the book, the SIP becomes useless
        mySemanticInterpretationTable.
   removeSemanticInterpretationPrinciple(this);
      }
      return result;
   }
}
```

### *12.4.2 SPECIALIZING GENERIC SIPS*

Some of the generic SIPs provided with the JSA framework can be easily customized to fit application needs. In a similar manner to that described in the previous section, such customization simply consists of creating an instance of the proper SIP adapter, which may be found in the `jade.semantics.interpreter.sips.adapters` package.

The most useful adapters are probably those that customize the Belief and Intention Transfer principles. Indeed, they make it possible to accurately control the beliefs and goals adopted by the semantic agent depending on both the beliefs or goals to adopt themselves and the external agent originating them. They both work on the same principle. First of all, their constructor expects a pattern of belief or goal to adopt and a pattern of an external agent as second and third arguments (the first argument being, as usual, a reference to the Semantic Capabilities the SIP adapter belongs to).

Next the `doApply()` abstract method must be defined. The first argument of this method gives the result of the matching between the pattern of belief or goal (previously passed to the constructor) and the concrete belief or goal to adopt. Similarly, the second argument gives the result of the matching between the pattern of agent and the concrete external agent originating the belief or the goal to adopt. This method must return:

- `null`, if the Belief or Intention Transfer SIP adapter is not applicable to the concrete belief or intention and the external agent;
- the `acceptResult` given as the third argument of the `doApply()` method, if the semantic agent accepts to adopt the belief or intention originated by the external agent;
- the `refuseResult` given as the fourth argument of the `doApply()` method, if the semantic agent does not accept to adopt the belief or intention originated by the external agent.

When the decision to accept a belief or a goal cannot be made directly in the `doApply()` method (e.g. because a request to another trusted agent and its reply are needed), this method has to 'install' the proper delayed computation algorithm on the semantic agent (e.g. by adding proper behaviours) and return an empty `ArrayList` (to 'absorb' the input SR). Additionally the installed algorithm, when finished, must interpret the `acceptResult` argument if the belief or intention has to be eventually adopted by the semantic agent, or the `refuseResult` argument otherwise (using the `interpret()` method of the `SemanticCapabilities`).

For example, the `BelievePriceSIP` of the `BookSellerAgent` prevents a seller agent from changing the selling price of his books upon a request from an external agent:

```
class BelievePriceSIP extends BeliefTransferAdapter {
  public BelievePriceSIP() {
    super(BookSellerCapabilities.this,
          SLPatternManip.fromFormula("(selling_price ??isbn ??price
             ??seller)"),
          SLPatternManip.fromTerm("??agent"));
```

```
    }

    protected ArrayList doApply(MatchResult matchFormula, MatchResult
       matchAgent, ArrayList acceptResult, ArrayList refuseResult) {
      return getAgentName().equals(matchFormula.term("seller") ?
             // If the selling price to adopt is mine, refuse it
            refuseResult :
             // Else adopt it
             acceptResult;
    }
}
```

All other SIP adapters work very similarly. Some useful ones make it possible to control the replies to a call for proposal (upon receipt of a CFP message) or to a proposal (upon receipt of a Propose message). Some others make it possible to catch the receipt of various messages such as an Inform stating a given action has been done (e.g. following a request), a Failure, a Cancel, etc. The complete list of available adapters can be found in the JSA documentation.

## 12.5 CUSTOMIZING BELIEF HANDLING

The belief base is a central component of the JSA framework since it is used to store the internal state of semantic agents as a collection of beliefs. These beliefs describe the complete representation of the world of an agent. It is very important to understand that an agent's beliefs include its beliefs about objective properties of the world (e.g. the title of a book), his beliefs about the beliefs of other agents (e.g. another agent believes that a particular book is for sale), and more generally any kind of mental attitude (a special case of beliefs according to the underlying theory). Mental state also include all of the agent's uncertainties and intentions.

More precisely, an agent's belief base stores the facts that the agent believes to be true. In other words, if the fact "`p`" belongs to a belief base owned by John, this means that the formula "`(B John p)`" is true. On the contrary, if the fact "`p`" does not belong to `B` it means that "`(not (B John p))`" is true. If, additionally, "`(not p)`" explicitly belongs to the belief base, the formula "`(B John (not p))`" is also true. Thus, it is possible to distinguish between three states of belief with respect to a fact: either the agent explicitly believes the fact (first case, the fact belongs to the base), or he believes its contrary (third case, the negation of the fact belongs to the base), or he does not believe anything about it (second case, neither the fact nor its negation belong to the base).

The belief base component is accessed through the `Kbase` interface. This interface defines the basic methods to modify and query the agent's beliefs. The main method to modify the agent's beliefs is `assertFormula()` which expects an argument consisting of a FIPA-SL formula to assert into the belief base. Asserting a fact "`p`" necessarily entails the truth of the formula "`(B myself p)`" ("`myself`" representing the agent owning the belief base), such that it is equivalent to assert "`(B myself p)`". The method `retractFormula()` allows a fact to be removed from the belief base. Actually, this method is provided for convenience, because removing a fact "`p`" is equivalent to asserting "`(not (B myself p))`" into the belief base, as explained above.

*Important note:* Although the `KBase` interface provides intuitive methods to assert and retract formulas into/from an agent's belief base, it is highly recommended to use the `interpret()` method of the `SemanticCapabilities` instead. This method creates an internal event described by the formula to assert (given as a parameter, just as for the `assertFormula()` method) and processes it through the overall interpretation algorithm (see Section 12.2). Thus, asserting a formula in this way additionally triggers all relevant actions specified in the interpretation chain of the agent. If none of these actions consumes this event, the formula will be finally asserted into the belief base.

The main method available to query the agent's belief base is query(). This method expects a FIPA-SL formula as an argument and returns a ListOfMatchResults object indicating whether the given formula is believed by the agent. If the method returns a non-null (respectively null) value for a formula "p", this means that "(B myself p)" is true (respectively false). Interestingly, the queried formula may include meta-references (see Section 12.1.2). In this case, if a non-null value is returned, it contains the list of all bindings of the meta-references that make the formula true. For example, if John believes "(title "ISBN 0439784549" "Harry Potter and the Half-Blood Prince")" and "(title "ISBN 0618343997" "The Lord of the Rings")", implying that he knows two books, then querying his belief base for "(title ??ISBN ??TITLE)" will return a list of two MatchResult objects. The first one binds the meta-references "ISBN" and "TITLE" respectively to ""ISBN 0439784549"" and ""Harry Potter and the Half-Blood Prince"", and the second one binds them respectively to ""ISBN 0618343997"" and ""The Lord of the Rings"".

The queryRef() method allows the handling of queries expressed using an Identifying Referential Expression (IRE, see Section 12.1.1.1). It expects as an argument an IdentifyingExpression object and returns as a result a ListOfTerm object containing all the objects of the world (expressed as FIPA-SL terms) denoted by the given IRE, or null if no object fits the IRE. Continuing with the previous example, calling this method with "(any ?isbn (exists ?title (title ?isbn ?title)))" returns a MatchResult object binding the variable "isbn" to the term ""ISBN 0439784549"". The same IRE with the "any" quantifier substituted for "iota" returns null because the objects fitting the description (i.e., ""ISBN 0439784549"" and ""ISBN 0618343997"") are not unique.

As Kbase is a Java interface, developers may use their own belief base implementation. However, the JSA framework provides a 'built-in' implementation, which should, for most cases, be a good trade-off between FIPA-SL expressiveness and efficiency. The assertion and query operations can be easily customized by a filter-based architecture. In particular, it provides fine control of the management of application-specific facts. For example, for efficiency reasons, it is generally not reasonable to store large amounts of data directly within the belief base, particularly if this data is already available on an appropriate medium (e.g. the byte representation of a picture). To avoid this, a simple assertion filter may catch assertions of such facts and store the corresponding data into the medium (e.g. a file or a database). Conversely, a query filter may intercept queries about these facts to properly retrieve the queried data from the actual storage medium.

Another typical use of such filters is to control the consistency of facts which depend on one another. For instance, a book implicitly has a unique selling price. Thus, an assertion filter on the "selling_price" predicate may easily retract all previous facts about this predicate in order to replace the previous price value with the new asserted one.

More precisely, the provided built-in belief base actually implements the FilterKBase interface, which provides the add/removeKBAssertFilter() and add/removeKBQueryFilter() methods to manage customized assertion and query filters on the belief base. The basic way to define application-specific filters is to inherit the KBAssertFilter or KBQueryFilter abstract classes by defining the apply() method. When a filter has to apply to a given pattern of formula, a more convenient way to define it is to inherit the KBAssertFilterAdapter or KBQueryFilterAdapter classes and to override the doApply() method. The apply() methods of these latter classes checks that the formula to assert or query matches the pattern given to their constructors, before calling the doApply() methods.

When asserting a formula F, the assertFormula() method of the provided FilterKBase implementation calls the apply() method of each assertion filter (in their order of creation with the addKBAssertFilter() method) and forwards its result to the following filter. Finally, the result of the last assertion filter is asserted into the belief base (see Figure 12.2). Thus, each assertion filter may modify the formula to assert depending on its needs. For example, if the apply() method of a filter returns the "true" formula, this prevents the actual assertion of the initial formula into the belief base.

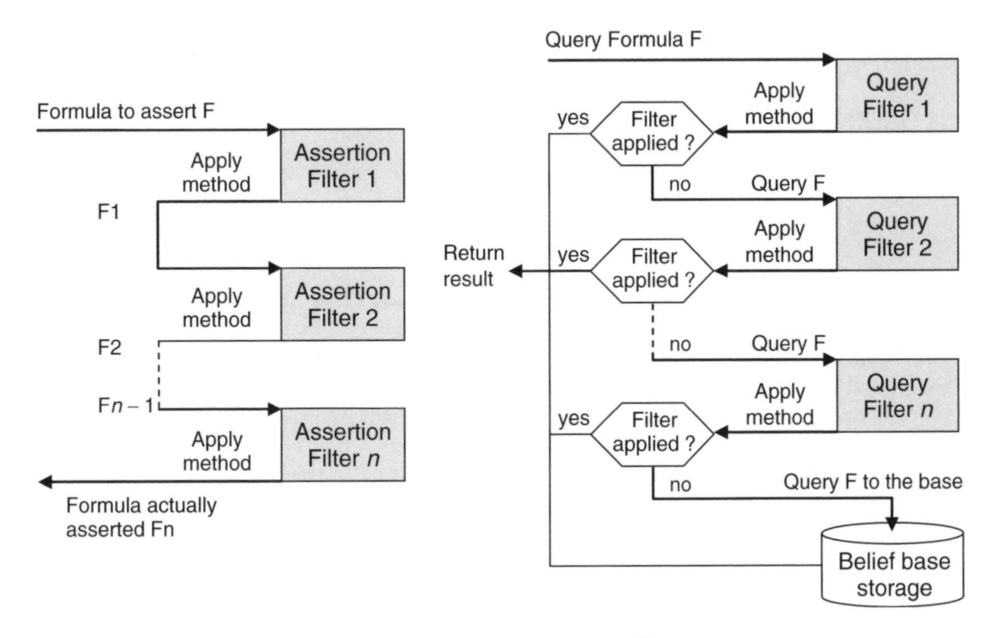


**Figure 12.2** Assertion and query filters

Similarly, when querying for a formula F, the `query()` method calls the `apply()` method of each query filter. If the result of this method indicates the filter was applicable then it is returned as the final result, otherwise the `query` method of the following filter is called with the same formula F. If no filter could be applied, then the belief base is finally searched for the queried formula F and the corresponding result is returned (see Figure 12.2). Thus, each query filter may 'bypass' the actual belief base and rely on external computation depending on its needs.

The `KBAssertFilterAdapter` and `KBQueryFilterAdapter` classes of the JSA framework ease the implementation of assertion and query filters to customize the built-in filter-based belief base. As recommended in Section 12.3, such declarations should take place within the `setupKbase()` method of the `SemanticCapabilities` class. In the `BookSellerAgent-Capabilities`, shown in the following example, the first filter retracts all the facts of a book when it is no longer for sale. The second filter retracts the current selling price when asserting a new price. The last filter is an example of a generic filter to check if a string is a substring of another one (e.g. may be used to retrieve books the title of which contains a given keyword).

```
class BookSellerAgentCapabilities extends SemanticCapabilities {

  protected KBase setupKbase() {
    FilterKBase kb = (FilterKBase) super.setupKbase();
    // 1. Assertion filter to handle the not_for_sale predicate
    kb.addKBAssertFilter(
      new KBAssertFilterAdapter("(B ??myself (not (for_sale ??isbn
         ??seller)))") {
        public Formula doApply(Formula formula, MatchResult match) {
          Term isbn = match.term("isbn");
          myKBase.retractFormula(ISBN_FORMULA.instantiate("isbn",
  isbn));
```

```
          myKBase.retractFormula(TITLE_FORMULA.instantiate("isbn",
   isbn));
          myKBase.retractFormula(SELLING_PRICE_FORMULA.instantiate
   ("isbn", isbn));
          myKBase.retractFormula(SELLING_DELAY_FORMULA.instantiate
   ("isbn", isbn));
          ...
          return new TrueNode();
      }});
    // 2. Assertion filter to handle the selling_price predicate
    kb.addKBAssertFilter(
      new KBAssertFilterAdapter("(B ??myself (selling_price ??isbn
         ??price ??seller))") {
        public Formula doApply(Formula formula, MatchResult match) {
          Term isbn = match.term("isbn");
          myKBase.retractFormula(SELLING_PRICE_FORMULA.instantiate
   ("isbn", isbn));
          return formula;
      }});
    // 3. Query filter to check if a string is a substring of
   another one
    kb.addKBQueryFilter(
      new KBQueryFilterAdapter("(B ??myself (zsubstr ??str
         ??substr))") {
        public MatchResult doApply(Formula formula, MatchResult
           match) {
          // The parameters of the zsubstr predicate must be
   instantiated
          String str = ((Constant)match.term("str")).stringValue();
          String substr =
             ((Constant)match.term("substr")).stringValue();
          return ( str.indexOf(substr) != -1 ) ? match : null;
      }});
    return kb;
  }
}
```

To take a step further, the whole interpretation algorithm of JSA-based agents exclusively accesses the belief base through the `KBase` interface, without making any assumption on its implementation. Consequently, developers may freely use other implementations than the `FilterKBase` provided. For example, one could think of implementing a belief base over a robust SQL database or a powerful reasoning engine (contributions are welcome!). However, particular attention must be paid to follow the storage principles given in the introduction of this section and to maintain a constant consistent state of the belief base.

## 12.6 HANDLING ACTIONS

To perform a particular action or simply to reason about it, a semantic agent needs an explicit representation of the action. Such a representation is handled in the JSA framework by the `SemanticAction` class. The semantic actions include application-specific actions, which represent the

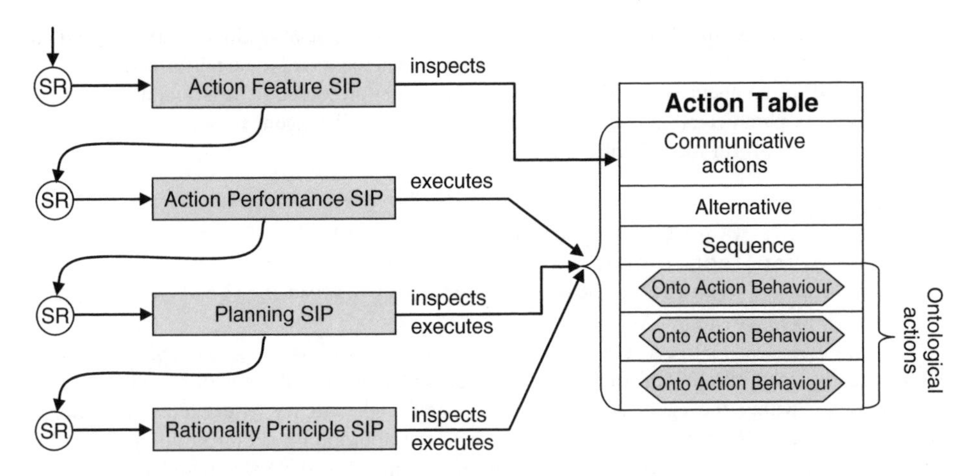


**Figure 12.3** Relationships between SIPs and the semantic action table

specific knowledge of an agent, all communicative actions specified by FIPA-ACL, and also the sequence and alternative operators defined by FIPA-SL.

More precisely, a `SemanticAction` instance acts as a prototype which makes it possible to inspect this action and also to create and execute clones of it. Prototypes are used for example to determine the intended effect and the preconditions of actions, including communicative actions. To conform to the JADE behaviour mechanism, the execution of an action consists of retrieving a behaviour from its prototype and adding this behaviour to the agent.

The prototypes of all actions usable by an agent are stored into the semantic action table of this agent (see the `SemanticActionTable` class). Generally, new action prototypes are not directly defined as instances of the `SemanticAction` class, but rather of a subclass of it, namely the `OntologicalAction` class. Introducing new action prototypes in the semantic action table can be achieved by redefining the `setupSemanticActions()` method of the `SemanticCapabilities` class, as shown in Section 12.3. By calling the `setupSemanticActions()` method onto the super object, all the predefined action prototypes (FIPA-ACL communicative actions and FIPA-SL operators) are automatically added into this table.

During the interpretation process, four kinds of SIPs usually deal with the semantic actions prototypes: Action Features, Action Performance, Rationality Principle and Planning (see Figure 12.3).

An Action Feature SIP only inspects a prototype to retrieve the characteristics of a communicative action that has just been carried out. These characteristics are used to produce SRs which will be interpreted in turn. Suppose, for example, that an Inform message with the following content has just been received by an agent called ‘buyer’ from another one called ‘seller’:

```
(title "ISBN 0439784549" "Harry Potter and the Half-Blood Prince")
```

Consequently, the Action Feature SIP of the buyer agent inspects the Inform prototype to retrieve the effect intended by the seller, then produces the following SR stating the buyer believes the seller intends him to believe this fact:

```
(B (agent-identifier :name buyer)
   (I (agent-identifier :name seller)
      (B (agent-identifier :name buyer)
         (title "ISBN 0439784549" "Harry Potter and the Half-Blood
            Prince"))))
```

The three other kinds of SIPs generally inspect a prototype but also clone it to carry out the corresponding action. For example, when a particular action is requested to be performed, the Action Performance SIP of an agent clones the corresponding action prototype (if it is found according to its name), sets its parameters and runs it. Consider that our seller agent now receives a Request from the buyer to perform the sale of a book:

```
(REQUEST
      :sender   (agent-identifier :name buyer)
      :receiver (set (agent-identifier :name seller))
      :content  "((action (agent-identifier :name seller) (SELL_BOOK
         ...)))")
```

Consequently, the Action Performance SIP of the seller agent will look for a SELL_BOOK action prototype within the agent action table, and, if found, will try to execute this action. A Rationality Principle SIP works similarly, but selects the appropriate action prototype according to the expected effect. Such a SIP can, for example, be applied to the SR hereunder by the seller agent:

```
(I (agent-identifier :name seller)
   (B (agent-identifier :name buyer)
      (title "ISBN 0439784549" "Harry Potter and the Half-Blood
         Prince")))
```

In this case, the intended effect matches the effect of an Inform act, which is to make the buyer know about the title of a particular book. Therefore, the Rationality Principle SIP selects and instantiates the corresponding Inform prototype and executes it. Finally, if no primitive action makes it possible to reach the intended effect, a Planning SIP can be used. Such a SIP combines primitive actions (based on their preconditions and effects) and returns a behaviour that realizes the corresponding course of actions.

We will now explain how to build an ontological action. Here is the skeleton of the `OntologicalAction` class, focused on its major items.

```
public class OntologicalAction implements SemanticAction {

    public OntologicalAction(    SemanticCapabilities
       myCapabilities,
                                 Term actionPattern,
                                 Formula postconditionPattern,
                                 Formula preconditionPattern) {...}
    ...
    public void perform(OntoActionBehaviour behaviour) {
        behaviour.setState(SemanticBehaviour.SUCCESS);
    }
}
```

An ontological action is mainly defined in terms of four elements, three of which are specified directly using the constructor while the last one is the `perform()` method. The former is a pattern that defines the general form of the FIPA-SL term that represents the action. For example, the following pattern represents the SELL_BOOK action. It expects three parameters: the buyer identifier, the ISBN code of the book to buy and the agreed price for the book.

```
(SELL_BOOK :buyer ??buyer :isbn ??isbn :price ??price)
```

This pattern matches FIPA-SL terms that can be embedded into FIPA-SL action expressions. The following example denotes a SELL_BOOK action the author of which is our seller agent:

```
(action (agent-identifier :name seller)
        (SELL_BOOK :buyer (agent-identifier :name buyer)
                   :isbn "ISBN 0439784549"
                   :price 9.99))
```

Such an action expression can serve as the content of a Request act or it can be embedded in a more complex FIPA-SL expression. The following expression, for example, states the buyer agent intends the seller agent to perform the SELL_BOOK action:

```
(I (agent-identifier :name buyer)
   (done (action (agent-identifier :name seller)
         (SELL_BOOK :buyer (agent-identifier :name buyer)
                    :isbn "ISBN 0439784549"
                    :price 9.99)) true))
```

The two next elements involved in an ontological action definition are two patterns of FIPA-SL formulas that represent the feasibility precondition and the postcondition of the action. Before executing an action, the interpretation engine always checks that its feasibility precondition holds. This is achieved with a special behaviour called `OntoActionBehaviour`. If the feasibility precondition does not hold, the behaviour state is set to `FEASIBILITY_FAILURE`. If the action is feasible and its execution succeeds then the behaviour state is set to `SUCCESS` and the postcondition is asserted into the belief base of the agent.

When no special precondition or postcondition is required, it is possible to use the FIPA-SL True formula (see `TrueNode`) instead of a complex one. This can be done because a True precondition always holds and a True postcondition is never asserted. To exemplify this, here is a partial definition of the SELL_BOOK action. The action pattern, together with a postcondition and a precondition, are specified using the constructor of `OntologicalAction`:

```
class BookSellerCapabilities extends SemanticCapabilities {
  class SellBookAction extends OntologicalAction() {

    public SellBookAction() {
        super(BookSellerCapabilities.this,
              "(SELL_BOOK :buyer ??buyer :isbn ??isbn :price
                 ??price)",
              "(not (for_sale ??isbn ??actor))", // postcondition
              "(for_sale ??isbn ??actor)");      // precondition
    }

     public void perform(OntoActionBehaviour behaviour) {...}
  }

  protected SemanticActionTable setupSemanticActions() {
    SemanticActionTable t = super. setupSemanticActions();
    t.addSemanticAction(new SellBookAction());
    return t;
  }
}
```

**Table 12.1** `OntoActionBehaviour` principles

| Precondition | Perform method | `OntoActionBehaviour` | | Postcondition |
|---|---|---|---|---|
| | | State | Ends | |
| Does not hold | Not executed | `FEASIBILITY_FAILURE` | Yes | Not asserted |
| Holds | Executed | `EXECUTION_FAILURE` (set during execution of perform) | Yes | Not asserted |
| Holds | Executed | `SUCCESS` (set during execution of perform) | Yes | Asserted |
| Holds | Executed | Other, like RUNNING (set during execution of perform) | No | Not asserted |

The last element involved in an ontological action definition is the `perform()` method which must be written by the programmer, but is automatically called by the `action()` method of the `OntoActionBehviour` when the precondition holds. Like the `action()` method of a classical JADE behaviour, the `perform()` method may be called several times to achieve a single execution of an action. Before the first call, the state of the behaviour is automatically set to `START`. It is then up to the 'perform' method to update this state at each execution. Finally, the behaviour ends its execution when its state equals `FEASIBILITY_FAILURE`, `EXECUTION_FAILURE` or `SUCCESS`. Table 12.1 summarizes the various `OntoActionBehaviour` principles.

To complete this section, here is a possible implementation of the `perform()` method of our SELL_BOOK action (the enclosing `SellBookAction` class is defined as an inner class of the one derived from `SemanticCapabilities`):

```
Term isbn;
Term buyer;
Term price;

public void perform(OntoActionBehaviour behaviour)
{
  try {
    switch (behaviour.getState()) {
      case OntoActionBehaviour.START:
        isbn = getActionParameter("isbn");
        buyer = getActionParameter("buyer");
        price = getActionParameter("price");
        behaviour.setState(OntoActionBehaviour.RUNNING);
        break;
      case OntoActionBehaviour.RUNNING:
        System.out.println("Bingo! " + getAgentName() + " sells the
           book " + isbn +
                                " to " + buyer + " at " + price +
                                   " euros ");
        behaviour.setState(OntoActionBehaviour.SUCCESS);
        break;
    }
  }
  catch (Exception e) {
        behaviour.setState(OntoActionBehaviour.EXECUTION_FAILURE);
  }
}
```

If the state of the behaviour is START, then the perform method simply takes into account the action parameters. Otherwise, it prints a message and sets the behaviour state to SUCCESS. Last, if an exception occurs, the behaviour state is set to EXECUTION_FAILURE.

## 12.7 SYNTHESIZING STANDARD AND ADVANCED USE OF THE JSA

To complete and summarize this section about the JSA framework, we provide some guidelines as to how to improve and rationalize the programming of semantic agents. As has become clear, programming semantic agents differs somewhat from the programming of classical JADE agents. In particular, a JSA programmer never explicitly handles the receipt of messages. However, the typical steps for building an agent remain the same: implementing the specific interpretation capabilities of the agent, its specific beliefs and its specific know-how.

***Implementing the Specific Interpretation Capabilities***: To achieve this it is first necessary to identify the specific interpretation capabilities. For example, a semantic agent requires specific interpretation capabilities to not accept beliefs or intentions from some agents (e.g. BelievePriceSIP), to adopt a goal (e.g. PriceProposalSIP) or to trigger a particular reaction on a particular belief (e.g. updating a GUI). To implement these capabilities proper SIPs must be introduced into the SIP table of the agent (see Section 12.4). The provided adapters can be used to do this (see the jade.semantics.interpreter.sips.adapters package).

***Implementing the Specific Beliefs of an Agent***: To achieve this it is first necessary to identify which facts can be asserted into the belief base of an agent. Such facts are always predicates (e.g. ISBN_FORMULA, TITLE_FORMULA), which, for convenience, can be defined as patterns within a shared class (see Section 12.1.2). For each kind of fact it must be determined whether it must be stored in the belief base as is, or in a special way (e.g. using objects in memory, files, relational databases). In the former case there is nothing to do, otherwise specific assert and query filters must be implemented to handle the corresponding facts (see Section 12.5).

***Implementing the Know-how of an Agent***: To achieve this it is first necessary to identify which ontological actions comprise the know-how of an agent (e.g. SELL_BOOK). For each action it must be determined which feasibility precondition(s) must hold before performing the action and which postcondition(s) must be asserted if the action succeeds. The OntologicalAction class must then be specialized to implement this action (e.g. SellBookAction, see Section 12.6).

A more sophisticated mechanism to run a complex course of actions is to use a planning algorithm. Whereas the JSA framework does not provide any planning facility, it provides the proper hook to plug in an external planner (through the Planning SIP entry of the SIP table). It is up to the programmer to define an application-specific planner.

## 12.8 CONCLUSIONS

The JADE Semantics add-on provides a particular BDI-like model for programming JADE agents. This model is exactly that used by the FIPA specifications to formalize the semantics of the FIPA Agent Communication Language. Consequently, the design of JSA-based agents relies on high-level abstractions (mainly beliefs, intentions and actions), which are automatically handled by a semantic interpretation engine (whose rules – or 'SIPs' – can be customized). Such a framework should therefore result in at least three concrete benefits from the developer perspective:

- Using high-level abstractions eases the coding of agents (e.g. there is no need to receive and analyse messages or to manage interaction protocols).
- The developed agents intrinsically benefit from the advanced features of the FIPA Agent Communication Language (e.g. queries, subscriptions or propagations are automatically handled).
- Consequently, semantic agents are naturally more flexible and open.

As for the software design of the JSA framework, a significant effort has been made to ensure it executes efficiently on the J2SE and J2ME/pJava profiles (the J2ME/MIDP profile is currently not supported) and providing extension points to fit various needs.

For future versions, it is being considered whether to make it possible to interface the belief base with a relational database, or indeed replacing it with a specialized reasoning engine. It would also be useful to provide some proper ways to use content languages other than FIPA-SL in exchanged messages, and support external ontologies (such as OWL-based ones).

# 13

# A Selection of Other Relevant Tools

This section describes some other important tools related to JADE. Many more exist, but are not mentioned in this book due to space limitations. A full list is available from the *add-ons* and *third-party software* areas of the JADE website together with a short description and the links required to download the software.

## 13.1 THE BEAN GENERATOR

The Bean Generator is a tool that supports agent engineers in creating message content ontologies compliant with the JADE support for managing content expressions described in Section 5.1. The tool is a plug-in for Protégé, which is a commonly used ontology editor that enables engineers to graphically model ontologies. Furthermore, additional functionality and storage formats can be 'plugged in' to the system.

Another advantage of the Protégé tool is that other ontologies can be imported. Repositories of existing ontologies ranging from biological domains to market place product and service descriptions, can be found at the Protégé community page and at the DAML site[1]. The languages used to represent these ontologies can be XML, RDF, DAML-OIL, XMI, SQL or UML.

The ontology model of Protégé consists of classes, slots and slot facets. Classes are concepts in the domain of discourse with which a taxonomic hierarchy can be constructed. Slots describe properties or attributes of these classes. A slot in itself is a frame that has a type. This can be a primitive class, such as String, Integer and Float, or an instance of another class. Furthermore, a slot has a value. Slot facets, listed in Table 13.1, describe properties (or constraints) of slots.

In order to create a message content ontology, we make use of a simple upper ontology as depicted in Figure 13.1. Several concepts are defined and an AgentAction defines the speech act, i.e. the action within a message to, for example, buy, or sell. The AID (Agent IDentifier) represents an agent.

After defining an ontology the Bean Generator will translate it using Protégé into Jade-compliant Java classes. This takes place as follows: every class in the Minimal Ontology, i.e. **Concept** and **AgentAction**, is the basis for the generation of a Java class. The taxonomic structure (i.e. inheritance relations) of the domain model is mapped on the inheritance capabilities of Java. Therefore if **S1** is a super-schema of **S2** then the class **C2** associated to schema **S2** will *extend* the class **C1**. Slots

[1] See http://protege.stanford.edu/ontologies.html and www.daml.org/data

*Developing Multi-Agent Systems with JADE* Fabio Bellifemine, Giovanni Caire, Dominic Greenwood

Table 13.1 Slot facets

| Facet | Description | Examples |
|---|---|---|
| *Cardinality* of a slot, 0,1,n | The number of values the slot can have | Class *person* can only have one father ($c = 1$)<br>class *father* can have multiple children ($c = N$) |
| *Allowed values* | Restriction of the value type of a slot | Integer, String, Instance of a class |
| *Numeric boundaries* | The minimum and maximum value for a numeric slot | The slot *age* is between 0 and 150 |
| *Required or optional* | Whether a slot is required or not | The slot *name* is required for the class *person* |

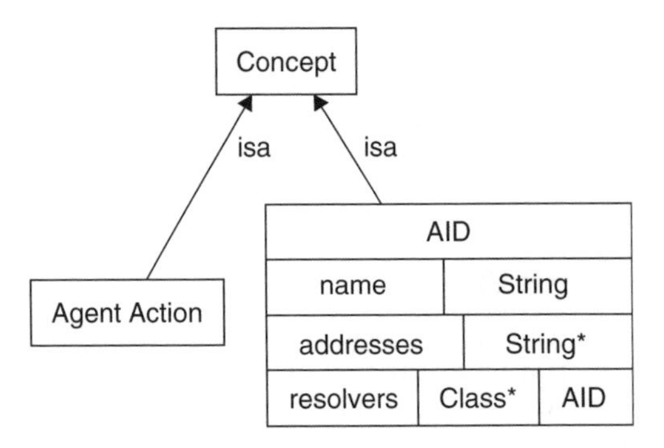


**Figure 13.1** Simple JADE Abstract Ontology

of a class are associated with data members of the Java Bean associated with the class. If the type of the slot is a primitive class, such as *String*, *Integer* or *Float*, then the Bean Generator maps them onto their Java equivalents, otherwise the member of the class is defined as an instance of the corresponding Java class. If the cardinality is higher than one, a class of type Collection[2] is used.

### *13.1.1 INSTALLING THE BEAN GENERATOR*

To use the Bean Generator tool Protégé must be properly installed. After that the following steps must be followed:

- Copy the file 'beangenerator.jar' into the *plugin* directory of Protégé (this file contains the code of the plug-in).
- Copy the file 'beangenerator.properties' into the root directory of Protégé (this file contains favourite settings for the plug-in).
- Copy the file 'protegeprojects.zip' into the project directory (this file contains the SimpleJADE-AbstractOntology project, the MusicShopOntology project, the OWLSimpleJADEAbstractOntology project and the OWLMusicShopOntology project).

[2] The class **Collection** is only an interface. The class **java.util.ArrayList** actually implements the class **Collection**.

### *13.1.2 WORKING WITH THE BEAN GENERATOR*

To illustrate the process of creating the required classes of a message content ontology, we define an example in an electronic commerce (business to consumer) domain. In this example agents represent parties that want to do business, such as buying and selling books. We define a selling agent and a buying agent. Both the buying agent and the selling agent will try to negotiate in order to get the best deal. An example of a negotiation is bargaining for the price of a book in a bookshop. There are four steps in constructing a message content ontology using the Bean Generator:

1. *Create a Protégé project.* This step implies opening a new or existing Protégé project including the *SimpleJADEAbstractOntology* into the project for a *regular* Protégé project (via project->manage included projects) or the OWLSimpleJADEAbstractOntology for an *OWL* project. Next go to the project>configure menu and select the OntologyBeanGeneratorTab checkbox.
2. *Add the ontological elements to the project.* This step implies creating concepts by making subclasses of the Concept class, creating agent actions by making subclasses of the Concept class and creating predicates by making subclasses of the Predicate class. The Concept, AgentAction and Predicate classes are included in the SimpleJADEAbstractOntology.

   At the end of this step, considering the same concepts, agent actions and predicates described in Section 5.1.3.1 for the book-trading case study, the Protégé GUI should appear as shown in Figure 13.2.
3. *Configure the ontological elements.* This step implies declaring the proper slots and facets for all the concepts, agent actions and predicates defined in the previous step.
4. *Generate the ontology and the bean classes.* This step implies selecting the Ontology Bean Generator tab, specifying (or selecting from the combobox) a *package name*, e.g. bookTrading.onto, specifying (or selecting from the combobox or from the [...] button) a *location* where the Java files are to be created (make sure the directory exists), specifying (or selecting from the combobox) a *ontology name*: e.g. BookTrading, and pressing the [*Generate Beans*] button (old generated files will be overwritten).

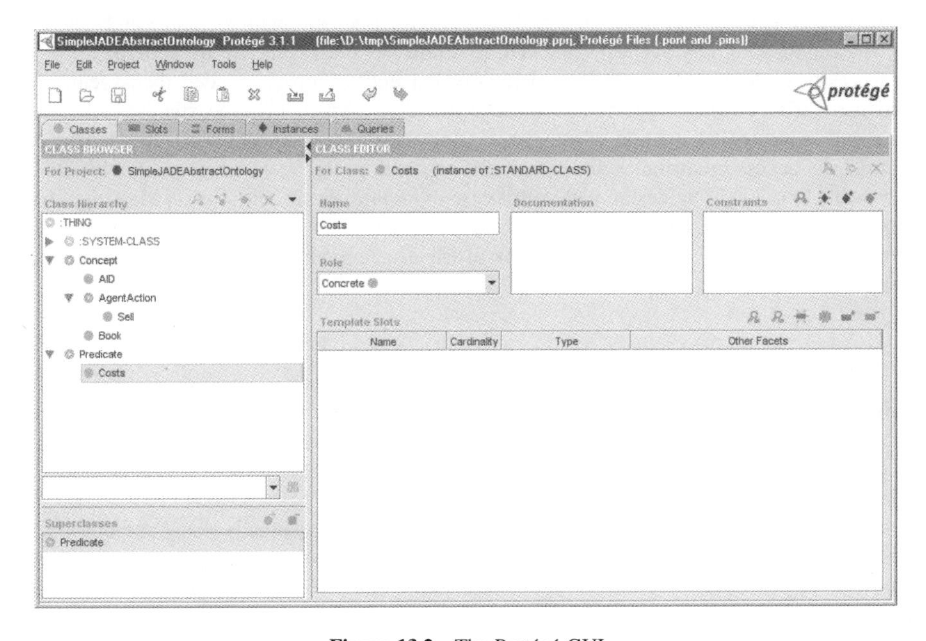


**Figure 13.2** The Protégé GUI

Both the ontology definition class as described in Section 5.1.3.1 and the ontological element classes as described in Section 5.1.3.2 are automatically generated.

### 13.1.3 ACKNOWLEDGEMENTS

The Bean Generator was originally developed within the IBROW project (http://hcs.science.uva.nl/projects/ibrow/home.html). At a later stage it was improved by Jamie Lawrence–Media Lab Europe (http://www.medialabeurope.org/), RWTH Aachen (http://www-i4.informatik.rwth-aachen.de/), Decis Lab (http://www.decis.nl) and Y'All (http://www.yall.nl). The source code has been donated to the JADE project.

## 13.2 JADEMX

This section describes the **jademx** JADE add-on, which provides two major capabilities: the ability to interface JADE agents with Java JMX (Java Management Extensions) and the ability to unit test JADE agents using JUnit.

Jademx is available for download from the third-party software area of the JADE website. Everyday, useful software systems rarely exist in isolation. Indeed, one of the strengths of JADE is that the full capabilities of the Java environment are available when creating a software agent application. JMX is the Java technology for management and monitoring of software systems; it was originally part of the Java EE enterprise platform (formerly known as J2EE), but as of Java 5 it is available as part of the standard J2SE environment. Furthermore, unit testing is an important technique for the development of robust software and JUnit is a standard methodology for the unit testing of applications written in Java. **Jademx** was developed for an industrial software agent effort requiring management using Java EE and to be unit-testable.

### 13.2.1 CODING A JADEMX AGENT

No code change is required for simple agent control. **Jademx** can obtain the name and state of any agent and can also kill any agent if the JADE platform has been launched by **Jademx**.

If access to custom JMX attributes, operations and notifications is needed then agents should extend the `jade.jademx.agent.JademxAgent` class, which in turn extends `jade.core.Agent` and is an implementation of `javax.management.DynamicMBean`. In addition to the standard JMX access to attributes and operations exposed by the agent as a DynamicMBean, it is also possible to use a proxy design pattern for ease of coding. To do this a proxy class that extends `jade.jademx.agent.JademxAgentProxy` and proxies the attributes and operation actions of the real agent must be created. The **jademx** distribution contains agents in its unit tests that can be used as examples of this approach.

### 13.2.2 CONFIGURING JADEMX AGENTS

A **jademx** agent can be configured either programmatically or by using XML. If agents are configured with XML, the configuration is usually named `jademx-config.xml`, which may be read from either a resource or URL. The definition of this file is available in XML Schema format from the `jademxconfig.xsd` file in the **jademx** distribution.

It is also possible to configure **jademx** applications programmatically using the class `jade.jademx.config.jademx.onto.JademxConfig`. This is particularly convenient if configurations must be generated dynamically. It is worth noting that `JademxConfig` implements `jade.content.AgentAction` and can be passed around as content in ACL messages; see the **jademx** tutorial for a reference to an example of this.

### *13.2.3 DEPLOYING JADEMX AGENTS*

A set of agents deployed by **jademx** may run under J2SE in either a Java servlet context, or as a JBoss service. Servlets are defined by the Java EE environment and JBoss is a popular open-source Java EE application server.

To use **jademx** under J2SE, an application must:

1. Instantiate a `JadeMXServer` to define the MBeanServer to use;
2. Construct a `JadeFactory`, which will instantiate the MBeans;
3. Obtain a `JadeRuntime` (which corresponds to a JADE `Runtime`) from the factory; and
4. Use the run-time to create `JadePlatform` MBeans.

Afterwards, the application should call `JadeRuntime.shutdown()` to shut down cleanly.

If JADE is to be run in an environment managed by a Java EE application server, then deploying **jademx** in a servlet context is particularly convenient. **Jademx** provides the `jade.jademx.server.JadeServletContextListener` class to create and destroy a **jademx** run-time when the corresponding servlet context is created and destroyed. To use this method, a `web.xml` file must be created in a Java EE `.war` file. **Jademx** is distributed with files already prepared that can be used to deploy under JBoss and WebLogic. Otherwise programmers can create their own for other application servers.

The JBoss application server has its own concept of *service*. If the JADE application is to be deployed as a JBoss service, refer to the **jademx** documentation for details and an example. **Jademx** provides the files `jboss-app.xml` and `jboss-service.xml` to help with creating an `.sar` file.

### *13.2.4 TESTING JADE AGENTS USING JUNIT AND JADEMX*

**Jademx** easily enables the execution of a JADE agent unit test under JUnit; the unit level is an ACL message. To do this, the following steps must be taken:

1. The class for the agent-under-test must extend `jade.jademx.agent.JademxAgent`.
2. Set up the agent under test and a dummy agent to receive the generated message.
3. Wait to make sure that the agent as a dynamic MBean has bound to the MBean server.
4. Set the message that to be injected into the agent under test (*important: make sure that the sender and receiver agents use local names, i.e. without the* **`@platform`** *part*).
5. Set the message that is expected to be received back.
6. Set which strings in the messages can vary to avoid spurious differences for things such as the agent platform or reply-id.) There are two ways to do this:
   (a) by setting 'variables' to show values that must be isomorphic or ignored;
   (b) by turning off comparison of message slots (see the tutorial for a reference to an example of this).
7. Assert that the expected message is received after the agent under test receives the injected message.

The **jademx** distribution contains further test documentation and executable example JUnit tests.

## 13.3 THE JAVA SNIFFER

The Java Sniffer is a stand-alone Java application, developed by Rockwell Automation, Inc., that can remotely connect to running JADE systems and is intended as an alternative to JADE's built-in sniffer. The tool receives messages from all agents in the system, reasons about the information, and presents it from different points of view.

The tool is able to visualize messages as a low-level UML sequence diagram and provides a high-level view via dynamically created traceable workflow diagrams. XML, Lisp and BitEfficient message encodings are supported according to the FIPA-ACL specifications and SL, XML, JDL (job description language) content languages are primarily supported. It offers visualization of statistical information, message and agent filtering, automatic log file creation, etc. The Java Sniffer is freeware with detailed licensing information available with the distribution.

The main visualization screen is divided into five sections (see Figure 13.3). Each section provides the observer with information at a different level of detail.

1. *List of Messages.* The window located at the bottom left corner displays a list of messages sent among agents as a UML sequence diagram similar to JADE's built-in sniffer. A scrollable list of agents is located at the top of this window and agents can be sorted by name, by the number of sent or received messages, and so on. Each row represents one message sent from one agent to another, displayed as an arrow pointing from the sender of the message to the receiver. The message is numbered, time-stamped, colour coded by message type, and an overall description constructed from key information of message content is also added. If a message has more than one receiver, it is displayed accordingly (see message 430). If the source and/or the target agent are not visible then the information about source and/or target agent is also added.
2. *Message Detail view.* The window located at the top left corner provides information about the content of the selected message. The format of the information displayed is dependent on the communication language used. It is possible to show the whole message as a tree or focus on some part of the message such as envelope, FIPA-ACL, and SL, XML or JDL content.
3. *List of Work Units.* The window located at the top right corner shows the list of work units requested by the agents. Each description of a work unit consists of work unit identification, work name, the number of subsequent messages belonging to this work unit, and the original requester. Also, the status of each work unit is pictographically visualized showing 'in progress', 'successfully' or 'unsuccessfully' finished.

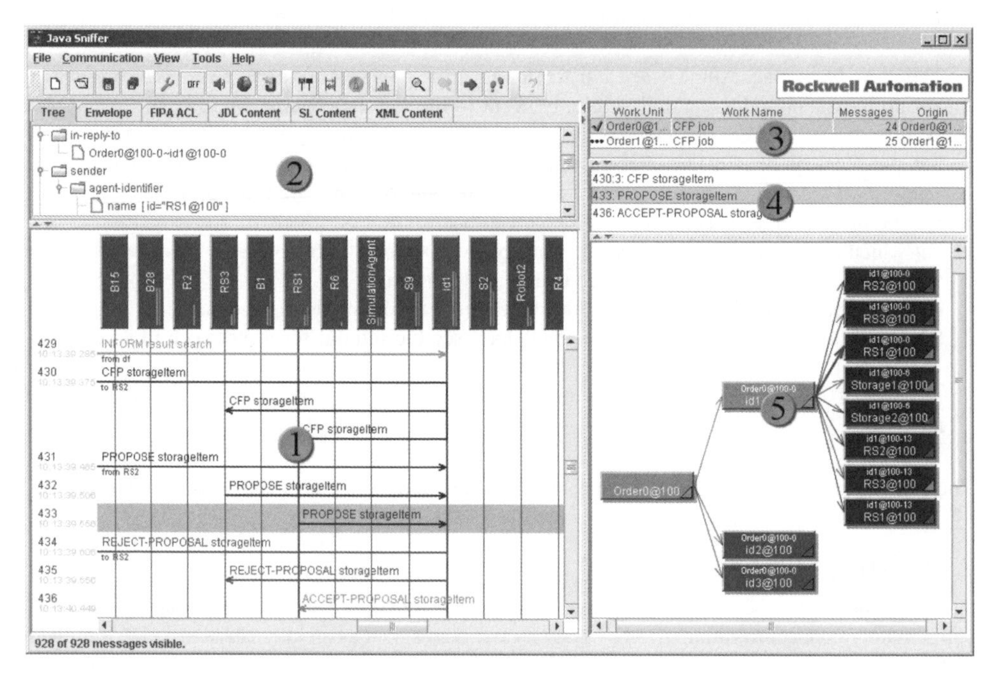


Figure 13.3 Main window of Java Sniffer application

4. *Workflow Detail view*. This shows details of either a selected agent or selected arrow in the Workflow view. If the list is filled with messages that are associated with the selected arrow the user may select one message from the list. The selected message will be automatically shown in the UML view and the Message Detail view will be also updated.
5. *Workflow view*. The window located at the bottom right corner displays a dynamically created tree of a workflow that belongs to the work unit selected in the list of work units. Any message from a FIPA CFP protocol or any request for, and reply to, JDL planning, commitment or execution of the work is visualized as an arrow pointing from the parent (creator of the work) to a child (solver) agent. The arrow represents all messages belonging to a particular part of the work. The Workflow view shows the whole tree automatically constructed from all parts of the work unit. It is possible to trace a workflow message by message, i.e. replay its dynamic construction.

All parts of the visualization screen are interconnected. For example, it is possible to identify some problem in the workflow window, select the appropriate part of the conversation among the agents and receive the list of messages involved in the conversation. Any message in the list can be selected and automatically displayed in the Message Detail view. Also, upon selecting a message in the list of messages, the appropriate workflow and conversation in this workflow are automatically selected. Therefore, the user can seamlessly switch between low-level and high-level visualization views.

Various filters can be used to reduce the number of messages that are logged or displayed. An agent filter can be used to suppress appearance of selected agents from the UML view. A message filter can be used to suppress visualization of selected message types. It is also possible to set up automatic periodic saving of messages to a file after a given number of messages. The user can also limit the number of messages kept in the memory so that upon arrival of a message that would exceed the selected limit, the oldest message is removed and all views are refreshed to reflect this change.

The user can display statistical information about communication in the system based on various settings (see Figure 13.4). Each series can have a set of agents assigned and consider a combination of incoming, outgoing or internal messages. It is possible to choose the time interval for the whole

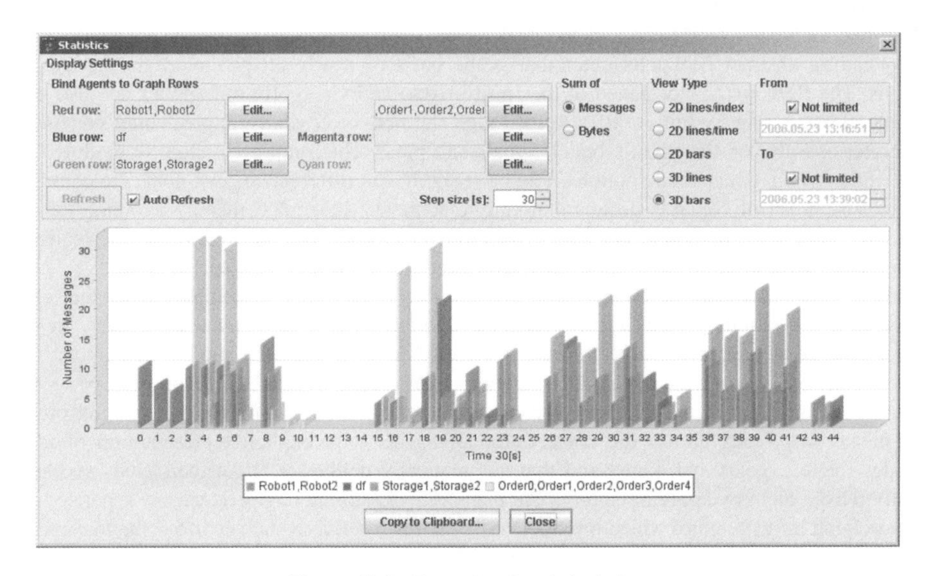


**Figure 13.4** Example of statistical view

graph and step sizes for columns. A variety of 2D and 3D views can be used and the resulting diagram can be stored to the system clipboard.

## 13.4 JADEX – ENGINEERING GOAL-ORIENTED AGENTS

In previous sections of the book agents have been considered as software artefacts that differ from objects mainly in their capability to autonomously execute tasks and to decide for themselves if and how something will be done. Following this view it is possible to design autonomous entities that communicate to solve application problems by cooperation or negotiation. An even more abstract view can be obtained if, besides autonomy, one explicitly supports the *proactiveness* of agents. Proactiveness basically means that agents can have their own goals and have behavioural freedom in which ways the goals are pursued, i.e. which means are used to fulfil the goals.

Most software systems and therefore also software agents, are goal-oriented in the way that they pursue the design goals laid down by the developer of the system. In this model, software artefacts only have implicit goals and are not aware of them. In contrast, proactiveness is closely related to the explicit representation of goals and other mental attitudes. By using explicit representations, agents can be made aware of their goals and have the opportunity to reason about them, e.g. to find alternative ways to achieve goals, giving up unreachable goals or adopting new ones given the right opportunities.

Explicit representations of goals and other mental attitudes have been advocated by many well-known researchers such as Dennett (1987), McCarthy (1979) and Newell (1990) as they drastically affect the human ability to understand and predict the behaviour of systems. This provides a natural abstraction layer on which software agents can be built.

Several approaches exist that propose different kinds of mental attitudes and their relationships. One prominent approach is the BDI model that was originally conceived by Bratman (1987) as a philosophical theory of practical reasoning explaining human behaviour with the attitudes: beliefs, desires and intentions. The basic assumption of the BDI model is that actions are derived in a two-step process called *practical reasoning*. In the first step – (goal) *deliberation* – it is decided which set of desires should be pursued in the current situation represented in the agent's beliefs. The second step – *means–end reasoning* – is responsible for determining how such concrete desires produced as a result of the previous step can be achieved by employing the means available to the agent (Wooldridge, 2000).

Rao and Georgeff (1995) adopted the BDI model for software agents by introducing a formal theory and an abstract BDI interpreter that is the basis of nearly all historical and current BDI systems. The BDI systems designed in the spirit of Rao and Georgeff's interpreter are also called Procedural Reasoning Systems (PRS), termed after the first successfully implemented system. The interpreter operates on the agent's beliefs, goals and plans which represent slightly modified concepts of the original mentalistic notions. The most significant difference is that goals are considered as a consistent set of concrete desires that can be achieved altogether, thereby avoiding the need for a complex goal deliberation phase. The main task of the interpreter therefore is the realization of the means–end process by selecting and executing plans for a given goal or event.

This section describes one of these interpreters – the Jadex BDI reasoning engine that allows development of rational agents using mentalistic notions in the implementation layer. In other words, it enables the construction of rational agents following the BDI model. In contrast to all other available BDI engines, Jadex fully supports the two-step practical reasoning process (goal deliberation and means–end reasoning) instead of operationalizing only the means–end process. This means that Jadex allows the construction of agents with explicit representation of mental attitudes (beliefs, goals and plans) and that automatically deliberate about their goals and subsequently pursue them by applying appropriate plans. The reasoning engine is clearly separated from its underlying infrastructure, which provides basic platform services such as life-cycle management and communication. Hence, running Jadex over JADE combines the strength of a well-tested agent middleware with the abstract BDI execution model.

For the programming of agents, the engine relies on established techniques, such as Java and XML and, to further simplify the development task, Jadex includes a rich suite of run-time tools that are based upon the JADE administration and debugging tools. It also includes a library of ready-to-use generic functionalities provided by several agent modules (capabilities).

### *13.4.1 JADEX ARCHITECTURE*

In Figure 13.5 an overview of the abstract Jadex architecture is presented. Viewed from the outside, an agent is a black box capable of sending and receiving messages. The interpreter handles *practical reasoning* via two components responsible for *goal deliberation* and *means–end reasoning*. The goal deliberation is mainly state-based and has the purpose of selecting the current non-conflicting set of goals.

Incoming messages, as well as internal events and new goals, serve as input to the means–end reasoning component that dispatches these events to plans selected from the plan library for further processing. Running plans may access and modify the belief base, send messages to other agents, create new top-level or subgoals, and cause internal events. The interpreter represents the only global component within Jadex. All other components are contained in reusable modules called capabilities. In the following sections a short introduction to these programming concepts is provided; for more detailed material the reader can refer to Braubach *et al.* (2005). The internal mode of operation, which is quite different from traditional BDI systems, has been formally explained in Pokahr *et al.* (2005).

### *13.4.2 AGENT PROGRAMMING CONCEPTS*

Contrary to most other BDI systems, Jadex intentionally does not promote a new agent programming language but instead relies completely on established software engineering techniques such as Java and XML. As a BDI agent consists of structural as well as behavioural parts, Jadex chooses a hybrid approach for defining and programming agents. The structural part comprises the agent's

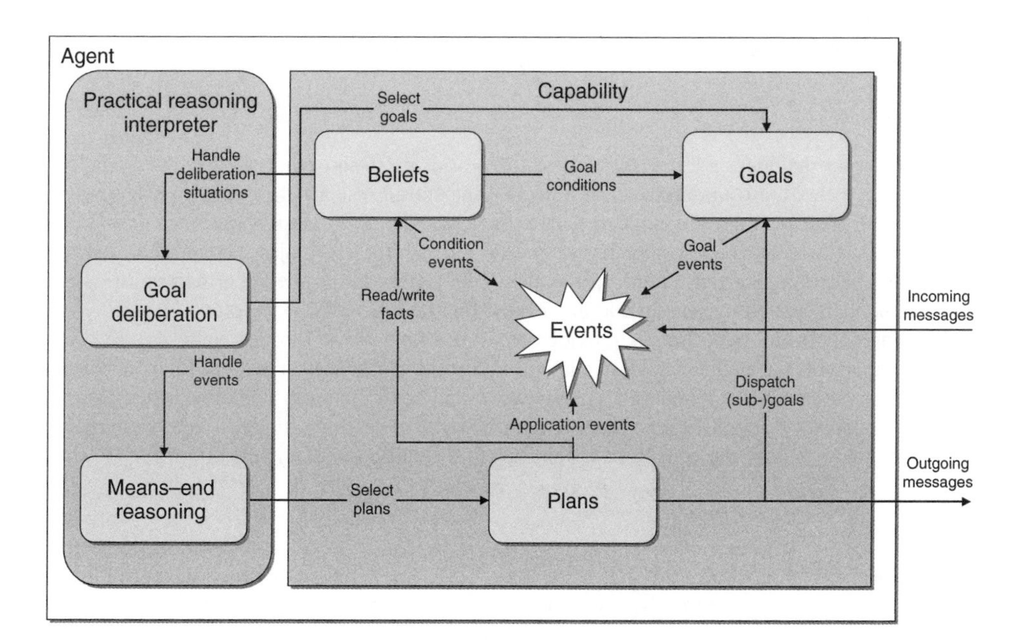


**Figure 13.5** Jadex abstract architecture

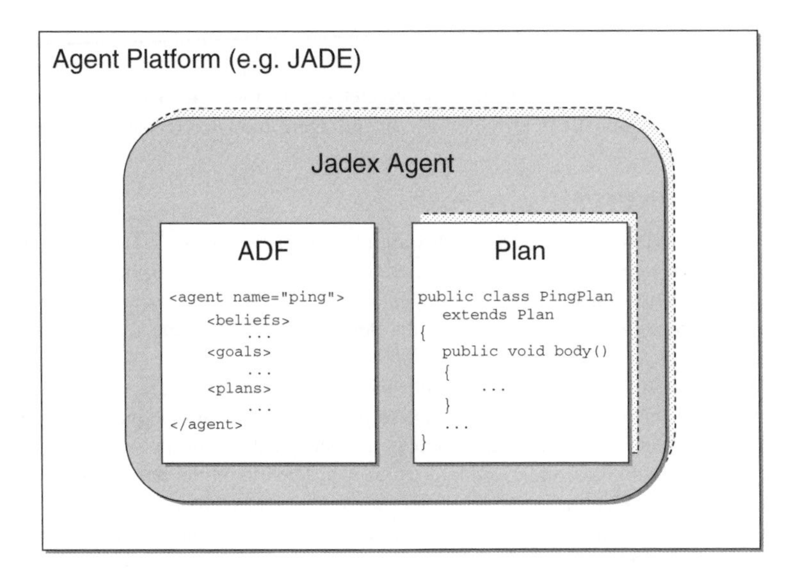


**Figure 13.6** Jadex agent specification

static design composed of elements such as belief, goal and plan types and additionally the agent's initial state consisting of e.g. goals and plans that will be pursued once the agent is activated. These structural aspects are specified in the Jadex XML language following an XML-Schema defining the BDI metamodel. On the other hand, the dynamic agent behaviour needs to be encoded also. All such procedural knowledge is contained in the Jadex plans and is described using plain Java.

Figure 13.6 illustrates the two conceptually different parts of a Jadex agent. On the left-hand side is the agent definition file (ADF) for the agent type and on the right-hand side are the procedural plans. The connection between these parts is established by an API enabling the Java plan classes to access BDI aspects such as reading the beliefs or issuing new subgoals.

#### 13.4.2.1 Beliefs

Beliefs represent the agent's knowledge about the world, itself and other agents. The belief representation in Jadex allows arbitrary Java objects to be stored instead of relying on a logic-based representation. This facilitates integration with existing software, e.g. classes generated by ontology modelling tools or database mapping layers can be directly reused. Objects are stored as named facts (called beliefs) or named sets of facts (called belief sets). Using the belief names, the 'belief base' can be manipulated by setting, adding or removing facts. In addition, a more declarative way of accessing beliefs and belief sets is provided by OQL-like queries. Besides being a passive data store for the agent, the belief base also plays a vital role in the reasoning processes as changes in beliefs are automatically monitored by the engine and conditions can be defined, referring, for example, to domain-relevant belief values. Hence belief changes may trigger a goal's creation or drop condition, or render the context of a plan invalid leading to a plan abort.

#### 13.4.2.2 Goals

Goals are the motivational force driving an agent's actions. They come in different flavours allowing various attitudes of an agent to be expressed. Jadex currently supports four application-relevant goal types: perform, achieve, query and maintain goals. *Perform goals* express an agent's wish to directly engage in actions which are already satisfied if something else has been achieved. Such behaviour is suitable for abstractly modelling activities such as driving around with a car just for fun. In contrast to such activity-centric goals, *achieve goals* are associated with a desired world state. An example of this would be having the car parked close to its driving destination. In this case the state and position of a car are relevant for the goal's fulfilment state and hence express the goal's target world state. Therefore, achieve goals clearly emphasize the decoupling of what is to be achieved and how it will be brought about. *Query goals* instead can be used for information retrieval; their outcome is not defined by some target condition but by a query that needs to be answered. Depending on whether the agent's current knowledge plans may or may not be executed, it means that if the information is readily available no additional work will be necessary. An example for a query goal is finding out the way to drive a car to an intended destination. Finally, *maintain goals* are used to describe situations that should be preserved by the agent under all circumstances. Strictly speaking, it means that whenever a specified situation is violated the agent will activate any applicable means to re-establish the desired world state. Ensuring that a car continues to function properly is an example for such a kind of goal. More information about the goal representation in Jadex can be found in Braubach *et al.* (2005).

When describing real-world scenarios with goals the problem often arises that not all of the individual agent goals can be pursued at the same time. Such goal interferences are an important design aspect that is directly addressed by the goal deliberation facilities of Jadex. With the built-in *Easy Deliberation* strategy (Pokahr *et al.*, 2005), goal cardinalities and inhibition links between goals can be modelled at design time, with the system ensuring that during run-time only valid goal subsets are active. If a goal set contains conflicting goals, the system will exploit the defined inhibition links to delay less important goals while executing the more important ones. Whenever goals are finished, the system considers the reactivation of currently inhibited goals.

#### 13.4.2.3 Plans

Means–end reasoning is performed with the objective of determining suitable plans for pursuing goals or handling other kinds of events such as messages or belief changes. Instead of performing planning from first principles, PRS systems such as Jadex use the plan-library approach to represent the plans of an agent. A plan consists of two distinct parts: the plan head and the plan body. The plan head contains information about the situations in which the plan will be used. Most importantly it includes the events and goals the plan can handle and conditions that are used to restrict the applicability. Using conditions it is also possible to abort a running plan immediately if the current situation demands it. The plan body represents the recipe of actions that will be performed if the plan is chosen for execution. Depending on the plan's purpose, its degree of abstractness varies continuously between very concrete and fully abstract. Concrete plans are fully specified at design time and consist of directly executable actions, whereas fully abstract plans are specified in terms of subgoals only.

Plan programming in Jadex requires the definition of the plan head in the ADF and the programming of the plan body in a pure Java class. The body is implemented by extending an existing Jadex framework class. This enables plan classes to access agent and BDI-specific functionality such as sending messages, accessing beliefs or dispatching subgoals. Errors that may occur in BDI processing, such as a failed subgoal or a timeout when waiting for a reply message, are mapped to BDI exceptions, causing the plan to fail if not intentionally caught by the programmer. Besides the main body, a plan may include special methods for clean-up operations needed to properly complete a plan in the case of either failure or success.

#### 13.4.2.4 Capabilities

A capability (Busetta *et al.*, 2000) results from the packaging of specific functionality into a module with precisely defined interfaces. An agent can be composed of an arbitrary number of capabilities that themselves may include any number of subcapabilities. A capability description is defined in a separate XML document similar to the ADF and also consists of beliefs, plans and goals needed to generate intended behaviour. Per default all capability elements have local scope and thus cannot be seen or used in other capabilities or in the agent. This ensures a maximum degree of encapsulation and avoids any kind of unintended interference between different capabilities. Those parts of the capability that constitute the capability's interface need to be explicitly exported.

Jadex contains several generic plans and predefined capabilities in the package `jadex.planlib`. Basic platform features can be accessed by using the AMS and DF capabilities. The AMS capability offers goals for agent management such as creating new agents or destroying existing ones, whereas the DF capability can be used for accessing yellow pages services such as registering agents or searching specific services via goals. Furthermore, from the *protocols capability* several well-known FIPA interaction protocols such as Request and Contract-Net are available via several goals. The rationale behind the goal-oriented view provided also for protocols is that it allows adoption of a more abstract viewpoint that concentrates not on message flows but rather on the domain activities within a protocol.

### *13.4.3 SUMMARY*

The Jadex BDI reasoning engine allows the development of rational agents using mental notions. Significant features of the engine include full support for the goal deliberation and means–end reasoning phases of the reasoning process, and the explicit representation of mental attitudes, such as beliefs, goals and plans.

The engine is specifically designed to build on established software engineering principles and practices as an independent layer that can be flexibly deployed on middleware platforms such as JADE. In terms of agent programming, the engine relies on established techniques, such as Java and XML, allowing easy development of agents in mature state-of-the-art environments such as Eclipse and IntelliJ IDEA. To further simplify development, the Jadex distribution includes a rich suite of run-time tools for administration and debugging purposes, and a library of ready-to-use generic functionalities provided by several agent modules (capabilities).

The Jadex BDI reasoning engine enables the construction of *complex real-world applications* by exploiting the ideas of intentional systems going back to Denmett and McCarthy. The high-level intentional view of the system-to-be can be preserved in the Jadex implementation, leading to easily understandable and effectively manageable solutions.

# Appendix A

# Command Line Options

This appendix lists the set of options that can be passed from the JADE command line. It also provides a description of how programmers can extend this list and retrieve command line parameters directly from their code.

## A.1 SYNTAX

The following is the EBNF syntax of the JADE command line; common rules apply for the token definitions:

```
CommandLine      = "java jade.Boot" Option* UserDefOption*
   AgentSpecifier*

Option           = ParameterName ParameterValues?

ParameterValues = ParameterValue
                 | ParameterValue ( ";" ParameterValue)*

ParameterValue   = Word | Number | String

UserDefOption    = ParameterName ParameterValue

AgentSpecifier   = AgentName ":" FullyQualifiedClassName ("("
   Argument* ")")?

AgentName        = Word | Number

FullyQualifiedClassName = (PackageName ".")* ClassName

PackageName      = Word

ClassName        = Word

Argument         = Word | Number | String
```

*Developing Multi-Agent Systems with JADE* Fabio Bellifemine, Giovanni Caire, Dominic Greenwood

### A.1.1 LEXICAL DEFINITIONS

```
ParameterName   =   "-" Word
```

The usual lexical definitions apply for Word, Number and String.

*Remember that the* `"-"` *symbol must always precede the parameter name and that symbol should not be considered part of the name.*

The command line can be extended with any user-defined option; notice that they only differ from the JADE options in that they must always be followed by a parameter value.

The AgentSpecifier part of the command allows the specification of a list of agents to be launched. Each AgentSpecifier should be separated by a space and is composed of three parts: the first substring (delimited by the ':' colon symbol) identifies the agent name; the second substring (delimited by an ending space or by an open parenthesis) identifies the fully qualified class name of the class that implements the agent (i.e. the class that extends `jade.core.Agent`); the final substring is an optional sequence of space-delimited arguments collected in parenthesis, to be passed to the agent. Example:

```
prompt> java jade.Boot Tom:jade.core.Agent(arg1 "argument 2" 3)
                       Bob:jade.core.Agent
```

The following tables list all available parameters. When nothing is specified for the parameter value, the option has no parameter and defaults to `true` when the option is passed on the command line, `false` otherwise.

## A.2 OPTIONS TO LAUNCH CONTAINERS AND MAIN CONTAINERS

| Parameter name | Parameter value | Description |
|---|---|---|
| `container` | | Launches a JADE container that, must join to, and register with, a main-container. The Host and Port parameters specify which main container to join (defaults to a main container) |
| `host` | host name | Host name of the device hosting the main container with which to register. This option can also be used when launching the main container in order to override the value of localhost and, for instance, pass the full domain of the host to allow the main container to be contactable from outside the local domain: e.g. `-host kim.cselt.it` when the localhost would have returned just `'kim'` (defaults to localhost) |
| `port` | port number | Port number where the RMI registry of the main container is listening to or, in case a main container is being launched, where the RMI registry should listen to (defaults to 1099) |
| `local-host` | hostname | Host name where this instance of the container can be contacted. As for the -host option above, it allows overriding of the default localhost value |
| `local-port` | port number | Port number where this container can be contacted (defaults to 1099) |

| Parameter name | Parameter value | Description |
|---|---|---|
| name | name | Name to be assigned to this platform. This option is ignored when launching a non-main container. By default, a globally unique name is generated from the concatenation of host number and port number (<hostname>:<portNumber>/JADE). Note that usage of this option is strongly discouraged since it might result in non-unique agent names |
| container-name | name | Name of the container (defaults to container-<number>) |
| jade_imtp_rmi_RMIIMTPManager_enablermilog | [true\|false] | Enable logging of all Java RMI operations (defaults to false) |
| nodeport | port number | TCP port where the local RMI node is to be exported (defaults to 1099) |

## A.3 GENERAL OPTIONS

| Parameter name | Parameter value | Description |
|---|---|---|
| gui | | Launches a GUI, i.e. an instance of the RMA (Remote Monitoring Agent) (defaults to false) |
| version | | Print JADE versioning information on standard output (defaults to false) |
| help | | Print JADE help on standard output with a summary of the most used options (defaults to false) |
| conf | file name | Read the JADE configuration parameters from the specified file. If the option is used without specifying any file name, then a GUI is launched that allows composition of configuration parameters, loading and saving from/to a file (defaults to false) |
| file-dir | dir name | Name of the directory where all the files produced by JADE will be generated (defaults to the working directory) |
| aclcodec | list of acl codec classes | List of fully qualified class names of ACL Codecs additional to the default String-based ACLCodec. Usage of these codecs is activated through the proper value of the field *aclRepresentation* of the Envelope of the sent/received ACLMessages. Look at the FIPA specifications for the standard names of these codecs (defaults to jade.lang.acl.StringACLCodec) |

## A.4 OPTIONS OF THE JADE KERNEL-LEVEL SERVICES

| Parameter name | Parameter value | Description |
|---|---|---|
| services | list of service classes | List of classes, each implementing a JADE kernel-level service, that must be loaded and activated during start-up. Class names must be separated by a semicolon, and must be fully qualified. The special messaging and management services are always activated and they do not need to be specified in this option. If the option is not used, by default the mobility and event notification services are started (default to jade.core.mobility.AgentMobilityService;jade.core.event.NotificationService) |

The following eight services are available with the JADE distribution, while others can be downloaded as add-ons or third-party software:

| Name | Fully qualified class name | Description |
|---|---|---|
| Messaging | jade.core.messaging.MessagingService | ACL message exchange and MTP management. Always active |
| Agent-Management | jade.core.management.AgentManagementService | Basic management of agent life cycle, container and platform life cycle. Always active |
| Agent Mobility | jade.core.mobility.AgentMobilityService | Support for agent mobility. Active by default |
| Notification | jade.core.event.NotificationService | Support for platform-level event system notifications. This is required to support the Sniffer and Introspector tools. Active by default |
| Persistent-Delivery | jade.core.messaging.PersistentDeliveryService | Allows buffering and persistent storage of undelivered ACL messages. Inactive by default |
| Main-Replication | jade.core.replication.MainReplicationService | Support for replicating the main container for fault tolerance purposes. The service must be activated on each node hosting a main container (see Section 9.1 for details). Inactive by default |
| Address-Notification | jade.core.replication.AddressNotificationService | Support for distributed notification of the list of active Main Containers (see 9.1 for details). Inactive by default |
| UDPNode Monitoring | jade.core.nodeMonitoring.UDPNodeMonitoringService | Support for platform nodes monitoring through UDP protocol instead of default TCP protocol. Inactive by default |

*A.4.1 OPTIONS OF THE MESSAGING SERVICE*

| Parameter name | Parameter value | Description |
|---|---|---|
| `jade_core_messaging_MessagingService_attachplatforminfo` | [true\|false] | When this option is specified, the JADE version, Java version, Operating System version and Operating System name are automatically attached to the Envelope of all messages directed towards agents residing outside the platform (defaults to false) |
| `jade_core_messaging_MessagingService_cachesize` | number | Specifies the size of the cache of the Global Descriptor Table kept by the Messaging Service to know where agents live without contacting the Main Container each time as described in Section 3.3 (defaults to 100) |
| `jade_core_messaging_MessageManager_poolsize` | number | The size of the pool of threads used by the Messaging Service to asynchronously deliver messages to recipients (defaults to 5) |
| `jade_core_messaging_MessageManager_maxqueuesize` | number | The size in bytes of the Messaging Service internal queue where messages are buffered before asynchronous delivery (defaults to 10 000 000, i.e. 10 Mbytes) |

*A.4.2 OPTIONS OF THE AGENT MANAGEMENT SERVICE*

| Parameter name | Parameter value | Description |
|---|---|---|
| `jade_AgentManagement_agentspath` | Path name | The path where the code locator will search for agent jar files (defaults to the current directory) |
| `accept-foreign-agents` | [true\|false] | Used to trigger the platform to accept foreign agents, i.e. agents whose name is not in the form <local-name>@<platform-name>. This is useful when the inter-platform mobility service described in Section 6.3 is used (defaults to false) |

*A.4.3 OPTIONS OF THE AGENT MOBILITY SERVICE*

| Parameter name | Parameter value | Description |
|---|---|---|
| `nomobility` | | This option disables the mobility and cloning support in the launched container for security concerns. In this way the container will not accept requests for agent migration or agent cloning. Notice that a platform can include both containers where mobility is enabled and containers where it is disabled. In this case an agent that tries to move from/to the containers where mobility is disabled will terminate due to a Runtime Exception. By default, mobility and cloning are supported |

### A.4.4 OPTIONS OF THE PERSISTENT DELIVERY SERVICE

| Parameter name | Parameter value | Description |
|---|---|---|
| persistent-delivery-filter | className | Fully qualified class name of the application-specific class that will be asked to filter all the undelivered ACL Messages |
| persistent-delivery-sendfailureperiod | Number | The Number parameter represents how often (in milliseconds) the Persistent Delivery Service will try to send all buffered, previously undelivered ACL messages (defaults to 60 000, i.e. 1 minute) |
| persistent-delivery-storagemethod | className | Fully qualified class name of the class that is used by the service to persist undelivered ACL messages. The class must implement the `jade.core.messaging.MessageStorage` interface. The class `jade.core.messaging.FileMessageStorage` is available to store messages into files. Defaults to `jade.core.messaging.PersistentDeliveryManager$DummyStorage` which persists messages in memory and not to files |
| persistent-delivery-basedir | dirName | Name of the directory where files with undelivered ACL messages should be saved/loaded from/to the persistent delivery service (defaults to the working directory) |

### A.4.5 OPTIONS OF THE MAIN REPLICATION SERVICE

| Parameter name | Parameter value | Description |
|---|---|---|
| backupmain | | Used to launch a backup-type main container (defaults to false) |
| smhost | host name | For a backup-type main container, this option must be used to select the host name where the Service Manager is to be exported (typically it is the local host). This option is only useful on nodes where the Main Replication service has been activated |
| smport | port number | For a backup-type main container, this option must be used to select the TCP port where the Service Manager is to be exported. This option is only useful on nodes where the Main Replication service has been activated |
| smaddrs | list of addresses | This option lists the addresses of all the backup-type instances of main container so that the container can select which one to register with or which one to re-register when the current one becomes unavailable. The order of selection is firstly the main container specified by the (-host, -port) couple of parameters (as usual), then those specified into the `smaddrs` parameters.<br>This option is an alternative to activating the Address Notification service: instead of setting up an update protocol, one simply provides a fixed list of well-known addresses for main container backups |

### A.4.6 OPTIONS OF THE UDP-BASED NODE MONITORING SERVICE

| Parameter name | Parameter value | Description |
|---|---|---|
| jade_core_nodeMonitoring_UDPNodeMonitoringService_port | portNumber | Specifies the port number where the main container will listen for UDP pings (defaults to 28 000) |
| jade_core_nodeMonitoring_UDPNodeMonitoringService_pingdelay | Number | Defines the time interval (in milliseconds) between two UDP pings sent by a peripheral container to the main container (defaults to 1000) |
| jade_core_nodeMonitoring_UDPNodeMonitoringService_pingdelaylimit | Number | Defines the maximum time (in milliseconds) the main container will wait for incoming ping messages before considering a peripheral container 'Unreachable' (defaults to 3000) |
| jade_core_nodeMonitoring_UDPNodeMonitoringService_unreachablelimit | Number | Defines the maximum time (in milliseconds) a peripheral container can be temporarily unreachable until it is removed from the platform (defaults to 10 000) |

It should be noted that all options related to the UDPNodeMonitoringService are only specified on the main container. The main container automatically propagates them to peripheral containers. Peripheral containers only need to be configured so that the UDPNodeMonitoringService is active by means of the `-services` option.

## A.5 OPTIONS RELATED TO MTPS

| Parameter name | Parameter value | Description |
|---|---|---|
| mtps | list of MTPs | List of Message Transport Protocols to be activated on this container. By default the HTTP MTP is activated only on main containers. Each MTP must be identified by the fully qualified class name of the class that implements the `jade.mtp.MTP` interface; optionally an address can be passed, between brackets, to each MTP to indicate the preferred listening address for incoming connection; each element of the list must be separated by a ';'. See Section 3.6 for an example |
| nomtp | | Used to indicate that no MTP should be activated. This has precedence over the `-mtp` option. By default the HTTP MTP is activated on main containers and no MTP on non-main containers |

### A.5.1 OPTIONS OF THE HTTP MTP

| Parameter name | Parameter value | Description |
|---|---|---|
| jade_mtp_http_port | port number | Local port number to listen for incoming HTTP connections (defaults to 7778) |
| jade_mtp_http_outPort | port number | Local port number to be used for all outgoing HTTP connections By default, a random port is used |
| jade_mtp_http_numKeepAlive | number | Maximum number of persistent connections that the MTP is allowed to keep. A value of 0 inhibits use of keep-alive connections (defaults to 10) |
| jade_mtp_http_timeout | number | The Server part of the HTTP MTP keeps alive incoming connections. This timeout (specified in milliseconds) is the maximum idle time for these connections intended to improve memory consumption. A value of 0 means infinite time (defaults to 60 000, i.e. 1 minute) |
| jade_mtp_http_proxyHost | host name | Host name and port number of the HTTP Proxy to redirect the platform outgoing messages to. If these parameters are not specified, then connections will be opened directly to the remote platforms |
| jade_mtp_http_proxyPort | port number | |
| jade_mtp_http_parser | saxClass name | Specification of which XML Parser to use. The fully qualified class name of the SAX XML Parser must be indicated. This option is mandatory with JDK 1.3 or earlier (defaults to org.apache xerces.parsers.SAXParser) |
| jade_mtp_http_https_keyStoreFile | fileName | Indicates the name of the file of the Java keystore file of the JADE platform |
| jade_mtp_http_https_keyStorePass | password | Indicates the password to access the keystore file above |
| jade_mtp_http_https_needClientAuth | [true\|false] | When using the HTTPS protocol, this option indicates whether platforms attempting to send messages to the local platform must first be authenticated. This feature restricts the set of platforms allowed to communicate with each other (defaults to false) |
| jade_mtp_http_https_friendListFile | fileName | Indicates the name of the file that contains the list of trusted certificates |
| jade_mtp_http_https_friendListPass | password | Indicates the password for the keystore containing the list of trusted certificates |

## A.6 OPTIONS TO CONFIGURE THE YELLOW PAGE DF SERVICE

| Parameter name | Parameter value | Description |
|---|---|---|
| jade_domain_df_autocleanup | [true\|false] | When set to true, indicates that the DF will automatically clean up registrations as soon as agents terminate (defaults to false) |
| jade_domain_df_maxleasetime | number | Indicates the maximum lease time (in milliseconds) that the DF will grant for agent description registrations (defaults to infinite) |
| jade_domain_df_maxresult | number | Indicates the maximum number of items found in a search operation that the DF will return to the requester. This will only be applied if there are no explicit search constraints specified (defaults to 100) |
| *Storing the knowledge base of the DF* | | |
| jade_domain_df_kb-factory | className | This parameter allows specification of the name of the factory class which will be used by the DF to create a knowledge base object for storing its catalogue. The specified class must be a subclass of `jade.domain.DFKBFactory`, which is also the default value of this parameter |
| jade_domain_df_db-default | [true\|false] | If set to true, indicates that the DF will store its catalogue into an HSQLDB database, which is started in the same JVM as JADE. This is the easiest way to persist the DF catalogue. Apart from using this parameter, only the HSQLDB class files must be added to the Java CLASSPATH. Any further configuration is done automatically by JADE. HSQLDB is not part of the JADE distribution but can be downloaded from the HSQL Database Engine project at the website: http://sourceforge.net/projects/hsqldb (defaults to false) |
| jade_domain_df_db-url | URL | Defines the JDBC URL of the database other the DF will store its catalogue. With this parameter the DF can be configured to use databases than HSQLDB |
| jade_domain_df_db-cleantables | [true\|false] | If set to `true`, indicates that the DF will clean the content of all pre-existing database tables, used by the DF. This parameter is ignored if the catalogue is not stored in a database (defaults to false) |
| *Configuring the DB Access* | | |
| jade_domain_df_db-driver | className | Indicates the JDBC driver to be used to access the DF database (defaults to the ODBC-JDBC bridge) |
| jade_domain_df_db-username | username | Indicates the user name to be used to access the DF database |
| jade_domain_df_db-password | password | Indicates the password to be used to access the DF database |

## A.7 OPTIONS SPECIFIC TO THE JADE-LEAP PLATFORM

| Parameter name | Parameter value | Description |
|---|---|---|
| icps | list of class names | This option allows specification of the ICP (Internal Communication Peer) to be activated by the singleton Command Dispatcher of the LEAP IMTP as described in Section 8.2.2 (defaults to jade.imtp.leap.JICP.JICPPeer) |
| proto | protocol name | This option must be specified when the JADE LEAP main container has to be contacted by means of a protocol different than the default JICP (e.g. http) (defaults to jicp) |
| jade_imtp_leap_ JICP_JICPServer_ acceptmediators | [true\|false] | This option enables/disables of the capability of a LEAP ICP to act as mediator and accept CREATE_MEDIATOR requests from front-ends (defaults to false) |
| *Split container (front-end) start-up* | | |
| host | host name | The host name or address of the mediator (defaults to the local host) |
| port | port number | The port where the mediator is accepting connections from front-ends (defaults to 1099) |
| exitwhenempty | [true\|false] | When setting this option to true, a split container will automatically terminate as soon as there are no more agents on it |
| connection-manager | class name | This option allows specification of a class implementing the jade.core.FEConnectionManager interface to be used to manage (front-end side) the connection between the front-end and the back-end (see Section 8.5.5.1) (defaults to jade.imtp.leap.JICP.BIFEDispatcher) |
| mediator-class | class name | This option allows specification of a class implementing the jade.imtp.leap.JICP.JICPMediator interface to be used to manage (back-end side) the connection between the front-end and the back-end (see Section 8.5.5.1) (default: depends on the value of the connection-manager option) |
| msisdn | telephone number | This option allows simulation of a system able to detect the telephone number of the device where the front-end is starting. As a result the newly born split container will be named using this telephone number. This option is ignored when a real system able to retrieve the device's telephone number is connected to the mediator by means of the PDPContextManager interface (see Section 8.5.5.3) (default none) |

| Parameter name | Parameter value | Description |
|---|---|---|
| owner | string | When specified, this property is propagated both to the back-end and (if present) to the PDPContext manager. It may be used to build an application-specific authentication mechanism (default none) |
| max-disconnection-time | number | The maximum amount of time (expressed in milliseconds) the front-end will attempt to re-establish the connection with the back-end when it goes down (defaults to 600 000, i.e. 10 min) |
| reconnection-retry-time | number | The time (expressed in milliseconds) between each reconnection retry attempted by the front-end when it is trying to re-establish the connection with the back-end (defaults to 10 000, i.e. 10 sec) |
| keep-alive-time | number | The time (expressed in milliseconds) between two successive keep-alive packets exchanged between the front-end and the back-end to keep the connection alive. A value of −1 can be specified to disable the keep-alive mechanism (defaults to 60 000, i.e. 1 min). |
| *Back-End Management Service configuration* | | |
| jade_imtp_leap_nio_BEManagementService_local-port | port number | This option allows specification of the port used by the BEManagementService to listen to incoming connections from front-ends (defaults to 2099) |
| jade_imtp_leap_nio_BEManagementService_poolsize | number | This option sets the size of the pool of threads used by the BEManagementService to read data from all the connections (defaults to 5) |
| jade_imtp_leap_nio_BEManagementService_leap-property-file | file name | This option specifies the name of a property file including configuration options to be applied to all back-ends that will be created by the BEManagementService (defaults to ./leap.properties) |

## A.8 EXTENDING THE COMMAND LINE WITH USER-DEFINED OPTIONS

The class `jade.core.Agent` provides a simple, yet powerful mechanism to access command line options and parameters directly from the programmer's agent code with the method:

```
public String getProperty(String key, String defaultValue)
```

where the first argument is the parameter name and the second argument is the default value returned if that parameter was not passed on the command line.

Note that the method applies to both the JADE command line options described in this appendix and to any user-defined options. For instance, if the following command line was used to launch JADE:

```
prompt> java jade.Boot -gui -myParam myValue
```

then the execution of the method `getProperty("myParam", null)` on any agent running in that container would return the String `"myValue"` and the execution of the method `getProperty("gui", null)` would return the String `"true"`.

# Appendix B

# List of Symbols and Acronyms

| | |
|---|---|
| ACC | Agent Communication Channel |
| ACL | Agent Communication Language |
| ACTS | Advanced Communications Technologies and Services (EU) |
| ADF | Agent Definition File |
| AID | Agent Identifier |
| AMS | Agent Management System |
| ANS | Address Notification Service of JADE |
| AP | Agent Platform |
| API | Application Programming Interface |
| ASCML | Agent-Society Configuration, Manager, and Launcher |
| AUML | Agent Unified Modeling Language |
| AWT | Abstract Window Toolkit |
| BDI | Belief Desire Intention |
| BE | Back-End |
| CA | Communicative Act |
| CDC | Connected Device Configuration |
| CFP | Call For Proposal |
| CLDC | Connected-Limited Device Configuration |
| CNP | Contract Net Protocol |
| CPU | Central Processing Unit |
| CT | Container Table |
| DAML | Darpa Agent Markup Language |
| DB | Data Base |
| DCE | Distributed Computing Environment |
| DF | Directory Facilitator |
| DTD | Data Type Definition |
| EBNF | Extended Backus-Naur Form |
| ENS | Event Notification Service of JADE |
| EU | European Union |
| FAB | FIPA Architecture Board |
| FE | Front-End |
| FIFO | First in First out |
| FIPA | Foundation for Intelligent Physical Agents |
| FP | Feasibility Precondition |

*Developing Multi-Agent Systems with JADE* Fabio Bellifemine, Giovanni Caire, Dominic Greenwood

| | |
|---|---|
| FSM | Finite State Machine |
| GADT | Global Agent Descriptor Table |
| GGF | Global Grid Forum |
| GPL | Gnu Public Licence |
| GPRS | General Packet Radio System |
| GSM | Global System for Mobile communication |
| GUI | Graphical User Interface |
| GUID | Globally Unique Identifier |
| HTTP | Hyper Text Transfer Protocol |
| HTTPS | Hyper Text Transfer Protocol Secure |
| ICP | Inter-Container Communication Peer |
| IDE | Integrated Development Environment |
| IDL | Interface Definition Language |
| IEEE | Institute of Electrical and Electronics Engineers |
| IETF | Internet Engineering Task Force |
| IIOP | Internet Inter-ORB Protocol |
| IMTP | Internal Message Transport Protocol |
| IP | Internet Protocol. When referred to FIPA and agents, in the book it is also used to refer Interaction Protocol |
| IPMS | Inter Platform Mobility Service of JADE |
| IRE | Identifying Referential Expression |
| ISBN | International Standard Book Number |
| ISO | International Organization for Standardization |
| IST | Information Society Technologies (EU) |
| J2EE | Java 2 Enterprise Edition |
| J2ME | Java 2 Micro Edition |
| J2SE | Java 2 Standard Edition |
| JAD | Java Application Descriptor |
| JADE | Java Agent DEvelopment Framework |
| JAR | Java Archive File |
| JAS | Java Agent Services |
| JAXB | Java Architecture for XML Binding |
| JCC | Jadex Control Centre |
| JCP | Java Community Process |
| JDBC | Java DataBase Connectivity |
| JDL | Job Description Language |
| JICP | JADE Inter-Container Protocol |
| JMS | Java Message Service |
| JMX | Java Management Extension |
| JSA | JADE Semantics Add-on of France Télécom |
| JSP | Java Server Pages |
| JSR | Java Specification Request |
| JVM | Java Virtual Machine |
| JWT | J2ME Wireless Toolkit |
| KB | Knowledge Base |
| KIF | Knowledge Interchange Format |
| KQML | Knowledge Query Manipulation Language |
| LADT | Local Agent Descriptor Table |
| LEAP | Lightweight Extensible Agent Platform. It is an add-on of JADE |
| LGPL | Library Gnu Public Licence |
| LRU | Least Recently Used |

| | |
|---|---|
| MAS | Multi-Agent System |
| MCRS | Main Container Replication Service of JADE |
| MIDP | Mobile Information Device Profile |
| MTP | Message Transport Protocol |
| MTS | Message Transport Service |
| NAT | Network Address Translation |
| O2A | Object to Agent |
| OIL | Ontology Interchange Language |
| OMG | Object Management Group |
| OQL | Object Query Language |
| OSI | Open Systems Interconnection |
| OWL | Web Ontology Language (W3C) |
| PC | Personal Computer |
| PDA | Personal Digital Assistant (electronic hand-held information device) |
| PRS | Procedural Reasoning System |
| RDF | Resource Description Framework |
| RE | Rational Effect |
| RMA | Remote Monitoring Agent |
| RMI | Java Remote Method Invocation |
| SAX | Simple API for XML |
| SIG | Special Interest Group |
| SIM | Subscriber Information Module |
| SIP | Semantic Interpretation Principle |
| SL | Semantic Language |
| SOAP | Simple Object Access Protocol |
| SQL | Structured Query Language (database query language) |
| SR | Semantic Representation |
| SSL | Secure Socket Layer |
| TC | Technical Committee |
| TCP | Transmission Control Protocol |
| UAB | Universitat Autonoma de Barcelona |
| UDDI | Universal Description, Discovery and Integration |
| UI | User Interface |
| UML | Unified Modelling Language |
| UMTS | Universal Mobile Telecommunications System |
| URL | Uniform Resource Locator |
| W3C | World Wide Web Consortium |
| WAP | Wireless Application Protocol |
| WG | Work Group |
| WLAN | Wireless Local Area Network |
| WSDL | Web Service Description Language |
| WSIG | Web Services Integration Gateway |
| XML | Extensible Markup Language |
| XMPP | Extensible Messaging and Presence Protocol |
| XP | eXtreme Programming |

# Bibliography

## References


Agentcities website, http://www.agentcities.org

Aksit, M., Wakita, K., Bosch, J., Bergmans, L. and Yonezawa, A. Abstracting Object Interactions Using Composition Filters. In *Proceedings of the ECOOP 1993 Workshop on Object-Based Distributed Programming*, pp. 152–184, Springer-Verlag, 1993.

Albert, M., Laengle, T., Woern, H., Capobianco, M. and Brighenti, A. Multi-agent Systems for Industrial Diagnostics. In *Proceedings of 5$^{th}$ IFAC Symposium on Fault Detection, Supervision and Safety of Technical Processes*, pp. 483–488, Washington, DC, June 9–11 2003.

ANT, The Apache ANT Project tool, http://ant.apache.org

Aspect-Oriented Software Development, http://aosd.net

Bellifemine, F., Poggi, A. and Rimassa, G. Developing Multi Agent Systems with a FIPA-Compliant Agent Framework. In *Software–Practice and Experience*, John Wiley & Sons, Ltd, vol. 31, pp. 103–128, 2001.

Bordini, R.H., Hübner, J.F. and Vieira, R. Jason and the Golden Fleece of Agent-oriented Programming. In Bordini, R., Dastani, M., Dix, J. and Seghrouchni, A. (eds), *Multi-Agent Programming*, Kluwer, 2005.

Bordini, R.H., Braubach, L., Dastani, M., Seghrouchni, A.E.F., Gomez-Sanz, J.J., Leite, J., O'Hare, G., Pokahr, A. and Ricci, A. A Survey of Programming Languages and Platforms for Multi-agent Systems. *Informatica*, **30**(1): pp. 33–44, 2006.

Bratman, M. *Intention, Plans, and Practical Reason*. Harvard University Press, 1987.

Braubach, L., Pokahr, A., Lamersdorf, W. Jadex: a BDI-Agent System Combining Middleware and Reasoning. In Walliser, M., Brantschen, S., Calisti, M. and Hempfling, T. (eds), *Whitestein Series in Software Agent Technologies*, Birkhäuser-Verlag, Springer Science+Business Media, Berlin, New York, 2005.

Brooks, R. Intelligence without Representation. *Artificial Intelligence*, **47**: pp. 139–159, 1991.

Brown, P. and Rossak, W. *Mobile Agents*. Morgan Kaufmann Publishers and dpunkt.verlag, 2005.

Buckle, P., Moore, T., Robertshaw, S., Treadway, A., Tarkoma, S. and Poslad, S. Scalability in Multi-agent Systems: the FIPA-OS Perspective. In d'Inverno, M., Luck, M., Fisher, M. and Preist, C. (eds), *Proceedings Foundations and Applications of Multi-Agent Systems*, vol. 2403 of *LNCS*, pp. 110–130, Springer, 2002.

Buhler, P.A. and Vidal, J.M. Towards Adaptive Workflow Enactment Using Multiagent Systems. *Information Technology and Management*, **6**(1): pp. 61–87, 2005.

Busetta, P., Howden, N., Rönnquist, R. and Hodgson, A. Structuring BDI Agents in Functional Clusters. In *Intelligent Agents VI, Agent Theories, Architectures, and Languages* (ATAL'99), LNCS 1757, pp. 277–289, Springer, 2000.

Bussmann, S. and Muller, J. A Negotiation Framework for Co-operating Agents. In Deen, S.M. (ed.), *Proceedings of CKBS-SIG*, pp. 1–17, Keele, UK, 1992.

Camarinha-Matos, L. and Afsarmanesh, H. 2001. Virtual Enterprise Modeling and Support Infrastructures: Applying Multi-agent System Approaches. In Carbonell, J. and Siekmann, J. (eds), *Multi-agents Systems and Applications*, pp. 335–364, Sanibel Island, FL, 2001.

Dastani, M., van Riemsdijk, M.B. and Meyer, J.C. Programming Multi-agent Systems in 3APL. In Bordini, R., Dastani, M., Dix, J. and Seghrouchni, A. (eds), *Multi-Agent Programming*, Kluwer, 2005.

Davies, N.J., Fensel D. and Richardson, M. The Future of Web Services. *BT Technology Journal*, **22**(1): pp. 76–82, January 2004.



*Developing Multi-Agent Systems with JADE* Fabio Bellifemine, Giovanni Caire, Dominic Greenwood

Decker, K. and Sycara, K. Intelligent Adaptive Information Agents. *Journal of Intelligent Information Systems*, **9**(3): pp. 239–260, 1997.

Dennett, D. *The Intentional Stance*. Bradford Books, 1987.

d'Inverno, M., Kinny, D., Luck, M. and Wooldridge, M. A Formal Specification of dMARS. In Singh, M.P., Rao, A.S. and Wooldridge, M. (eds), *Intelligent Agents IV: Proceedings of the Fourth International Workshop on Agent Theories, Architectures, and Languages*, pp. 155–176, Springer, 1998.

Durfee, E. Distributed Problem Solving and Planning. In Weiß, Gerhard, (ed.), *Multiagent Systems: a Modern Approach to Distributed Artificial Intelligence*, pp. 121–164, MIT Press, Cambridge, MA, 1999.

Durfee, E. and Victor, L. Using Partial Global Plans to Coordinate Distributed Problem Solvers. In *Proceedings of the 10th International Joint Conference on Artificial Intelligence*, pp. 875–883, Milan, August 1987.

Estlin, T.A., Gaines, D., Fisher, F. and Rebecca Castaño, R. Coordinating Multiple Rovers with Interdependent Science Objectives. In *Proceedings of the 4th International Joint Conference on Autonomous Agents and Multiagent Systems*, pp. 879–886, Utrecht, The Netherlands, 2005.

Ferguson, I.A. Towards an Architecture for Adaptive, Rational, Mobile Agents. In Werner, E. and Demazeau, Y. (eds), *Decentralized AI 3 – Proceedings of the Third European Workshop on Modelling Autonomous Agents and Multi-Agent World*, pp. 249–262, Elsevier, Amsterdam, The Netherlands, 1991.

FIPA, Foundation for Intelligent Physical Agents, website: http://www.fipa.org

Fricke, S. Bsufka, K., Keiser, J., Schmidt, T., Sesseler, R. and Albayrak, S. *Communications of the ACM*, **44**(4): pp. 43–48, 2001.

Garson G. Quantification in Modal Logic. In *Handbook of Philosophical Logic*, Vol. II: *Extensions of Classical Logic*, pp. 249–307, D. Reidel Publishing Company, 1984.

Genesereth, M.R. and Ketchpel, S.P. Software Agents. *Communications of the ACM*, **37**(7): pp. 48–53, 1994.

Georgeff, M., Communication and Interaction in Multi-Agent Planning. In *Proceedings of the 3rd National Conference on Artificial Intelligence*, pp. 125–129, Washington, DC, 1983.

Georgeff, M., A Theory of Action for Multi Agent Planning, *Proceedings of the 4th National Conference on Artificial Intelligence*, pp. 121–125, Austin, TX, 1984.

Georgeff, M. and Lansky, A. Reactive Reasoning and Planning: an Experiment with a Mobile Robot. In *Proceedings of the 7th National Conference on Artificial Intelligence*, pp. 677–682, Seattle, WA, 1987.

GPL-FAQ, Frequently Asked Questions about the GNU GPL, http://www.gnu.org/copyleft/gpl-faq.html

Goldberg, D., Cicirello, V., Dias, M., Simmons, R., Smith, S., Smith, T. and Stentz, A. A Distributed Layered Architecture for Mobile Robot Coordination: Application to Space Exploration. In *Proceedings of the 3rd International NASA Workshop on Planning and Scheduling for Space*, Houston, TX, 2002.

Greenwood, D., Buhler, P. and Reitbauer, A. Web Service Discovery and Composition using the Web Service Integration Gateway. In *Proceedings of the IEEE International Conference on e-Technology, e-Commerce and e-Service*, pp. 789–790, Hong Kong, China, 2005.

Greenwood, D., Vitaglione, G., Keller, L. and Calisti, M. Service Level Agreement Management with Adaptive Coordination. *Proceedings of the International Conference on Networking and Services* (ICNS'06), Silicon Valley, USA, 2006.

Hayzelden, A.L. and Bourne, R.A. *Agent Technology for Communication Infrastructures*, John Wiley & Sons, Ltd, London, UK, 2001.

Hendler, J., Berners-Lee, T. and Miller, E. Integrating Applications on the Semantic Web. *Journal of the Institute of Electrical Engineers of Japan*, **122**(10): pp. 676–680, 2002.

Howden, N., Ronnquist, R., Hodgson, A. and Lucas, A. JACK Intelligent Agents – Summary of an Agent Infrastructure. In *Proceedings of the 5th International Conference on Autonomous Agents*, 2001.

HTTP, RFC2616, Hypertext Transfer Protocol – HTTP/1.1. Request for Comments, 1999. http://www.ietf.org/rfc/rfc2616.txt

Huber, M. JAM: A BDI-Theoretic Mobile Agent Architecture. In *Proceedings of the 3rd International Conference on Autonomous Agents*, pp. 236–243, New York, NY, 1999.

Hudson, D. and Cohen, M. Use of Intelligent Agents in the Diagnosis of Cardiac Disorders. *Computers in Cardiology*, pp. 633–636, Memphis, Sept. 23–25, 2002.

IIOP OMG Internet Inter-ORB Protocol Specification, Common Object Request Broker Architecture 2.2. Object Management Group, 1999.

JADE, Java Agent Development Framework, website: http://jade.tilab.com

JCF, Java Collection Framework, http://java.sun.com/j2se/1.4.2/download.html#docs

Jennings, N. The Archon System and its Applications. In *Proceedings of the 2nd International Working Conference on Cooperating Knowledge Based Systems* (CKBS-94), pp. 13–29, Dake Centre, University of Keele, UK, 1994.

Jennings, N. and Wooldridge, M. Applications of Intelligent Agents. In *Agent Technology: Foundations, Applications, and Markets*, pp. 3–28, Secaucus, NJ, Springer-Verlag, Berlin, 1998.

Jennings, N., Faratin, P., Johnson, M.J., Norman, T.J., O'Brien, P. and Wiegand, M.E. Agent-based Business Process Management. *International Journal of Cooperative Information Systems* **5**(2-3): pp. 105–130, 1996.

JiBX website: http://jibx.sourceforge.net/

JSR82, Java Agent Services, Java Specification Request no. 82, http://www.jcp.org/en/jsr/detail?id=87

JSR 222, Java Architecture for XML Binding (JAXB) v. 2.0, http://www.jcp.org/en/jsr/detail?id=222

Klusch, M. Information Agent Technology for the Internet: a Survey. *Data Knowledge Engineering* **36**(3): pp. 337–372, 2001.

Labrou, Y. Agents and Ontologies for e-Business. *Knowledge Engineering Review*, **17**(1): pp. 81–85, 2002.

Labrou, Y., Finin T., and Peng, Y. Agent Communication Languages: the Current Landscape. *IEEE Intelligent Systems*, **14**(2): pp. 45–52, 1999.

Lange, D. and Oshima, M. *Programming and Deploying Java™ Mobile Agents with Aglets™* . Addison-Wesley, 1998.

Lanzola, G. and Boley, H. *Experience with a Functional-logic Multi-agent Architecture for Medical Problem Solving*, pp. 17–37, Idea Group Publishing, Hershey, PA, 2002.

Laukkanen, M. and Helin, H. Composing Workflows of Semantic Web Services. In *Proceedings of the 1st International Workshop on Web Services and Agent Based Engineering*, Sydney, Australia, July 2003.

LEAP website http://leap.crm-paris.com

Ljungberg, M. and Lucas, A. The Oasis Air-traffic Management System. In *Proceedings of the 2nd Pacific Rim International Conference on Artificial Intelligence*, Vol.II, pp. 1183–1189, Seoul, Republic of Korea, Sept. 15–18, 1992.

Louis V. and Martinez T., The JADE Semantic Agent: Towards Agent Communication Oriented Middleware, *AgentLink News*, **18**, pp. 16–18, http://www.agentlink.org/newsletter/18/AL-18.pdf, August 2005a.

Louis V. and Martinez T. An Operational Model for the FIPA-ACL Semantics. *Proceedings of the AAMAS'05 workshop on Agent Communication*, pp. 101–113, Utrecht, the Netherlands, July 2005b.

McCarthy, J. Ascribing Mental Qualities to Machines. In *Philosophical Perspectives in Artificial Intelligence*, pp. 161–195, Humanities Press, 1979.

Mayfield, J., Labrou, Y. and Finin, T. Evaluating KQML as an Agent Communication Language. In Wooldridge, M., Müller, J.P. and Tambe, M. (eds), *Intelligent Agents II* (LNAI 1037), pp. 347–360. Springer-Verlag, Heidelberg, 1996.

Mir, J. *Protocolos criptográficos para canales de comunicación anónimos y protección de itinerarios en agentes móviles*. PhD thesis, Escola de postgrau, 2004.

Moreno, A. and Nealon, J. *Applications of Software Agent Technology*. Whitestein, 2003.

Muller, J.P., Pischel, M. and Thiel, M. Modelling Reactive Behaviour in Vertically Layered Agent Architectures. In Wooldridge, M. and Jennings, N.R. (eds), *Intelligent Agents: Theories, Architectures, and Languages* (LNAI 890), pp. 261–276, Springer-Verlag, Heidelberg, 1995.

Neagu, N., Dorer, K., Greenwood, D. and Calisti, M. LS/ATN: Reporting on a Successful Agent-Based Solution for Transport Logistics Optimization. *Proceedings of the IEEE 2006 Workshop on Distributed Intelligent Systems* (WDIS'06), Prague, 2006.

Negri, A., Poggi, A., Tomaiuolo, M. and Turci, P. Dynamic Grid Tasks Composition and Distribution through Agents. *Concurrency and Computation: Practice and Experience*, **18**(8): pp. 875–885, 2006.

Newell, A. *Unified Theories of Cognition*. Harvard University Press, 1990.

Nwana, H.S., Lee, L. and Jennings, N.R. Coordination in Software Agent Systems. *BT Technology Journal*, **14**(4): pp. 79–88, 1996.

Nwana H.S., Ndumu D.T., Lee L.C. and Collis J.C. ZEUS: a Toolkit for Building Distributed Multiagent Systems. *Applied Artificial Intelligence*, **13**(1–2): pp. 129–185, 1999.

Parunak, H. Manufacturing Experience with the Contract Net. In Huhns, M. (ed.), *Distributed Artificial Intelligence*, pp. 285–310, Pitman, London, 1987.

Picco, G.P. Understanding Code Mobility (Tutorial Session). In *ICSE '00: Proceedings of the 22nd International Conference on Software Engineering*, pp. 834, ACM Press, New York, 2000.

Pokahr, A., Braubach, L. and Lamersdorf, W. A Flexible BDI Architecture Supporting Extensibility. In Skowron, A., Barthes, J.-P., Jain, L., Sun, R., Morizet-Mahoudeaux, P., Liu, J. and Zhong, N. (eds), *2005 IEEE/WIC/ACM International Conference on Intelligent Agent Technology* (IAT-2005), pp. 379–385, IEEE Computer Society, 2005.

Protégé, http://protege.stanford.edu/

Rao, A.S. and Georgeff, M. BDI Agents: from Theory to Practice. In *Proceedings of the 1st International Conference on Multi-Agent Systems*, pp. 312–319, San Francisco, CA, 1995.

Richards, D., van Splunter, S., Brazier, F. and Sabou, M. Composing Web Services Using an Agent Factory. In *Proceedings of the 1st International Workshop on Web Services and Agent Based Engineering*, Sydney, Australia, July 2003.

RMI, Java RMI Specifications, part of the Sun JDK documentation at docs\guide\rmi\spec\rmiTOC.html

Russell, S.J. and Norvig, P. *Artificial Intelligence: a Modern Approach*, 2nd edn. Prentice Hall, 2003.

Sadek, M.D. Attitudes Mentales et Interaction Rationnelle: Vers une Théorie Formelle de la Communication. Thèse de Doctorat Informatique, Université de Rennes I, France, 1991.

Searle, J. *Speech Acts*, Cambridge, MA, Cambridge University Press, 1969.

Silva, N., Rocha, J. and Cardoso, J. E-Business Interoperability through Ontology Semantic Mapping. In *Proceedings of the Processes and Foundations for Virtual Organizations*, pp. 315–322, Lugano, Switzerland, 2003.

Smith, R. and Davis, R. The Contract Net protocol: High Level Communication and Control in a Distributed Problem Solver. *IEEE Transactions on Computers*, **29**(12): pp. 1104–1113, 1980.

SOAP, Simple Object Access Protocol specifications, http://www.w3.org/TR/soap/

Suguri, H. Comtec Agent Platform, 1998.

Sycara, K., Paolucci, M., Ankolekar, A. and Srinivasan, N. Automated Discovery, Interaction and Composition of Semantic Web Services. *Journal of Web Semantics*, **1**(1): pp. 27–46, 2003.

Tanenbaum, A.S. and Van Steen, M. *Distributed Systems: Principles and Paradigms*. Prentice Hall PTR, Upper Saddle River, NJ, 2001.

Thielscher, M. FLUX: a Logic Programming Method for Reasoning Agents. *Theory and Practice of Logic Programming*, **5**(4–5): pp. 533–565, 2005.

Tsalgatidou, A. and Pilioura, T. An Overview of Standards and Related Technology in Web Services. *Distributed and Parallel Databases*, **12**(2–3): pp. 135–162, 2002.

UDDI, Universal Description, Discovery, and Integration, specifications, http://www.uddi.org/specification.html

W3C Web Service Architecture Working Group. *Web Service Architecture Recommendation*. http://www.w3.org/TR/2004/NOTE-ws-arch-20040211/, 2004.

W3CSem, W3C. *The Semantic Web*. http://www.w3.org/2001/sw, 2004.

WAP, Wireless Application Protocol Specification Version 1.2. WAP Forum, 1999. http://www.wapforum.org/what/technical.htm

Weikum, G. Special Issue on Infrastructure for Advanced E-services. *IEEE Data Engineering*, **24**(1), 2001.

White, J.E. Telescript Technology: Mobile Agents. In Bradshaw Jeffrey, (ed), *Software Agents*, AAAI Press/MIT Press, 1996.

Winikoff, M. JACK Intelligent Agents: an Industrial Strength Platform. In Bordini, R., Dastani, M., Dix, J. and Seghrouchni, A. (eds), *Multi-Agent Programming*, pp. 175–193, Kluwer, 2005.

Wooldridge, M. *Reasoning about Rational Agents*. The MIT Press, 2000.

Wooldridge, M.J. and Jennings, N.R. Intelligent Agents: Theory and Practice. *Knowledge Engineering Review*, **10**(2): pp. 115–152, 1995.

WSDL, Web Service Description Language specifications, http://www.w3.org/TR/wsdl

## FIPA Specifications

The following is the current set of Standard FIPA specification. These, and others not referenced in this text, are available for download from http://www.fipa.org.

FIPA1. Specification SC00001, FIPA Abstract Architecture Specification.

FIPA8. Specification SC00008, FIPA SL Content Language Specification.

FIPA14. Specification SI00014, FIPA Nomadic Application Support Specification.

FIPA23. Specification SC00023, FIPA Agent Management Specification.

FIPA26. Specification SC00026, FIPA Request Interaction Protocol Specification.

FIPA27. Specification SC00027, FIPA Query Interaction Protocol Specification.

FIPA28. Specification SC00028, FIPA Request When Interaction Protocol Specification.

FIPA29. Specification SC00029, FIPA Contract Net Interaction Protocol Specification.

FIPA30. Specification SC00030, FIPA Iterated Contract Net Interaction Protocol Specification.

FIPA33. Specification SC00033, FIPA Brokering Interaction Protocol Specification.

FIPA34. Specification SC00034, FIPA Recruiting Interaction Protocol Specification.

FIPA35. Specification SC00035, FIPA Subscribe Interaction Protocol Specification.
FIPA36. Specification SC00036, FIPA Propose Interaction Protocol Specification.
FIPA37. Specification SC00037, FIPA Communicative Act Library Specification.
FIPA61. Specification SC00061, FIPA ACL Message Structure Specification.
FIPA67. Specification SC00067, FIPA Agent Message Transport Service Specification.
FIPA69. Specification SC00069, FIPA ACL Message Representation in Bit-Efficient Specification.
FIPA70. Specification SC00070, FIPA ACL Message Representation in String Specification.
FIPA71. Specification SC00071, FIPA ACL Message Representation in XML Specification.
FIPA75. Specification SC00075, FIPA Agent Message Transport Protocol for IIOP Specification.
FIPA84. Specification SC00084, FIPA Agent Message Transport Protocol for HTTP Specification.
FIPA85. Specification SC00085, FIPA Agent Message Transport Envelope Representation in XML Specification.
FIPA88. Specification SC00088, FIPA Agent Message Transport Envelope Representation in Bit Efficient Specification.
FIPA91. Specification SI00091, FIPA Device Ontology Specification.
FIPA94. Specification SC00094, FIPA Quality of Service Specification.
FIPAHST. History of FIPA. Available online via http://www.fipa.org/subgroups/ROFS-SG.html.

# Index



*Developing Multi-Agent Systems with JADE* Fabio Bellifemine, Giovanni Caire, Dominic Greenwood